SECOND EDITION

FIRE BEHAVIOR AND COMBUSTION PROCESSES

Raymond Shackelford

Alfred J. Rager

Jeffery J. Zolfarelli

JONES & BARTLETT
LEARNING

World Headquarters
Jones & Bartlett Learning
25 Mall Road
Burlington, MA 01803
978-443-5000
info@jblearning.com
www.psglearning.com

National Fire Protection Association
1 Batterymarch Park
Quincy, MA 02169-7471
www.NFPA.org

International Association of Fire Chiefs
4025 Fair Ridge Drive
Fairfax, VA 22033
www.IAFC.org

Jones & Bartlett Learning books and products are available through most bookstores and online booksellers. To contact the Jones & Bartlett Learning Public Safety Group directly, call 800-832-0034, fax 978-443-8000, or visit our website www.psglearning.com.

Substantial discounts on bulk quantities of Jones & Bartlett Learning publications are available to corporations, professional associations, and other qualified organizations. For details and specific discount information, contact the special sales department at Jones & Bartlett Learning via the above contact information or send an email to specialsales@jblearning.com.

28401-0

Production Credits
Vice President, Product Management: Marisa R. Urbano
Vice President, Content Strategy and Implementation: Christine Emerton
Director, Product Management: Cathy Esperti
Director, Content Management: Donna Gridley
Director, Project Management and Content Services: Karen Scott
Product Manager: Janet Maker
Manager, Content Strategy: Kim Crowley
Content Strategist: Elena Sorrentino
Manager, Program Management: Kristen Rogers
Program Manager: Dan Stone
Senior Digital Project Specialist: Angela Dooley
Director, Marketing: Andrea DeFronzo
Sr. Product Marketing Manager: Elaine Riordan
Vice President, International Sales, Public Safety Group: Matthew Maniscalco
Director, Sales, Public Safety Group: Brian Hendrickson
Content Services Manager: Colleen Lamy
Senior Director of Supply Chain: Ed Schneider
Procurement Manager: Wendy Kilborn
Composition: S4Carlisle Publishing Services
Cover Design: Scott Moden
Text Design: Scott Moden
Media Development Editor: Faith Brosnan
Rights & Permissions Manager: John Rusk
Rights Specialist: Liz Kincaid
Cover Image (Title Page, Chapter Opener): © Stocktrek Images/Stocktrek Images/Getty Images
Printing and Binding: Lakeside Book Company

Library of Congress Cataloging-in-Publication Data
Names: Shackelford, Raymond, author. | Rager, Alfred J., author. | Zolfarelli, Jeffery J., author.
Title: Fire behavior and combustion processes / Raymond Shackelford, Alfred J. Rager, Jeffery J. Zolfarelli.
Description: Second edition. | Burlington, MA : Jones & Bartlett Learning, [2024] | Revised edition of: Fire behavior and combustion processes / Raymond Shackelford. c2009. | Includes bibliographical references and index.
Identifiers: LCCN 2023024007 | ISBN 9781284206562 (paperback)
Subjects: LCSH: Fire. | Combustion. | Fire extinction. | Fire extinguishing agents. | Wildfires. | Transportation--Fires and fire prevention. | Hazardous substances--Accidents. | Emergency management.
Classification: LCC TP265 .S523 2024 | DDC 363.37--dc23/eng/20230629
LC record available at https://lccn.loc.gov/2023024007

6048

Printed in the United States of America
27 26 25 24 23   10 9 8 7 6 5 4 3 2 1

Dedication

Dr. Ray O. Shackelford

Comments by Daniel P. Coffman (Dan) for the Second Edition of the textbook ***Fire Behavior and Combustion Processes***, originally authored by Dr. Ray O. Shackelford, on his contribution to the professional development of the fire service. May 13, 2023.

Dr. Ray Shackelford, the original author of *Fire Behavior and Combustion Processes*, left a tremendous legacy that not only influenced this generation of firefighters, but fire and emergency services for generations to come.

He was a visionary, leader, fire chief, educator, author, and businessman—a fire service giant with over 50 years of innovation, involvement, and influence. A labor leader with the Los Angeles County Firefighters IAFF Local #1014, he would establish many standards and practices common in labor/management relations today.

In the 1970s, he was a principal in the development of the Incident Command System and California's FIRESCOPE, the model for the National Incident Management System (NIMS), which is used around the globe today. As the Distinguished Professor and Chair of the Fire Protection Administration Program at the California State University, Los Angeles in the 1980s, he developed and implemented many of the management practices utilized by fire administrators today, such as Master Planning as a Fire Chief. He would write, research, teach, and influence hundreds of fire service professionals.

As an original and long-time participant of the U.S. Fire Administration's Fire and Emergency Services Higher Education Professional Development Initiative, he was instrumental in the development and proliferation of internet distance learning programs globally and furthering the professional development of the American fire service.

Chief Shackelford was the author of many books and papers. This book would become the most popular textbook for the *Fire Behavior and Combustion* course. The Second Edition is a tribute to his continuing contribution to aspiring firefighters and the fire service globally.

I had the great honor to call Dr. Ray Shackelford a friend and mentor for over 40 years. He was an impressive man who influenced me and so many of his students, firefighters who worked for and with him, and others, including yourself, who will use this book.

Daniel (Dan) P. Coffman, MSPA/CFO
Professor, Fire Technology/Fire Administration

Brief Contents

Contents

Preface

Fire Behavior and Combustion Processes was designed to provide a straightforward yet comprehensive resource for students enrolled in fire science degree programs, or as a refresher for active firefighters. It provides an understanding of the basic principles of fire chemistry, the processes of fire combustion, and fire behavior. The subject of fire behavior is often a complex one, and this book seeks to clarify theoretical concepts, explain their importance, and illustrate how they can be applied in a practical way when responding to emergency situations. There is also a special emphasis on safety, with each explanation drawing a connection between how a fire behaves and how it affects the safety of individual firefighters and their team.

In June 2001, the U.S. Fire Administration hosted the third annual Fire and Emergency Services Higher Education (FESHE) Conference at the National Fire Academy campus in Emmitsburg, Maryland. Attendees from state and local fire service training agencies, as well as colleges and universities with fire-related degree programs, attended the conference and participated in work groups. Among the significant outcomes of the working groups was the development of standard titles, outcomes, and descriptions for six core associate-level courses for the model fire science curriculum that had been developed by the group the previous year. The six core courses are *Fundamentals of Fire Protection, Fire Protection Systems, Fire Behavior and Combustion, Fire Protection Hydraulics and Water Supply, Building Construction for Fire Protection,* and *Fire Prevention.*

This textbook covers all the objectives and content for the FESHE course, *Fire Behavior and Combustion,* instituted by the National Fire Academy. FESHE is an organization of the secondary educational organizations that meet at the National Fire Academy annually to develop and approve a national curriculum that is standardized in scope, content, and outcomes. A correlation guide, which cross-references the FESHE course objectives with the content in this book, can be found in Appendix A.

About This Book

This book takes the complex layers of fire behavior and separates them out to help facilitate learning. Chapters 1 through 4 provide the background and theory involved in fire behavior. They introduce the student to the history and foundations of fire protection, fire chemistry, fire combustion processes, and fire extinguishment. Chapters 5 through 10 proceed to apply the theory to various incidents likely to be encountered by firefighters, and explain how firefighters can react safely for a successful outcome.

Chapter 1 sets up context for all the chapters that follow by providing a look back at pivotal fires in history and important lessons learned. This chapter also includes a section on firefighter injuries and fatalities in the United States, stressing the importance of safe practices in the fire service.

Chapter 2 highlights the basics of fire chemistry, including an explanation of basic physical and chemical properties of fire and how they relate to firefighter response.

Chapter 3 explains the combustion process, covering fire classifications, underlying causes, the various stages, and types of fire events.

Chapter 4 explores extinguishing agents that are available to firefighters and their various applications. A section that identifies the latest technological advances in fire extinguishment, such as compressed air foam and ultrafine water mist, keeps firefighters knowledgeable about systems they may encounter on the job.

Chapter 5 provides the foundation for the chapters that follow by introducing the concepts of firefighting strategies and tactics. In this chapter, the book begins to explore the various types of attack modes based on different types of fire scenarios.

Chapter 6 moves deeper into the discussion of firefighting strategies and tactics by presenting special concerns in firefighting activities. It includes a discussion of specific actions: prefire and postfire event planning, addressing utility concerns, working in confined enclosures, ventilation, and salvage and overhaul considerations.

Chapter 7 focuses specifically on high-rise building fires and the special actions and considerations involved in response to this fire scenario. This chapter clearly distinguishes high-rise fires from others and explains the unique challenges associated with fighting these types of fires.

Chapter 8 covers wildland firefighting for those areas of the country where wildland fires are a concern. Included is a section on special techniques for wildland fire response, as well as the various tools and resources required for fighting these fires. This offers students an opportunity to expand their knowledge of firefighting in different types of locales.

Chapter 9 proceeds to cover transportation fires and all related firefighting activities that are a consideration for response to these types of fires. It includes a look at fires in passenger vehicles, motor homes, buses, recreational vehicles, trucks, railcars, aircraft, and boats. The chapter also considers the latest technologies, such as electric-powered, hybrid, and hydrogen-powered vehicles, to ensure that firefighters understand the unique challenges associated with vehicle fire response today.

Chapter 10 concludes the book with a concentrated look at hazardous materials incidents and terrorist events that are a consideration for firefighters of today. A discussion of the National Incident Management System (NIMS) is also included to ensure that firefighters have an understanding of the logistics of this emergency response plan.

A recommended approach to using the text would be to carefully read Chapters 1 through 4 and answer all of the review questions at the end of each chapter before tackling the remaining chapters of the book. Once completed, the student should apply the basic chemistry and combustion principles found in the earlier chapters. For example, Chapter 5 applies the basic chemistry and combustion as well as the extinguishing principles to various occupancies that fire responders will encounter.

About the Authors

Alfred J. Rager began his career in public safety as an EMT working for Regional Ambulance on both BLS and ALS units. After a year of private-sector ambulance service, he was hired by the city of Alameda Fire Department and worked as a firefighter on its engine, truck, and ambulance services. There, he held the positions of Firefighter (1984–1992), Lieutenant (1992–1998), Captain (1998–2015), and Acting Battalion Chief (2000–2015). Al served a full and eventful career before retiring after 31 years with the City of Alameda Fire Department.

During his career with the City of Alameda, Al was in charge of the rescue rig and confined space programs. In this role, he taught on-duty personnel and many of the recruits how to safely and effectively utilize the tools and equipment on the Medium Rescue Rig assigned to his station as well as ran required quarterly drills. He updated the firefighter probation manual to reflect current standards and policies, as well as the latest safety practices. Al also sat on several firefighter and lateral entry-level oral boards.

While active in the fire service, Al served as a United States Marine Corps Reservist (1980–1986). In this role, he was a jet engine mechanic (intermediate maintenance) working on helicopter engines at NAS Alameda.

After retiring, Al began teaching at Chabot College as a Fire Science Professor. He teaches the FESHE classes FT-2, FT-3, FT-4, FT-5, and FT-6, as well as the Fire Academy and Pre-Fire Academy.

Al has an associate's degree in fire service technology from Chabot College and a bachelor's degree in fire administration from Cogswell Polytechnical College. He has also maintained his EMT license and has been a practicing EMT for the past 40 years.

Al was introduced to Jeff Zolfarelli in 2018 and they have been team teaching, upgrading, and further developing the fire behavior and chemistry class (FT-3) ever since.

Al has a wife and four children and, at this writing, one grandchild. In his spare time, he enjoys watching soccer matches and playing an occasional round of golf.

Jeffery J. Zolfarelli began his career in public safety as a police officer serving in the K-9 and SWAT divisions. After 10 years of service, he transferred over to the fire service. He began his career as a firefighter with the city of San Leandro, California, where he held the positions of Firefighter, Captain, and Acting Battalion Chief. With the San Leandro Fire Department, he also began a challenging career as a hazardous materials technician. From there, he moved to the Newark Fire Department, where he worked for 3 years answering initial responses to the city, as well as serving as Division Chief of Training. From Newark, Jeff moved on to the Livermore–Pleasanton Fire Department in his hometown of Pleasanton, California. There he served as a Division Chief for 5 years and was promoted to a Deputy Chief of Operations for 5 years before retiring. He served a full and event-filled 25 years in the fire service.

While working as a firefighter, Jeff began his teaching career with the Chabot–Las Positas Community College District. There he has taught fire chemistry and behavior for over 32 years. He created the current curriculum for the district's Fire Service Technology Program in the area of fire chemistry and behavior. With the many changes in the fire service, he has been true to holding the principles and practices of fire chemistry and physics as they apply to firefighting to help keep firefighters aware and safe in their careers.

In addition to serving as a United States Air Force Reservist, Jeff holds an associate's degree in administration of justice from Chabot College, a bachelor's degree in business management, and a master's degree in emergency planning and operations from the CSU Long Beach.

In his personal time, Jeff likes to take long-distance motorcycle rides, work on his model railroad layouts, and garden with his wife.

Acknowledgments

Jones & Bartlett Learning would like to thank all of the authors, contributors, and reviewers of *Fire Behavior and Combustion Processes, Second Edition.*

Contributors

Brian Centoni
Fire Captain
Alameda County Fire Department
California

Klaus Zalinskis
Retired Safety Inspector, North County Fire Authority
Chabot–Las Positas College District
Dublin, California

Lisa Zolfarelli
Graphic Designer
Rancho Cordova, California

Reviewers

Donald R. Adams, Sr., EdD

Robert Bartoli
Solano College
Fairfield, California

Michael R. Harper
Ret. Fire Captain—Adjunct Professor/Fire Instructor
Lincoln Park Fire Department/Schoolcraft College
Lincoln Park, Michigan

Mark E. Karp
Adjunct Instructor/Staff
Nicolet Area Technical College
Rhinelander, Wisconsin

Keith Kawamoto
Fire Technology Professor
College of the Canyons
Santa Clarita, California

Jack M. Minassian
Hawai'i Community College
Kona, Hawai'i

Andrew J. Naklick
A.A.S. Fire Protection Technology, SUNY Jefferson
Managing Fire Officer, National Fire Academy
City of Watertown Fire Department
Watertown, New York

Sean F. Peck, M.Ed., CFO, FM, NRP
Palomar College
San Marcos, California

Chuck Perry, MPA
Metropolitan Community College
Kansas City, Missouri

Michael Ripoll
Associate Professor
Central Piedmont Community College
Charlotte, North Carolina

John A. Sabia II
B.S. Organizational Management
M.S. Emergency Management
Seminole State College
Sanford, Florida

Dr. Christopher J. Schultz, DPA
Public Safety Administration Lecturer
Assistant Dean
Heavin School of Arts, Sciences, and Technology
Thomas Edison State University
Trenton, New Jersey

Lieutenant Michael A. Smith, MPA
Lynn Fire Department, Lynn, Massachusetts
Anna Maria College, Paxton, Massachusetts
North Shore Community College, Danvers, Massachusetts
Bunker Hill Community College, Boston, Massachusetts

Jeremy Stewart, MS
Volunteer State Community College Fire Science Director
Fire Rescue District Chief
Clarksville, Tennessee

Mark Strzyzynski
Fire Chief (Retired)
Henrietta Fire District
Henrietta, New York

Mike Van Bibber
Fire Chief (Retired)
Professor of Fire Sciences
Julian, California

Dennis Michael Van Tassell, B.S, MPA
City of Wyoming Fire Department
Wyoming, Michigan

David A. Yakowenko, B.S.
Fire Administration, Instructor III
Blackhawk Technical College
Rock County, Wisconsin

CHAPTER

1

American Fire Service: The Past, Present, and Future

LEARNING OBJECTIVES

After studying this chapter, you will be able to:

- Explain how history has shaped the American attitude toward fire prevention and fire control efforts.
- Explain the five major differences between the United States and other industrialized countries that contribute to higher fire losses in the United States.
- Describe the fire service of today, including its successes, its problems, and its efforts to improve.
- Describe new technologies and systems the fire service has used in recent years.
- Explain the challenges and opportunities open to the fire service in the 21st century.

CASE STUDY

The 1906 Earthquake and Fire

On April 18, 1906, at 5:12 AM, there was a large earthquake in California. From San Juan Bautista to Cape Mendocino, 296 miles (477 kilometers) of shift in the Earth's crust was felt, and devastating events occurred. The earthquake lasted only 1 minute and was felt as far north as Oregon, as far south as Los Angeles, and as far east as Nevada.

In the mid-1800s, the U.S. Congress passed the U.S. Swamp Land Acts. These acts enabled the farmers and connecting property owners in the San Francisco Bay Area to claim the wetlands that were touching their properties, fill them in, and build upon them.

The landfill provided more farmlands and land on which to build. However, filled-in wetlands proved to be unstable ground. This instability was one reason properties collapsed during the 1906 earthquake.

The earthquake caused the gas lines to break. The brick-and-clay water mains of the water delivery system also broke, leading to the loss of the water supply. This had a catastrophic effect when fires erupted after the earthquake.

Thirty fires broke out almost immediately and raged through the city for 3 days. At that time, many of the structures were Type 3 masonry, Type 4 heavy timber, or Type 5 wood construction, which burned quickly.

Military personnel from the Presidio stepped in and offered explosives to create a fire block. The fire block would remove the standing structures. The military's misunderstanding of a fire block caused further fire spread. It made more fuel by reducing the size of the flammable materials from buildings to splinters of rubble. By reducing the size of the material, the spread of fire increased.

1. What effects would the time of day have on the response to the earthquake and fire?
2. What were the infrastructure issues at this incident?
3. What impact did the loss of water supply have after the earthquake?
4. How could the responding military personnel have handled this incident better?

Introduction

The culture and history of the United States have shaped the American fire service into what it is today. By comparing the past and present-day fire service, we can see that much has been done to reduce fire deaths and losses. However, improvement is still needed. Reducing these deaths and losses remains the biggest challenge facing the fire service.

To understand the fire service of today, think about its history and how it has changed. This chapter includes a summary of the U.S. fire service's history, what the fire service is currently doing, and a look at its future challenges.

Fire Service of the Past

Since the colonial settlements in North America, fire has been a big part of U.S. history. These early settlements had several conflagrations and large fires. **Conflagrations** are fires with building-to-building flame spread over a great distance. In some cases, the fires consumed the settlements.

Before motorized vehicles, many fire departments used horse-drawn wagons. Horses were susceptible to illnesses. As an example, during the Great Boston Fire of 1872, the northeastern United States was experiencing epizootic equine flu. This disease weakened the horses and thus required firefighters to pull the fire engines, hoses, and ladders. During the mid-1920s, the Dodge Graham brothers invented the chemical pumper, the first motorized fire vehicle. The pumper allowed firefighters to use an additive in water to quickly put out fires. However, firefighters still had only a limited capacity for extinguishing agents due to the tank size. In the early 1960s, the LaFrance open-cab fire truck was invented. This truck was built on a larger chassis and could carry more firefighters and water to the scene (**FIGURE 1-1**).

The Industrial Revolution

At the end of the 1800s, there was an industrial revolution in the United States. The cities became crowded. More people worked in factories and lived in the cities instead of working and living on farms. This growth in building construction led to poorly built and unplanned

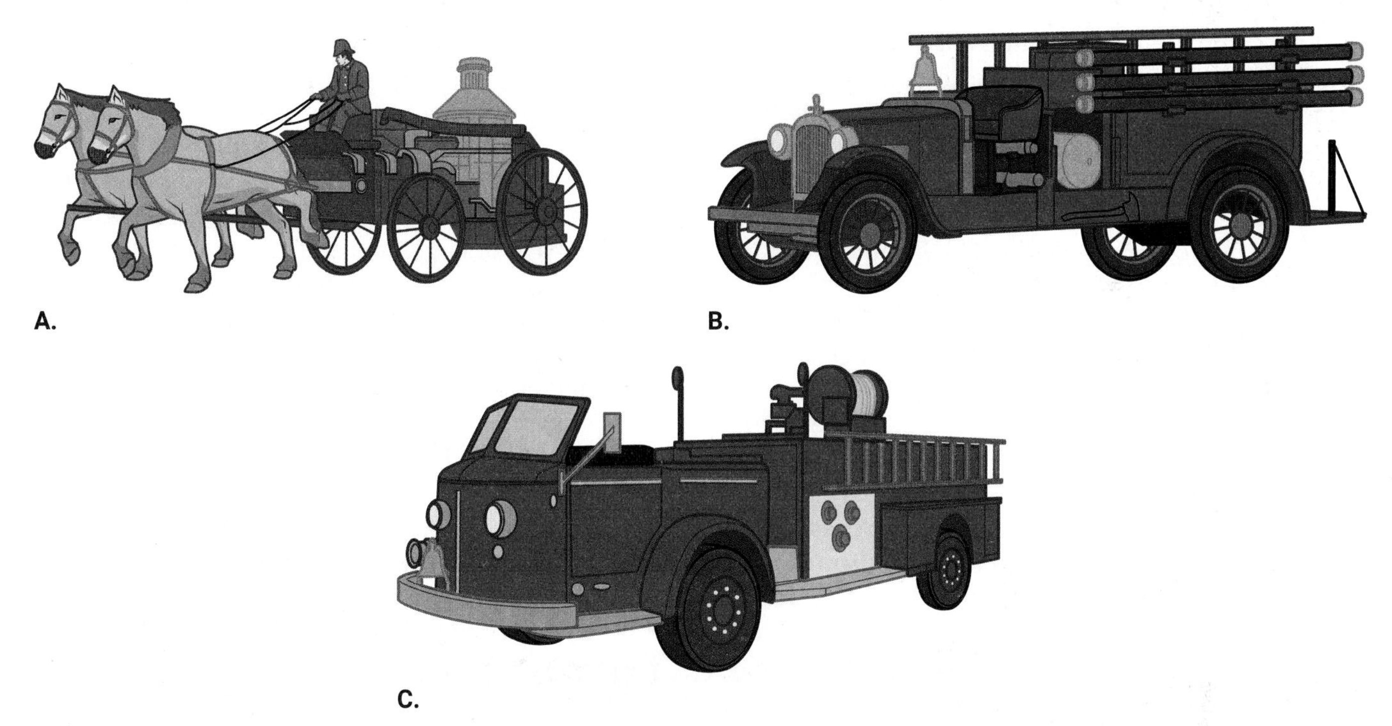

FIGURE 1-1 Some early-era American fire trucks. **A.** Horse-drawn wagon. **B.** Dodge Graham Brothers chemical pumper. **C.** LaFrance open-cab fire truck.

developments in the cities. Many structures were large compared to the smaller ones of earlier settlements. In many cases, cities ignored fire safety issues.

Talking about this time in the U.S. fire service, the editor of *Fire Engineering* magazine said, "Little control was exerted over construction methods and materials, fire loading and occupancy loads . . . and while there were many chances for the fire service to step up to the challenge there is little evidence that fire departments were quick to seize the opportunity" (Manning 1997).

Fire departments did not often offer input on new construction because building laws were controlled by the building departments. Building inspectors worked for the state or city building department. At that time, those agencies were often corrupt or did not see the importance of safe buildings. A top priority was building as fast and as cheaply as possible.

Corruption in building departments led to poor construction and little or no enforcement of fire codes. This resulted in large fires that destroyed whole cities. Some fires took the lives of the public as well as firefighters. The Industrial Revolution was challenging to the fire service because growth and expansion were valued over life safety.

Two Great Fires

On October 8, 1871, two fires raged out of control on the same day in the United States. One was a forest fire in northeastern Wisconsin. In this fire, 1,152 people died and 2,400 square miles of forested land burned. The second fire, known as the Great Chicago Fire, was a much more widely publicized fire. It attracted more media attention than the Wisconsin forest fire even though fewer lives were lost.

Because of the media attention, the Chicago fire was the subject of many songs, fables, and tales in U.S. history. There were rumors that the fire had been started by a cow kicking over a lantern. Now there is evidence that strong winds coupled with an earlier lumberyard fire resulted in a flying ember that started a fire in the straw in Mrs. O'Leary's cow barn.

The Chicago fire led to the deaths of 300 people and destroyed buildings in a 2.5 square mile area. It burned for 2 days before firefighters were able to stop it. Quick, poor-quality construction that used little or no fire-resistant materials was a factor in the widespread damage.

After the Great Chicago Fire, the city council approved a law that required new buildings to be built of stone or masonry. Unfortunately, the new law was not enforced. As a result, the fire insurance companies worried about the possibility of another large fire.

Insurance Companies Respond

In 1874, the fire insurance companies in the United States were being represented by the National Board of Fire Underwriters. This group was formed to make the Chicago city council enforce the laws created in response to the Great Chicago Fire. However, the city council did not respond. The insurance companies punished the city council by closing all insurance offices. The city council then agreed to enforce the laws. The insurance companies reopened the offices and increased the fire insurance rates by 20%.

The actions resulting from the Great Chicago Fire were the basis for today's Insurance Services Office. The **Insurance Services Office (ISO)** is an agency funded by insurance companies to apply a rating schedule to cities and fire departments. This system rates cities and fire departments from 1 to 10. The highest fire classification rating received by a public agency is 1, and the lowest is 10. Cities and fire departments may have different ratings based on this schedule. This rating is used to set the fire insurance costs for fire protection services. A method was needed to assess the city and the fire service's ability to prevent a large fire. The method became higher fire insurance costs for a poor fire rating. The goal of these ratings is to encourage fire departments and cities to improve their fire protection services. The rating schedule has changed over the years. However, the ISO still measures the fire service's ability to control large fires and adjusts the rate for fire insurance costs.

The Decade of Conflagrations

The years 1900 to 1909 have been called "the Decade of Conflagrations." Five of the most significant fires in the United States happened during this period. These fires brought to the attention of the public issues that added to the loss of lives and property. Areas of particular concern were:

- the need for fire-resistive construction materials in buildings;
- the need for a dependable water supply;
- the regular inspection of fire and life-saving equipment; and
- the safe storage of combustible and flammable materials.

As noted in the chapter case study, the 1906 earthquake and fire in San Francisco, California, brought the need for fire-resistive construction to the forefront of concern. The earthquake and fire were described as a double scourge because the earthquake broke gas lines and water mains, which then caused fires. An estimated 674 people died and 3,500 others were injured. The earthquake and fire destroyed 28,000 buildings that spanned 514 city blocks. As a result, the city of San Francisco installed a fire water system so that the domestic water was backed up in case of an earthquake. The fire water system uses cisterns that have been in place for over 110 years. This Decade of Conflagrations made it clear that there were grave fire problems.

The cycle of large fires and rebuilding has existed since the early 1900s. Despite growing awareness of fire problems, the fire service has made limited progress in safety. We still have large fires that devastate our towns and cities. For today's fire students, understanding combustion, the use of fire-resistant building materials, and fire technologies can help stop this cycle of conflagrations.

Fire Service of the Twentieth Century Through Today

The increasing size of buildings in the cities prompted the fire service to develop a way for firefighters to reach the upper floors of tall buildings. In the 1930s, the introduction of the turntable ladder allowed firefighters to reach heights of 150 feet (46 m). After World War II, a bucket was added to the top of the ladder. This crane and bucket became known as an aerial work platform, cherry picker, or elevated platform.

The 1960s brought modern fire trucks with newer water pumps, ladders, and cherry pickers. Today's fire trucks are designed so that they can perform different tasks based on the type of emergency response. Generally, fire trucks are in one of three categories: pumpers, turntable ladders, and specialized fire trucks (**FIGURE 1-2**).

Discovering the U.S. Fire Problem

During the early 1960s, the International Association of City/County Managers joined with the International Association of Fire Chiefs. Together they led and published several studies. Their goals were to judge and share the scope of the fire problem in the United States.

These studies found that despite the United States spending billions of dollars on fire protection, communities still had sizable losses of life and property. Because of these reports, the U.S. fire service recognized the need to understand what was causing the losses and how to change the trend.

The U.S. fire service also recognized the need for public support to control its fire problems. In 1966, fire service leaders held a conference at the Wingspread Conference Center in Racine, Wisconsin, to

A.

B.

C.

FIGURE 1-2 Today's standards in fire trucks. **A.** Fire truck with fully enclosed cab, 500-gallon water tank, and 1,500 gallons per minute (gpm) pumping capacity. **B.** Fire truck with tiller, 105-ft aerial ladder, and a large carrying capacity of ground ladders, smoke ejectors, and other tools. **C.** A smaller mini pumper that is used for wildland fires and can be driven off-road if needed.

A: © BD Images/Shutterstock; **B:** © HOT SHOTS/Alamy Stock Photo; **C:** Courtesy of SVI Trucks.

find answers to the nation's fire problems. Their report urged an examination of the fire problem in the United States. The result was the 1973 presidential committee report, ***America Burning***.

The *America Burning* report found several areas for improvement in the fire service. The suggestions called for improved enforcement of fire codes, data collection, and processing. It also called for improving the quality of training, equipment, technology, and education for the fire service. Over the next few years, the fire service addressed some of these suggestions, but progress was slow.

However, two of the report's suggestions led to the birth of the U.S. Fire Administration and the National Fire Academy. These agencies made fire data collection, analysis, and training national concerns. Improvement in these areas will continue as long as federal funding continues.

In 1987, the U.S. Fire Administration met with fire service leaders. The goal was to check the progress made on the original *America Burning* report suggestions. They found that some progress had been made on the goals set by *America Burning*. Other areas still needed to improve, such as fire prevention, public education, and changing the public's acceptance of fires destroying large areas and buildings every year.

The conference produced the 1987 report, ***America Burning Revisited***. This report urged the nation and the fire leaders to continue improving the fire service.

In 2000, the Federal Emergency Management Agency (FEMA) and the U.S. Fire Administration held another conference. The goal was to look again at the fire service and review its progress since the 1987 report. The conference led to the 2002 report, ***America Burning Recommissioned, America at Risk***. This report showed failures of the fire service on continuing U.S. fire problems. The report:

- Pointed to the frequency and severity of fires or the Nation's failure to adequately apply and fund known loss reduction strategies.
- Recommended that the fire service should apply more funding and resources to fire prevention.
- Determined that firefighters often expose themselves to unnecessary risks.
- Recommended more education and training to transition from firefighting to fire prevention activities (*America Burning Recommissioned*, 2002).

Because of the report, in 2000, the National Fire Academy held a conference for fire service educators. They came from across the country to make standards

FIGURE 1-3 National Professional Development Model.

Courtesy of FEMA (Federal Emergency Management Agency).

for the fire service. Their goal was to see if a national education program could be made for firefighters. After 4 years, the educators created a model (**FIGURE 1-3**). The model shows the criteria for fire department jobs and a framework for applying training and experience to the education system. A national education and training track for the fire service was viewed as a needed step to bring the fire service to a higher level of professionalism.

In 2004, at the Firefighter Life Safety Summit in Tampa, Florida, fire service groups met about the problem of firefighters dying in the line of duty (over 100 deaths annually). This led to the Everyone Goes Home program and 16 Firefighter Life Safety Initiatives. These initiatives are the basis for thousands of fire and EMS agencies to ensure the safety of their members.

Since 1966, fire service leaders have held conferences at the Wingspread Conference Center. They meet every 10 years to address critical issues facing the fire industry. In each conference report, leaders have included statements about how the fire service can reduce the loss of life, injuries, and property, while keeping firefighters safe. At the meeting in 2016, the attendees decided that they should meet every 5 years because the industry changes so quickly. The statements from the 2021 conference addressed the wildland fire crisis and the need for more research, education, and funding. However, the main request was for the federal government to add fire and emergency services to its critical infrastructure. As critical infrastructure, fire and emergency services would get the needed funding.

Comparing the U.S. Fire Problem

According to the U.S. Fire Administration, from 1990 to 2019, fire death rates per million population consistently fell throughout the industrialized world. The North American and Eastern European regions' fire death rates fell faster than those in other regions (U.S. Fire Administration/U.S. Fire Statistics).

During this same period, the fire death rate in the United States dropped by 41%. Although its standing is greatly improved, the United States still has one of the higher fire death rates in the industrialized world.

Original studies in the late 1980s showed five major areas in which the United States differed from its international partners. These statistics remain a factor today (Fire Rescue, 2021).

1. On average, less than 3% to 5% of the total fire department budget is spent on fire prevention.
2. Fire services receive little funding.
3. Wood is often used in construction.
4. The use of plastics has increased, which increases the heat output of a fire over wood products.
5. U.S. people allow uncontrolled fires to occur.

As noted above, U.S. fire departments spend, on average, only 3% to 5% of the total fire department budget on fire prevention. Most other countries spend an average of 5% to 15% of their budgets on fire prevention. The study recommended more funding for fire prevention.

These studies also pointed out that fire has been a local government responsibility, funded only by local money. This is often not enough. More recently, the federal government has approved some federal grant programs. These programs have funded personnel protection and other equipment, fire trucks, and self-contained breathing apparatus (SCBA). They have also given seed money for wellness programs and staff. Unfortunately, the funding is cyclical and is often provided after a great deal of pleading for federal support. This has led to an increase in grant proposals being written and submitted.

Building construction in the United States differs from that in other countries. The primary building material in the United States is wood. There is more wood construction than stone or brick construction. This has led to buildings made of wood that have little or no fire resistance. Compared to other countries that use steel and concrete with fire proofing added to the structure, wood construction creates a grave fire problem for the U.S. fire service.

Since the early 1960s, the use of plastics in furniture and building has increased the average amount of British thermal units. **British thermal units (BTUs)** are units of measure for the heat energy needed to raise the temperature of 1 pound of water 1 degree Fahrenheit. This is per square foot of floor space. When a fire on plastics occurs in a small space, there is a higher possibility of a flashover fire. A **flashover fire** is a fire that happens when all of the contents of a room or enclosed space reach their ignition temperature and result in a rapid increase in fire growth and intensity. Also, the heat from burning plastics leads to fires that quickly consume all combustible materials in an enclosed space (see Chapter 3).

Lastly, a fire event in the United States is socially accepted. Fire insurance and charity groups are on hand for those involved in a fire. In the United States, fire prevention focuses on climate change, droughts, and heavily wooded forests.

U.S. Firefighter Deaths and Injuries

In 2022, the National Fire Protection Association (NFPA) released its annual report *Firefighter Fatalities in the US in 2021* (**TABLE 1-1**). The report identified the five leading causes of firefighter deaths. The leading cause of death for on-duty firefighters was heart attack (46%), followed by trauma, such as a fatal head or internal injury (29%). The third highest cause of death was asphyxiation and burns (12%), with stroke as the fourth highest cause (4%). The fifth area of firefighter deaths is categorized as miscellaneous (8%). This category includes deaths and injuries that range from responding from and returning to quarters, training activities, station maintenance work, and other activities.

Over the past decade, there has been a consistent downward trend in firefighter deaths and injuries. The exceptions are the increased number of deaths in 2020 and 2021. Of the 135 firefighter deaths in 2021, 65 were due to coronavirus disease of 2019 (COVID-19). Many of

TABLE 1-1 Firefighter Fatalities, Fire Ground Injuries, and Total Injuries (2012–2021)

Year	Deaths	Fire Ground Injuries	Total Injuries
2012	81	31,490	69,400
2013	106	29,760	65,880
2014	91	27,240	63,350
2015	90	29,130	68,085
2016	89	24,325	62,085
2017	87	24,495	58,835
2018	82	22,975	58,250
2019	65	23,825	60,825
2020	102	22,450	64,875
2021	135	N/A	N/A

Data from Campbell, Richard. 2022. *Firefighter Injuries on the Fireground.* National Fire Protection Association. https://www.nfpa.org/-/media/Files/News-and-Research/Fire-statistics-and-reports/Emergency-responders/ospatterns.pdf; Fahy, Rita F., and Jay T. Petrillo. 2022. *Firefighter Fatalities in the US in 2021.* National Fire Protection Association. https://www.nfpa.org//-/media/Files/News-and-Research/Fire-statistics-and-reports/Emergency-responders/osFFF.pdf.

these cases were linked to medical emergencies, when firefighters were exposed to infected people. Other cases were tied to exposure at the fire station. The 2021 number of COVID deaths is an improvement over the 78 COVID deaths in 2020. Firefighter deaths from COVID should decline as the pandemic comes to an end.

Emergency Medical Services

There is a close link between the fire service and emergency medical care. In the United States in the 1930s, the fire service started providing first aid units for injured firefighters. This service was later offered to the public. It provided first aid to those with medical problems and injuries. This service continues to grow.

In the 1950s, radio systems improved, so two-way contact with hospitals and fire dispatch became possible. Firefighters trained as paramedics could talk directly with the hospital doctor. Directed by the doctor, the paramedic could save lives. Today, the use of smart devices, smart device applications, and live video feed allows for faster data transfer. Updates of the patient's condition go from the paramedic to the transport unit to the emergency room doctor in a rapid exchange. The fire service became vital to the community when it included emergency medical care and transport. Today, 65% of the requests for fire department help are for medical emergencies (NFPA 2020 report).

The fire department's emergency medical services allow them to meet changing community needs. In the future, improved communication and medical advances will help the fire service improve services.

Building and Fire Code Enforcement and Improvements

Although there have been great advances in building and fire codes, improvements are still needed. The fire and safety report issued after the World Trade Center terrorist attack in 2001 uncovered advances that would have prevented the buildings from collapse. The fires and building damage caused by the two jet airplanes flown into the buildings brought down both high-rise buildings. Over 3,000 people were killed. Before this event, many believed there were no more fire and safety problems in modern high-rise buildings. The 2002 World Trade Center study (Corley, Hamburger, and McAllister 2002) found issues that still needed to be addressed, such as:

- increasing the use of impact-resistant spaces around egress systems;
- examining the steel floor to make sure that truss bolting materials used to attach the floors to the walls are sturdy;
- grouping emergency exit stairways in the center of the building instead of scattering them throughout the buildings; and
- providing passive fire protection systems that protect buildings.

This report and its recommendations are important for firefighters to understand. Especially important are the relationships between building construction methods, fire-resistant materials, fire combustion processes, and fire behavior. Each of these can have an effect on firefighter safety.

In some cases, improvements in building construction led to problems for firefighters. For example, consider energy-saving windows. These windows play a role in the high heat conditions in fires today. They are designed to keep heat and cool air inside and keep unwanted air outside. They do not break easily, which shortens the time to a flashover fire. Firefighters trying to break the windows for ventilation find the plastic glass materials much more difficult to break.

The use of plastics has added not only more heat but also smoke to fires. This makes firefighting more dangerous, even though firefighters are now better able to handle these smoky fires. Plastics give off almost twice the number of BTUs as wood products during combustion. However, we are not reducing the use of plastics. Today, we see a projected increase of 3.9% for household furniture made of plastics, as reported in a study for the plastic furniture market (Grand View Research, 2019–2025).

The use of plastics in furniture has also raised cancer-related line-of-duty deaths. In a 2016 study by the Centers for Disease Control and Prevention (CDC) and the National Institute for Occupational Safety and Health (NIOSH), rates of cancer and deaths involving cancer among firefighters were higher than previously reported. Firefighters should reduce their risk of exposure to toxic fumes from burning plastics. To do this, the fire service needs to teach members about safe work practices, including using masks during all phases of firefighting.

Training and Education

Firefighters are better trained today because of improved procedures, requirements, and techniques. During the past 30 years, firefighting has demanded a higher level of training. As more toxic materials find their way into our lives, the fire service is now tasked with training firefighters to deal with them.

Training has become more complex, too. Today's firefighters use web-based training programs to imitate fire scenes. Computerized fire scenes are useful, offer diverse training in a safe setting, and may offer faster learning. These online programs are also used for ongoing education, such as the EMS and wildland refresher courses.

Not only is the training better for firefighters, but the requirements are strictly enforced. Before the internet, data collection and publication took months. With a web-based program, real-time reporting and data collection are quicker and more accurate. Online training systems hold students responsible for their own training. Programs such as OSHA's Near Miss and those from Vector Solutions allow students to record their training. Students can view their progress toward fire department standards. They can test for certification at the end of the program.

Higher levels of education are a must for daily firefighting activities. Before a firefighter can drive a fire truck, he or she must take driver-operator training. Those who want decision-making jobs in the fire service need a college degree or a certificate from a fire academy. They look for college degrees for promotion to captain and chief officer positions. Today, with supporting educational institutions, students passing an approved course can submit a FESHE application for credit from the National Fire Academy.

When firefighters are at the firehouse, they are training and preparing for the next fire. In the future, more training, education, and experience will be needed to meet the challenges.

Administration and Coordination

A great deal of time, effort, and resources was put into making the Incident Command System. The **Incident Command System (ICS)** is a management system used on the emergency scene to keep order. ICS follows set guidelines and tracks firefighters and their efforts to put out fires. Since 2004, the ICS has been known as the National Incident Management System. The **National Incident Management System (NIMS)** is a management system adopted by FEMA that combines resources from public and private agencies. NIMS is an all-hazards system (see Chapter 10).

One of the ways NIMS has been used is to address the California wildfires. For several years, California has had an expanding loss of life and property due to wildland/urban interface fires. **Wildland/urban interface** is the line, area, or zone where structures and other human developments merge with wildland or fuels from vegetation. These wildland/urban interface fires have damaged billions of dollars of land between the forest/wildland area and the coast.

There were minor issues in the fire service's response to these fires. However, one of the major problems has been the inability to combine different agencies' resources into one force to deal with the wildland fire. This includes the combined response on the initial dispatch to these fires.

Some wildland fires spread rapidly through four to six areas, sometimes even before all the resources could be brought together. This issue has been resolved through NIMS and dispatch aided by computers.

The ICS, as part of NIMS, can apply to all types of emergencies. Readers should review and become familiar with this system and how it works. It is used in every emergency, whether natural or made by humans.

Incident Command System: Goal and Key Parts

The goal of the ICS is to provide a complete command structure for one or many agencies. This system makes the command and control of an incident more effective. It provides a higher level of firefighter safety. According to FEMA Emergency Management Institute (2017), to do this, the system should contain the following key parts:

- An agency will be developed, with the functions of command, operations, planning, logistics, and finance/administration.
- The system will allow for use by federal, state, and local fire agencies. Thus, the language used in the system must be acceptable within all levels of government.
- The system must be used as the everyday operating system at all incidents. The constant operation allows for a shift from a small incident to a large incident or one that involves more than one agency. There would be few changes for participating agencies.
- The system for handling large fires must be thorough. The start-up or beginning phases must work with only a few people. It must have units to handle the most vital activities. As the incident grows or becomes more complex, management can allow people to keep a span of control over functional as well as geographic areas. A **span of control** is the number of people or resources that one supervisor can manage during an incident.
- The system must require that each agency's power over an area cannot be transferred to other agencies. Each agency will have full command authority over its area at all times. Helping agencies will work under the incident commander (IC). The IC

is chosen by the agency in charge of the area where the incident occurs unless the agencies agree to other plans.

- Incidents that cover more than one area are managed under a unified command. It includes one command post and one action plan. The plan is developed by and applied to all agencies present.
- The goal is to have the system staffed and operated by members of any agency. An incident could involve the use of workers from many agencies, each one working at different places within the system.
- The agency can expand or shrink based on the needs of the incident. Both size and complexity are measures for the size of the management system.

This system was built by and for the fire service. It is now accepted as the operating system for most of the fire service as well as many government agencies. Its continued use and improvement will prepare these agencies for emergencies and disasters well into this century.

Equipment and Personal Protection

Fire equipment has improved over the past few years with powerful diesel engines, better braking systems, and larger pumps. Larger-size hose, better-designed nozzles, and safer protective equipment and clothing have also all made the fire service more efficient and safer.

Larger and lighter fire hose made with synthetic materials is one example of how technology has helped the fire service provide more service with fewer people. Almost all small synthetic attack lines today are 1 3/4 inches (45 mm) in diameter. This allows for more water than the older 1 1/2-inch (38-mm) cotton lines could deliver. It also allows for a quicker attack because the hose is lighter and easier to handle than the cotton hose lines.

Nozzles also have a better design. Today's nozzles provide more water to the source of the fire with less water damage. They are more easily controlled and tend to leak less than earlier types.

These are just a few of the many advances in technology that have affected the fire service since the Industrial Revolution. One of the biggest issues facing the fire service is applying technology to best serve the fire service. We see big changes in our lives. We can expect that advances in technology will change the fire service now and in the future.

Protective Systems

Improved technology has made smoke and carbon monoxide (CO) detectors more reliable and less costly. Lower costs have made them available to more people. In addition to the requirements for smoke detectors, in 2018, states required CO monitors to be installed in homes. These warning devices have greatly reduced the number of fire deaths. They provide early warnings of dangerous conditions and give people time to escape. The creation of the quick-acting fire sprinkler heads and improved home sprinkler systems continue to decrease home fire losses.

The use of protective systems in homes and commercial buildings is another factor in reducing deaths and injuries from fires. The introduction of these systems into the mainstream has changed the response by fire departments. It will likely continue to do so. In addition, new construction and remodels are now required to install these systems during construction.

English or Customary System Versus the SI Measuring System

In the United States, the fire service uses the English or customary system to measure. Other countries use a form of the metric system called the International System of Units or SI. Americans working in laboratories and medicine and foreign companies also use this method. It was created in the 19th century to standardize how length, distance, weight, and volume are measured. The U.S. fire service should use the SI system to share information with fire services across the globe.

SI Measuring System

Firefighters are required to use numbers in their work. For example, a number system is used for the size of hose lines in a fire attack, which is expressed in relation to the diameter of the hose. **Diameter** is the distance from one side of a circle to the other. This is in relation to the **circumference**, the length around the circle, and **radius**, the length from the middle of the circle to the outside of the circle. For house fires, 1 1/2-inch (38-mm) or 1 3/4-inch (45-mm) hose lines are used. A single 5-inch (127-mm) hose line will be used to supply water, while the capacity of the engine pump is described as a 1,500 gpm (5,678 L/min) centrifugal pump.

For these numbers to make sense, they must use some unit of measure that describes what is being measured. So, we use three main units of measure: distance, mass, and time.

Distance, Mass, and Time. The basic unit for length in SI is the meter (m). It is also the basic unit of length in the metric system and is almost equal to the English yard. Whereas the meter is 39.7 inches in length, the English yard is 36 inches or 3 feet. From this base unit, area can be found in square meters (m^2) and volume in cubic meters (m^3). Measurements for speed can be derived from length and time and described in feet per second or meters per second (fps or m/s).

Because all units of surface and volume are developed using the meter standard, the relationship between all units is understandable as they are both divided and multiplied by factors of 10. In the metric system, cube subdivision is used to describe the relationship of length, area, and volume. The cube is one decimeter squared and contains 1,000 cubic centimeters.

Energy and Work. Energy has been defined as the capacity to perform work. Work occurs when a process is applied to an object over a distance. In the SI system, the unit for work is the joule (J). The joule is derived from the unit of kg m/s^2 and the distance in meters. In the English system, the unit of work is the foot-pound (ft lb).

Fire Service and Technology

The U.S. Fire Administration funds the fire service. More funding from other sources has helped research and develop equipment, protective gear, and fire trucks. From 2005 to 2015, the Fire Safety Research Institute (FSRI) received $11 million in grants for projects to make the public and firefighters safer.

FEMA, part of the Department of Homeland Security, offers grants for the fire service. These are Assistance to Firefighters Grants (AFG), Staffing for Adequate Fire and Emergency Response (SAFER), and Fire Prevention & Safety Grants (FP&S). Some of the new technologies and research afforded by these grants are briefly discussed.

Alternative Fuels

The use of alternative fuels for vehicles has been increasing in the United States. This is to reduce air pollution and reliance on oil from other countries. An **alternative fuel** is any energy source other than the hydrocarbon fuels. Some examples are compressed natural gas (CNG), liquefied natural gas (LNG), and liquefied petroleum (LP-Gas). Fueling vehicles with hydrogen is being researched. However, the use of hydrogen as a fuel for vehicles has not been widely accepted and vehicles are not in mass production.

Hybrid and electric vehicles are the most common alternative-fuel vehicles in the United States. Firefighters must be ready to put out fires at alternative-fuel storage/supplying sites. They must also be ready for fires in vehicles using these fuels. Every day, there are more charging stations. They appear at businesses, shopping centers, and schools. They offer easy access to drivers.

With the rise of electric vehicles, we are seeing changes to the design of our fire trucks. In recent years, fire departments have bought electric fire trucks. These vehicles work better and reduce emissions and the cost of energy (**FIGURE 1-4**).

Understanding fire chemistry and combustion will greatly improve firefighters' ability to put out fires caused by these alternative fuels (see Chapter 9). Especially with electric vehicles, there is a greater chance of a fire reigniting than with the other alternative fuels.

Thermal Imaging

Today, thermal imaging with portable cameras is a valuable tool. **Thermal imaging** uses infrared waves to find heat given off by a substance or body. The fire service can use this tool to find fire in a building, a grain storage silo or coal storage bin, or the active edge of a wildland fire.

These cameras have also been used in rescue missions where a building collapse has trapped people under piles of rubble. A probe can go through the piles and find heat from a body. Infrared cameras can also warn of upcoming flashover fires. To read these signs, firefighters must be skilled at using the camera.

Most thermal imaging cameras can go through only thin building materials. This limits the camera's ability to find a fire or person inside some buildings. In 2020, the U.S. military gave some of the latest in

FIGURE 1-4 LAFD gets a Rosenbauer electric fire truck.
Courtesy of Rosenbauer International AG.

thermal imaging equipment to local governments. This new equipment sees through materials better. Fires can be found behind thicker building walls. Improved thermal imaging shows heat differences and clear images. Possible victims can be found instantly. Firefighters can search what is in a room and find obstacles or risky areas. This is just one example of why firefighters must be better trained to understand and use this high-tech equipment.

Global Positioning Systems

The fire service uses global positioning systems. A **global positioning system (GPS)** is a navigation system. It uses satellites in space to map locations on Earth. This system helps the fire service to find the fire companies closest to an incident. It then directs that company to the fire.

The system connects to a dispatch system that uses computers. The computer can track companies sent to the incident. This tracking is vital because fire companies also provide medical services. This means that they are not as available for fires. This system offers real-time data. It allows for better coverage with fewer fire trucks.

GPS allows a computer system to apply maps for fire modeling. **Fire modeling** is a simulation used to figure out a fire's outcome. For wildland fires, one can program the computer with fuel, terrain, weather, and fire history data. These data are used for projection of fire behavior in real time or time intervals on a map. This helps the IC to plan and make decisions.

During an incident, with a GPS system in SCBA, firefighters can be tracked. They can track each other, even if someone is not moving. GPS with smart personal protective equipment (PPE) can gather an individual's biological data in real time, such as body temperature, heart rate, effort, and air level. Using Bluetooth technology, this information can go to the IC off-site. The IC can use the data to remove worn-out firefighters. They can be sent to rehab for fluids and rest. The IC can also send for backup if needed.

Ultrafine Water Mist

Halon 1301, a **halogenated extinguishing agent**, is an extinguishing agent that has one or more halogen atoms added to a hydrocarbon after removing the hydrogen atom from the hydrocarbon molecule. Halon 1301 was used in the 1960s and 1970s, but was found to be a substance that drains the ozone layer. There was a search for another extinguishing agent. Some alternative agents have been found, but their use may also have serious consequences. From an environmental viewpoint, ultrafine water mist-based systems seem to be a good alternative.

Ultrafine water mist is water dispensed under high pressure through very fine nozzle outlets that create nearly microsized droplets. The small size of the droplet improves the water's shift to steam. The droplets scatter like a gas, which helps them act as a total flooding agent.

The ability of water mist to put out fires depends on many things. It depends on the size and stability of the mist, how the mist travels, and how fast the droplets evaporate.

Firefighters can now carry these systems. Developed in 1971, this high-pressure water system may improve for future use.

Compressed Air Foam Systems

Class A and Class B foams with air under pressure is a technology used at the scene of structure fires, wildland fires, and flammable liquid fires. The **compressed air foam system (CAFS)** is a self-contained, stored-energy fire suppression unit that supplies compressed air foam. The foam is a Class A hydrocarbon-based surfactant. **Surfactant** is a soap material that reduces the surface tension of water. This allows the water to go through materials more easily and makes water a more effective agent to put out fires. This foam is made of water and additives formed around air bubbles, which makes it a lasting wet but light agent. Because of this trait, firefighters need less water to put out fires. It makes hose lines lighter and easier to manage. It also reduces water damage. The fire service is now supplying fire trucks with these materials. Improvements in this technology will result in wider use.

Improvements in Building Safety

Researchers are trying to reduce threats to buildings, building occupants, and first responders. The following are only some of the areas that are targeted for in-depth research.

1. **Increasing structural integrity. Structural integrity** is the engineering field that ensures that a building is designed and built for safe use under normal conditions and in an emergency. One can raise structural integrity through codes and standards, tools, and guidance to prevent building collapse. The collapse of the World Trade Center towers sparked a look at structural safety.

2. **Improving fire resistance.** Other countries use steel that is resistant to fire. The United States needs more testing of steel under building fire conditions to add fire-resistant steel to U.S. construction.
3. **Improving emergency exits and access.** The events of 9/11 showed the United States that we had to construct our buildings differently. Our models for the placement of doors, stairways, and elevators were inefficient. There will be studies to develop simulation tools. These tools will better capture the movement of people in a building under fire and other emergencies. Exit models have been created at the National Institute of Standards and Technology (NIST). These models help predict the movement of people in all types of situations.
4. **Cybernetic building systems.** Technology and rising costs of building construction will drive companies to develop building controls. These will improve performance and reduce costs.

Cybernetic building systems are control systems that combine building services, such as energy control (HVAC systems), fire and security systems, and building transport systems. These systems can also include earthquake detection and real-time checks of the building. Indoor air quality (IAQ) checks to find chemical, biologic, and radiologic agents inside the building are also included.

For the fire service, these systems provide on-site fire systems as well as faster fire department alerts. As these systems improve, so will the need for the fire service to understand and respond to these systems.

FIGURE 1-5 Fire service drone.

FIGURE 1-6 Robot designed for fire suppression.

Fire Service of the Future

The fire service must prepare for the challenges of this century. Climate change is increasing the number and strength of wildfires in the United States. The fire service needs technology to protect the lives of the public and fire service.

Emergency staff are training drone operators to work in disaster zones. These areas are too dangerous for firefighters to enter. Drones monitor the air, find locations, and report conditions. They can also take photos of the scene and map the area (**FIGURE 1-5**).

During natural disasters, roads may become so damaged that drones may be the best way to locate victims and gather data. Combined with thermal imaging cameras, drones can find victims and bodies. Combined with gas air monitors, the drones can check the quality of breathable air in the area as well as the presence of toxic air. To assess damage, drones can map locations and areas of concern. These data can be shared on websites. Evacuees can see their property without entering the disaster zone.

The use of robotics to improve firefighting is also being advanced. Today, we have robotics applied to fire prevention sprinkler systems. The U.S. Navy has developed a humanoid robot for combating shipboard fires, the Tactical Hazardous Operations Robot (THOR). It was created for their Shipboard Autonomous Firefighting Robot (SAFFiR) program. Today's robots, such as the Turbine Aided Firefighting machine (TAF 20), can enter small confinement danger zones. Others, such as Fire Ox, can carry their own water supply. These robots can scout the conditions of an area that would present a danger to firefighters. Sending robots into these areas lowers the risk to firefighters (**FIGURE 1-6**).

The Internet of Things (IoT) is another technology that is being used by the fire service. The IoT is the idea that almost anything can be connected to the Internet. Networks can transmit data from IoT sensors to aid in fire prevention and response.

IoT sensors in buildings can monitor electrical systems and find any increase in heat. An alert can then spark inspections. Some IoT sensors can signal their own upkeep and need for repair.

If a fire breaks out, IoT sensors can warn firefighters where the fire is. They can inform the firefighters of the fire's strength and spread. They can also tell if people in the building are near the fire. These data help firefighters decide on the needed equipment and approach. They can help streamline evacuation and save lives.

Companies are also adding IoT technology to firefighters' PPE. In 2021, a firefighter helmet with IoT that uses cellular and Wi-Fi technology was released. It has a thermal imaging camera in the helmet to give firefighters a better idea of where they are (**FIGURE 1-7**). Data are sent to a tablet for the IC to lead efforts to put out the fire. The cloud offers incident analysis, both in real time and after the incident.

FIGURE 1-7 Normal and enhanced vision of the area of origin.

Courtesy of Qwake Technologies.

Technology that was once science fiction, such as drones, robots, and futuristic helmets, now has a place in firefighting. The future of technology is only limited to our imagination.

CASE STUDY CONCLUSION

1. **What effects would the time of day have on the response to the earthquake and fire?**

   Time of day has an impact on any incident. In this situation, many people, including the military personnel, would have been asleep.

2. **What were the infrastructure issues at this incident?**

   At that time, many of the structures were Type 3 masonry, Type 4 heavy timber, or Type 5 wood construction, which burned quickly. The gas and water supply lines had ruptured.

3. **What impact did the loss of water supply have after the earthquake?**

   The loss of the water supply meant there was not enough water to fight the fires that had broken out throughout the city.

4. **How could the responding military personnel have handled this incident better?**

   If the military personnel had known how fuel size affected ignition, they would not have used explosives.

WRAP-UP

SUMMARY

- The culture and history of the United States have shaped the American fire service into what it is today.
- Since the colonial settlements in North America, fire has been a big part of U.S. history.
- The growth in building construction during the Industrial Revolution led to poorly constructed and unplanned developments in the cities.

- Corruption in building departments led to poor construction and little or no enforcement of fire codes, which resulted in large fires that destroyed whole cities.
  - 1871 fire in northeastern Wisconsin
  - 1871 Great Chicago Fire
- Five of the most significant fires in the United States happened during "the Decade of Conflagrations," the years 1900 to 1909.
- The 1906 earthquake and fire in San Francisco, California, brought the need for fire-resistive construction to the forefront of concern.
- Generally, fire trucks are in one of three categories: pumpers, turntable ladders, and specialized fire trucks.
- The recommendations from the 1973 presidential report, *America Burning*, resulted in the creation of the U.S. Fire Administration and the National Fire Academy.
- Over time, the fire service has applied new technology, improved the training and education of firefighters, improved firefighting equipment, and developed an Incident Command System.

KEY TERMS

alternative fuel Any energy source other than the hydrocarbon fuels.

America Burning The 1973 presidential committee report on the U.S. fire service.

America Burning Recommissioned, America at Risk The 2000 revisit of both the original *America Burning* report (1973) and the *America Burning Revisited* report (1987).

America Burning Revisited The 1987 revisit of the *America Burning* report to review the progress on the recommended improvements.

British thermal unit (BTU) Unit of measure for the heat energy needed to raise the temperature of 1 pound of water by 1 degree Fahrenheit.

compressed air foam system (CAFS) A self-contained, stored-energy fire suppression unit that supplies compressed air foam.

circumference The length around the circle.

conflagration A fire with building-to-building flame spread over a great distance.

cybernetic building system Control system that combines building services, such as energy control (HVAC systems), fire and security systems, building transport systems, earthquake detection, and real-time checks of a building.

diameter The distance from one side of a circle to the other.

fire modeling A simulation used to figure out a fire's outcome.

flashover fire A fire that happens when all the contents of a room or enclosed space reach their ignition temperature and explode in fire.

global positioning system (GPS) A navigation system that uses satellites in space to map locations on Earth.

halogenated extinguishing agent An extinguishing agent that has one or more halogen atoms added to a hydrocarbon after removing the hydrogen atom from the hydrocarbon molecule.

Incident Command System (ICS) A management system used on the emergency scene to keep order.

Insurance Services Office (ISO) An agency funded by the insurance companies to apply a rating schedule to cities and fire departments and set the rate for fire insurance costs.

National Incident Management System (NIMS) A management system adopted by FEMA that combines resources from public and private agencies.

radius The length from the middle of the circle to the outside of the circle.

span of control The number of people or resources that one supervisor can manage during an incident.

structural integrity The engineering field that ensures a building is designed and built for safe use under normal conditions and in an emergency.

surfactant A soap material that reduces the surface tension of water, allowing the water to go through materials more easily.

thermal imaging A tool that uses infrared waves to find heat given off by a substance or body.

ultrafine water mist Water dispensed under high pressure sent through very fine nozzle outlets that create nearly microsized droplets.

wildland/urban interface The line, area, or zone where structures and other human developments merge with wildland or fuels from vegetation.

REVIEW QUESTIONS

1. What are the four main public concerns shown during "the Decade of Conflagrations" that apply to today's fire service?
2. What are the five differences between the United States and other industrialized countries that add to higher fire losses in the United States?
3. Describe the differences between the Incident Command System (ICS) and the National Incident Management System (NIMS).
4. How did the World Trade Center study help to improve building design and fire codes?
5. How does a global positioning system help the fire service?

DISCUSSION QUESTIONS

1. How have the lessons learned from U.S. fires led to new training standards, policies, and procedures?
2. How has the advancement of fire trucks led to more efficient firefighting?
3. What building construction issues have led to stricter building and fire codes?
4. How have new technologies impacted the fire service?
5. What areas of the fire service do you think technology will impact in the future?

APPLYING THE CONCEPTS

Loma Prieta Earthquake, 1989

On October 17, 1989, at 5:04 PM, a 6.9 magnitude earthquake occurred on the San Andreas fault. This caused building collapse, water main breakage, and instability in the soil of the Marina District in San Francisco.

This earthquake affected the same areas as the 1906 earthquake and resulted in 63 deaths and 3,757 injuries. This incident also caused a collapse of the Cypress Street Viaduct in Oakland and stopped the World Series baseball game at Candlestick Park in south San Francisco.

Compared to the 1906 earthquake, the 1989 earthquake was of equal magnitude and destruction. This earthquake also resulted in water main breaks and fires, but the outcome was different because of how the incident was managed.

1. What were some of the factors that helped manage this incident that were not present during the 1906 earthquake and fire?
2. How did technology improve the water delivery in this incident?
3. With the ensuing fire, how did the updated building and fire codes improve the ability of people to survive this incident?
4. What lessons did we learn from the neighboring Oakland Cypress Street Viaduct collapse?

REFERENCES

Airbus/Secure Land Communications. "Fire and Rescue: Modern Communications for Fire and Rescue Services." n.d. Accessed December 22, 2022. https://www.securelandcommunications.com/segments/fire-and-rescue.

Angle, James S., Michael F. Gala, Jr, T. David Harlow, William B. Lombardo, Craig M. Maciuba. *Firefighting Strategies and Tactics*, 4th ed. Burlington, MA: Jones & Bartlett Learning, 2020.

Avsec, Robert. "Wave of the Future: Electric Fire Pumpers Are More Than Simply Green Technology." FireRescue1. August 2021. Accessed December 22, 2022. https://www.firerescue1.com/fire-products/fire-apparatus/articles/wave-of-the-future-electric-fire-pumpers-are-more-than-simply-green-technology-EMnQvN3wonDMVRfh/.

Baker, Christopher. "Wingspread Conference Reports." National Fire Heritage Center. April 8, 2020. Accessed December 22, 2022. https://fireheritageusa.org/archives-news/the-wingspread-conference-reports.

BehrTech. "3 Remarkable IoT Applications for Fire Safety." n.d. Accessed December 22, 2022. https://behrtech.com/blog/3-remarkable-iot-applications-for-fire-safety/.

Burner Fire Control. "Compressed Air Foam Systems (CAFS)." n.d. Accessed February 27, 2023. https://www.burnerfire.com/equipment/cafs-single-agent/.

Campbell, Richard, and Shelby Hall. *United States Firefighter Injuries in 2021*. National Fire Protection Association (NFPA). December 2022. Accessed April 7, 2023. https://www.nfpa.org/-/media

/Files/News-and-Research/Fire-statistics-and-reports/Emergency-responders/osffinjuries.pdf.

Centers for Disease Control and Prevention (CDC). *Findings from a Study of Cancer among U.S. Fire Fighters.* July 2016. Accessed December 22, 2022. https://www.cdc.gov/niosh/pgms/worknotify/pdfs/ff-cancer-factsheet-final-508.pdf.

Corley, Gene, Ronald Hamburger, and Therese McAllister. 2002. "Executive Summary." In *World Trade Center Building Performance Study: Data Collection, Preliminary Observations, and Recommendations,* edited by Therese McAllister, 1–7. Washington, DC: Federal Emergency Management Agency. https://www.fema.gov/pdf/library/fema403_execsum.pdf

Delmar Cengage Learning. *The Firefighter's Handbook: Firefighting and Emergency Response,* 3d ed. Clifton Park, NY: Delmar Cengage Learning, 2008.

Everyone Goes Home. "Firefighter Life Safety Initiatives." n.d. Accessed December 22, 2022. https://www.everyonegoeshome.com/16-initiatives/.

Fahy, Rita R., and Jay T. Petrillo. *Firefighter Fatalities in the US in 2021.* National Fire Protection Association (NFPA). August 2022. Accessed April 7, 2023. https://www.nfpa.org/-/media/Files/News-and-Research/Fire-statistics-and-reports/Emergency-responders/osFFF.pdf.

FEMA. *America Burning: The Report of the National Commission on Fire Prevention and Control.* Washington, DC: FEMA, 1973.

FEMA. *America Burning Revisited.* Washington, DC: FEMA, 1987.

FEMA. *America Burning Recommissioned, America at Risk: Findings and Recommendations on the Role of the Fire Service in the Prevention and Control of Risks in America.* Washington, DC: FEMA, 2000.

FEMA. "Assistance to Firefighters Grants Program." Last updated April 14, 2023. Accessed April 28, 2023. https://www.fema.gov/grants/preparedness/firefighters.

FEMA. "Fire Death Rate Trends: An International Perspective." Topical Fire Report Series 12, no. 8. (2011), 1–8. Accessed December 22, 2022. https://www.usfa.fema.gov/downloads/pdf/statistics/v12i8.pdf.

Federal Emergency Management Agency Emergency Management Institute. "National Incident Management System (NIMS) 2017 Learning Materials." FEMA Emergency Management Institute. 2017. https://training.fema.gov/nims/docs/nims.2017.instructor%20student%20learning%20materials.pdf

Fire Apparatus & Emergency Equipment/FDIC International. "SCBA Technology Gives Firefighters More Information, Greater Situational Awareness, Greater Protection." 2019. Accessed December 22, 2022. https://www.fireapparatusmagazine.com/ppe/scba-technology-gives-firefighters-more-information-greater-situational-awareness-greater-protection/#gref.

Firefighter Nation. "Wingspread V1 Conference Report." 2017. Accessed December 22, 2022. https://www.firefighternation.com/firerescue/wingspread-vi-conference-report/#gref.

FireRescue1. "Firefighter Safety Research Initiative Receives Federal Funding for 10th Consecutive Year." August 2015. Accessed December 22, 2022. https://www.firerescue1.com/fire-grants/articles/firefighter-safety-research-institute-receives-federal-funding-for-10th-consecutive-year-4XpnbA1fQk4ZYVNP/.

Gagnon, Robert M. *Design of Special Hazard and Fire Alarm Systems,* 2nd ed. Clifton Park, NY: Delmar Cengage Learning, 2008.

Grand View Research. "Plastic Furniture Market Size, Share, & Trends Analysis Report by Application (Residential, Commercial), by Distribution Channel (Online, Offline), by Region, and Segment Forecasts, 2019–2025." 2019. Accessed March 13, 2023. https://www.grandviewresearch.com/industry-analysis/plastic-furniture-market.

Havens, Nate. "The Exciting Technologies Revolutionizing Firefighting in 2022." The Last Mile. January 27, 2022. Accessed December 22, 2022. https://thelastmile.gotennapro.com/the-exciting-technologies-revolutionizing-firefighting-in-2022/.

International Fire Fighter. "Firefighter Fatalities: An Increase in Firefighter Fatalities over the Last Two Years Identified by NFPA and a Perspective from Around the Globe." June 2, 2023. Accessed April 7, 2023. https://iffmag.com/firefighter-fatalities-an-increase-in-firefighter-fatalities-over-the-last-two-years-identified-by-nfpa-and-a-perspective-from-around-the-globe/.

Manning, Bill (ed.). "Around the Fire Service, 1879–1889," *Fire Engineering* (February 1997), 57–61.

Manning, Bill (ed.). "Around the Fire Service, 1909–1919," *Fire Engineering* (June 1997), 74–79.

Manning, Bill (ed.). "Around the Fire Service, 1930–1939," *Fire Engineering* (July 1997), 79–84.

Manning, Bill (ed.). "Around the Fire Service, 1950–1959," *Fire Engineering* (September 1997).

Manning, Bill (ed.). "History of the Fire Service, the Last 120 Years," *Fire Engineering* (September 2000).

Midwest Fire. "The History of the Fire Truck." 2017. Accessed December 22, 2022. https://midwestfire.com/history-fire-truck/.

National Fire Protection Association (NFPA). *Emergency Medical Services NFPA 2020 Report.* 2020. Accessed January 9, 2023. https://www.nfpa.org/assets/files/AboutTheCodes/450/450_F2020_EMS_AAA_SD_SRstatements.pdf.

National Institute for Standards and Technology (NIST). "Virtual Cybernetic Building Testbed." May 16, 2014; last updated September 26, 2017. Accessed May 3, 2023 https://www.nist.gov/laboratories/tools-instruments/virtual-cybernetic-building-testbed.

National Park Service (NPS). n.d. *1906 Earthquake: The U.S. Army's Role.* Accessed December 22, 2022. https://www.nps.gov/goga/planyourvisit/upload/sb-1906-earthquake.pdf.

O'Mara, Kelly. "Large Parts of the Bay Area Are Built on Fill. Why and Where?" February 6, 2020. Accessed December 22, 2022. https://www.kqed.org/news/11799297/large-parts-of-the-bay-area-are-built-on-fill-why-and-where.

Patrascu, Daniel. "Fire Truck History." Autoevolution. May 30, 2009. Accessed December 22, 2022. https://www.autoevolution.com/news/fire-truck-history-7249.html.

Rackspace Technology. "The First Ever IoT-Enabled Firefighter Helmet Provides 'Superpowers' Through Augmented Reality." March 2021. Accessed December 22, 2022. https://www.rackspace.com/newsroom/first-ever-iot-enabled-firefighter-helmet-provides-superpowers-through-augmented-reality.

REFERENCES CONTINUED

Safe Piping Matters. "Building Materials Matter: Burdened by Facility Infrastructure." June 2022. Accessed December 22, 2022. https://safepipingmatters.org/building-materials-matter-firefighters-burdened-by-facility-infrastructure/.

Schaenman, Philip. *International Concepts in Fire Protection: Ideas That Could Improve U.S. Fire Safety.* Arlington, VA: Tri-Data Corporation, 1982.

Schaenman, Philip. *International Concepts in Fire Prevention.* Arlington, VA: Tri-Data Corporation, 1993.

Schaenman, Philip. *Profile of the Urban Fire Problem in the United States.* Arlington, VA: Tri-Data Corporation, 1999.

Society of Fire Protection Engineers. *Fire Protection Engineering.* https://www.sfpe.org/publications#Guides.

Wendt, Ronnie. "PPE Gets Smart." Industrial Fire World. March 2021. Accessed December 22, 2022. https://www.industrialfireworld.com/596880/ppe-gets-smart.

The following reports as well as many others are available to members of the fire service at *https:/www.usfa.dhs.govlapplications/publications/.*

Goldstein, Phil. "Fire Technology in Smart Cities and Beyond: How IoT Helps Fight Fires." StateTech. August 2020. Accessed December 22, 2022. https://statetechmagazine.com/article/2020/08/fire-technology-smart-cities-and-beyond-how-iot-helps-fight-fires-perfcon.

U.S. Naval Office of Information. "Shipboard Autonomous Firefighting Robot (SAFFIR)." October 20, 2021. Accessed February 27, 2023. https://www.navy.mil/Resources/Fact-Files/Display-FactFiles/Article/2160601/shipboard-autonomous-firefighting-robot-saffir/.

Vergun, David. "Department Uses Thermal Imaging to Detect COVID-19." U.S. Department of Defense/DOD News. May 2020. Accessed December 22, 2022. https://www.defense.gov/News/News-Stories/Article/Article/2178320/department-uses-thermal-imaging-to-detect-covid-19/.

Verisk. "Fire Suppression Rating Schedule (FSRS) Overview." n.d. Accessed May 5, 2023. https://www.isomitigation.com/ppc/fsrs/.

VSTEP Simulation. *The Benefits of Virtual Training for Firefighters.* Accessed December 22, 2022. https://www.vstepsimulation.com/assets/uploads/2020/04/White-Paper-The-benefits-of-virtual-training.pdf.

CHAPTER

2

Fire Chemistry

LEARNING OBJECTIVES

After studying this chapter, you will be able to:

- Understand and explain the basic structure of atoms.
- Explain how atomic structure determines the behavior of elements and compounds.
- Understand basic chemical and physical properties and concepts and how they influence the behavior of materials involved in fires and hazardous materials incidents.
- Correlate chemical structure with chemical names to predict some hazardous chemical behaviors.
- Understand key physical properties of chemicals and how these properties relate to fire protection.

CASE STUDY

Dixie Cold Storage 1984

At 2:45 PM on September 17, 1984, employees found an anhydrous ammonia leak at the Dixie Cold Storage plant in Shreveport, Louisiana. Anhydrous means without water. Anhydrous ammonia is used as a refrigerant and fertilizer. It is colorless and has a strong odor of ammonia. Its exact flash point has not yet been determined. It can be extremely flammable in the presence of an oxidizing agent. The cold storage building was a concrete tilt-up construction that stored refrigeration processing equipment and cold storage lockers.

The employees notified the fire department when the leak was discovered. The fire department sent units in to investigate. Two hazardous materials (hazmat) team members were sent to shut off the flow of anhydrous ammonia to the area. Other responding units remained outside.

At approximately 4:05 PM, the ammonia reached an ignition source and ignited. The explosion that followed killed one firefighter and the other had burns over 72% of his body.

1. In 1984, whom could the firefighters have consulted before entering the building?
2. What could the firefighters have done to protect themselves before entering the building?
3. Why is the relationship between the anhydrous ammonia and the conditions of the building of vital importance?

Introduction

This chapter reviews basic chemistry and physical processes. These impact the chemical and physical properties of materials. Firefighters are faced with an increasing use of hazardous materials. As a result, there are more chances for leaks, spills, and incidents involving these materials. Firefighters must understand these chemicals and their reactions in the fire combustion process.

Firefighters sometimes encounter a chemical reaction that leads to the release of energy or toxic vapors. Such incidents can threaten the lives of first responders and the public. This chapter prepares first responders by giving definitions, basic concepts, and descriptions of the physical and chemical processes of fires and chemicals that firefighters may encounter.

Atoms

Atoms are the basic building blocks and the smallest units of an element that takes part in a chemical reaction. Atoms of almost every element can combine with other atoms to form molecules. Molecules are two or more atoms tightly bound together by chemical bonds. For example, water (H_2O) is made up of two atoms of hydrogen and one atom of oxygen. They combine, creating a molecule that differs from the original atoms (hydrogen and oxygen).

Structure of the Atom

Atoms consist of three types of particles: protons, neutrons (except hydrogen), and electrons. The atom's identity as an element is defined as the number of protons in its nucleus. The **proton** is heavy and has a positive electrical charge. The **neutron** is equal in weight to the proton and has no electrical charge. Outside the nucleus, spinning around it in orbit, are the **electrons**. These are light and carry a negative charge. The structure of an atom is shown in **FIGURE 2-1**.

Bonding of Atoms

In a normal state, the electrical charge of the atom as a whole is zero. This means an atom contains the same

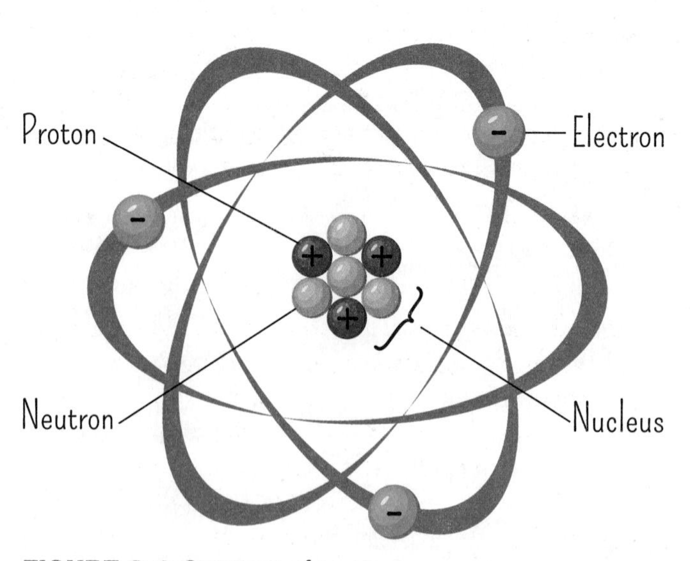

FIGURE 2-1 Structure of an atom.

number of electrons as protons. The atom is now said to be balanced. Electron orbits are arranged in layers, or shells. An atom can have an infinite number of shells. The inner shell is full when it contains two electrons and does not seek more. The number of electrons in the outer shell of an atom is important. Atoms combine with other atoms to fill their outer electron shells or rid themselves of extra electrons. Many large atoms have quite a few electrons, as many as 100 or more. Regardless of their size, the atom's behavior is guided primarily by the number of electrons in the outer shell. These electrons are called **valence electrons**. The valence is the outer orbiting shell that holds electrons. The number of electrons held in this valence is related to the column where the atom is found on the periodic table. The valence electrons can be found in columns 1 to 7. These electrons allow bonding with the other atoms in the periodic table.

The bonding of atoms follows the octet rule. The **octet rule** refers to the preference of atoms to have eight electrons in the outer valence shell. Atoms will share, gain, or lose electrons to obtain the same number of electrons as the noble gas near them in the periodic table. The noble gases have complete outer shells, meaning they have eight electrons in their outer valence shells. When atoms have fewer than eight electrons in their outer valence shells, they tend to react to form more stable compounds. Atoms that have eight electrons in their outer valence shells are very stable.

Sometimes forces outside an atom cause it to gain an extra electron. It now contains more negative electrons than protons, so the sum of the electrical charge is negative. The atom has become a **negative ion**, also called an anion. The atom is likely to combine with another atom that is missing electrons, a **positive ion**, also called a cation. It does this to return its overall charge to zero.

Atoms are held together with a chemical bond. A **chemical bond** is the attractive force that binds two atoms. This combination is stable at room temperature but becomes unstable at high temperatures. There are two main types of chemical bonds: ionic bonds and covalent bonds.

Ionic Bonds

An ionic bond occurs when an outer electron from an atom that is a nonmetal or salt is magnetically drawn away. This atom then attaches to that atom. An **ionic bond** is formed between two charged atoms, or ions, to create a molecule. Through a chemical process, one atom releases an electron from its outer valence, or shell, and another atom accepts the electron to its outer valence. This release and acceptance of electrons changes the atom to an ion. It is an ion because it no longer has an equal number of electrons and protons. One example is the combination of sodium (Na) and chlorine (Cl). Sodium releases an electron from its outer valence and becomes a positively charged ion (cation). Chlorine accepts the electron from sodium to its outer valence and becomes a negatively charged ion (anion). The compound formed is salt (NaCl) (**FIGURE 2-2**).

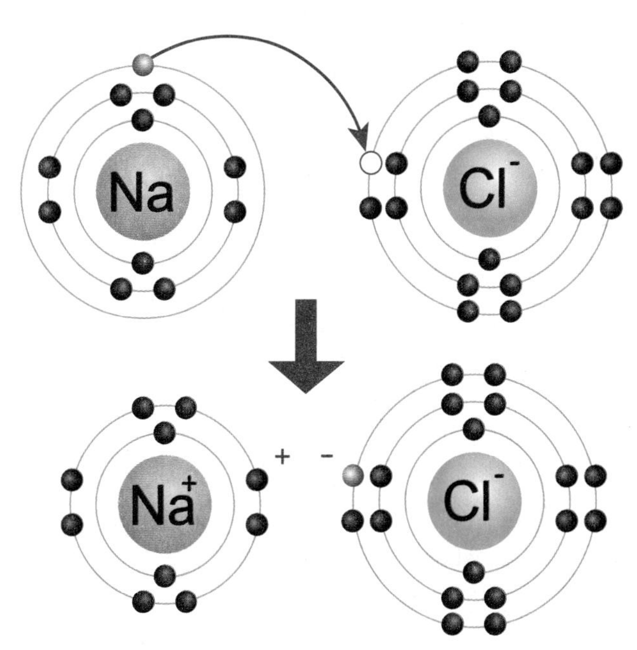

FIGURE 2-2 The bond between sodium and chlorine forms salt.

Covalent Bonds

A **covalent bond** is the sharing of electron pairs by combined atom(s). This type of bond occurs between two atoms of the same element or between elements near each other in the periodic table. This type of bonding usually occurs between nonmetals.

Examples of covalent bonds are water (H_2O) and methane gas (CH_4). Covalent bond types include single, double, and triple bonds (**FIGURE 2-3**).

- **Single Bonds**
  - A single bond is between two atoms in a molecule that share a single electron, such as a water molecule (H_2O), in which two hydrogen atoms each share one electron with an oxygen atom to form two single bonds.

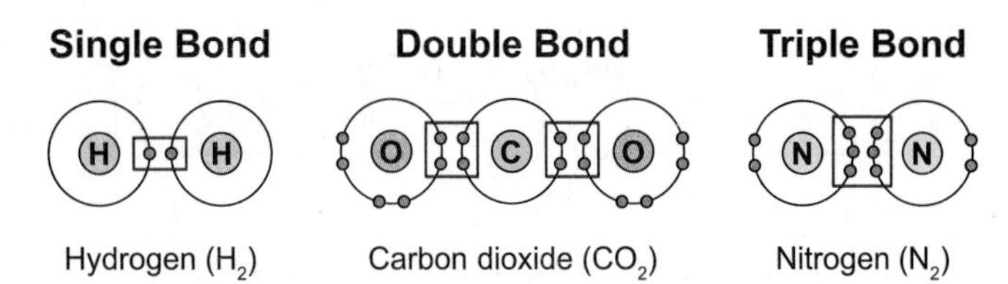

FIGURE 2-3 Single, double, and triple bonds.

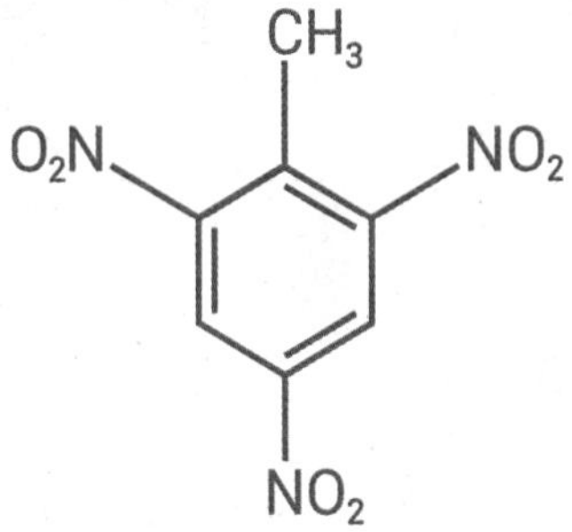

FIGURE 2-4 Chemical structure of trinitrotoluene (TNT), a type of resonant bond.

- **Double Bonds**
  - A double bond is formed when two atoms in a molecule share two electrons. Oxygen (O_2) is an example of a double bond.
- **Triple Bonds**
  - A triple bond is formed when two atoms in a molecule share three electrons. An example is cyanide (CN).
- **Resonant Bonds**
  - Another important bond in chemistry is called a resonant bond. This bond forms when an electron holds six separate atoms together. An example is trinitrotoluene (TNT) (**FIGURE 2-4**).

Elements

Elements are the simplest form of matter. Each element is made of one or more atoms. Each also has a unique number of protons in its nucleus, which helps define what it is. All of the naturally occurring elements (the synthetic ones are not included here) exist in one of three physical forms: solid, liquid, or gas. Most elements naturally occur as solids, with only two or three as liquids, and a handful as gases. Elements can be grouped into families by their similar properties. For example, atoms of the halogen elements may be attached to other atoms to form different compounds called halogenated compounds. **Halogenated compounds** are formed from any chemical reaction in which one or more halogenated atoms are combined with an existing compound. This changes the properties of the material. In this grouping of compounds, we find chlorine, bromine, and fluorine to be examples of halogenated compounds. These compounds have efficient and effective fire-extinguishing capabilities but have other traits that have resulted in the banning of their use. See Chapter 4 for further discussion of halogenated extinguishing agents and their replacements, the halocarbon extinguishing agents.

Periodic Table

The periodic table shows an arrangement of elements. It was introduced in 1869 by the Russian chemist Dmitri Mendeleev. Mendeleev wanted to classify the chemical properties of the elements he knew, so he listed them according to increasing atomic number. **Atomic number** is the number of protons in an element. The **atomic weight**, or mass, is the average number of protons and neutrons in atoms of a chemical element. Mendeleev was able to show how the elements could be arranged in a chart format based on blocks.

The individual blocks contain four main parts: the name of the element, the atomic number, the atomic weight or mass, and the element's symbol.

Element Symbols

Symbols are used to allow us to write out chemical formulas easily. Each chemical has been assigned a one- or two-letter abbreviation. The symbols are made up of a principal letter or letters from the name of the element. The element's symbol often represents the first letter of the element's name. For example, hydrogen is listed as H.

Some elements are identified with symbols representing the Latin or Greek names given to the element. For example, iron is listed as Fe. Fe stands for the word ferrum, which means iron in Latin. Note the following examples:

copper (cuprum)—Cu
gold (aurum)—Au
iron (ferrum)—Fe
lead (plumbum)—Pb
mercury (hydrargyrum)—Hg
potassium (kalium)—K
silver (argentum)—Ag
sodium (natrium)—Na
tin (stannum)—Sn

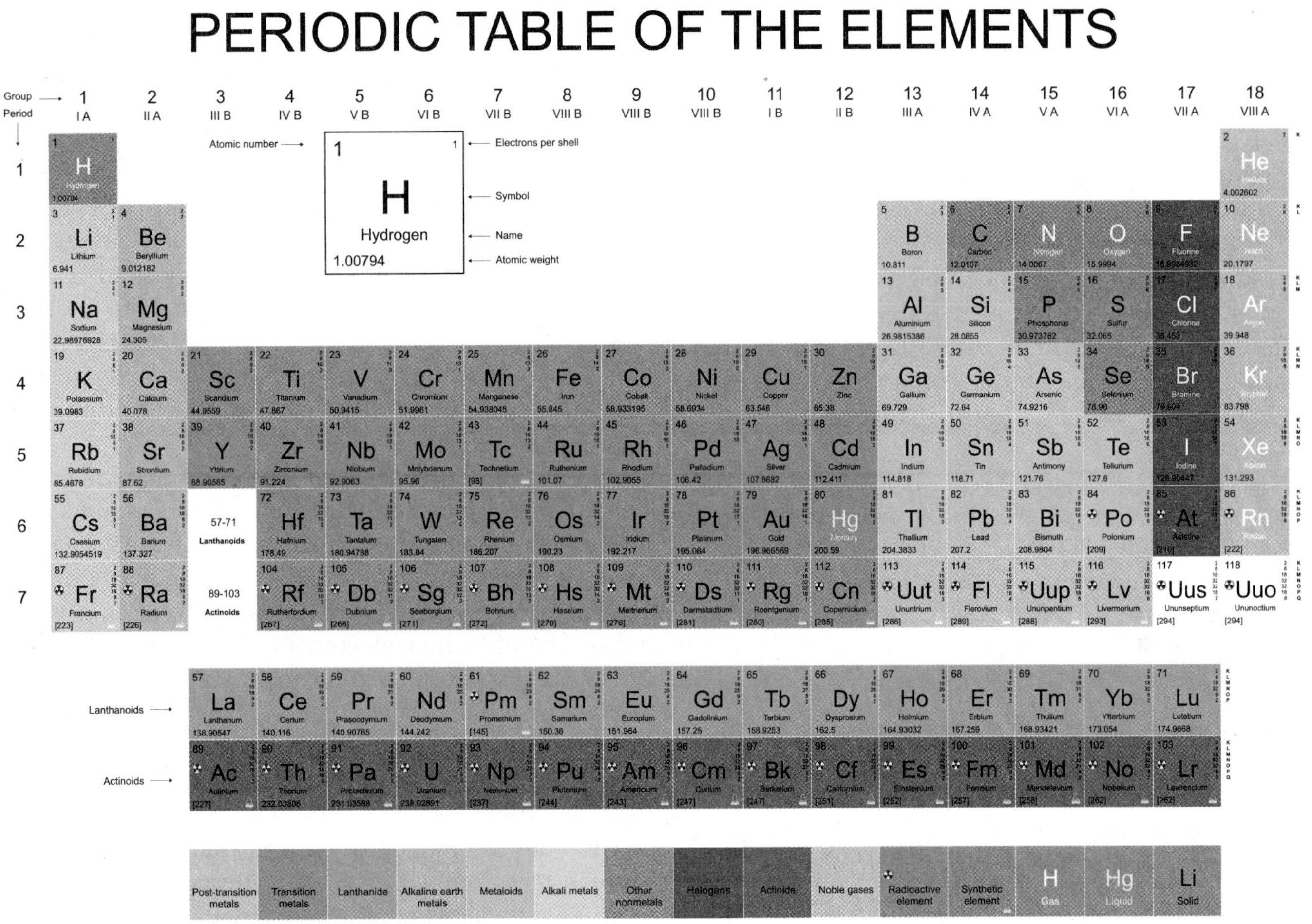

PERIODIC TABLE OF THE ELEMENTS

FIGURE 2-5 The periodic table.

It is important to note that the symbols almost always consist of an uppercase letter or a capital letter and lowercase letter. This helps identify the individual elements in a chemical formula.

The Organization of the Periodic Table

The periodic table shows 90 natural elements and 28 synthetic elements for a total of 118 elements (**FIGURE 2-5**). The elements listed in the table are found in nature. Those that are synthetic are listed below the table. The synthetic elements have a radioactive property not found in the natural elements. All matter found on Earth can be identified through the periodic table.

The table is organized into periods, groups, and families. A **period** is a horizontal row of elements that identifies how many orbits revolve around the nucleus of the atom. A **group** is a vertical column of elements. There are 18 groups (denoted by the numbers at the top of each column), and the elements in each group have the same number of valence electrons. As a result, elements in the same group often display similar properties and reactivity.

Families in the periodic table also represent groups of elements with like properties. The five main families include alkali metals, alkaline earth metals, transition metals, halogens, and noble gases. Many of these families represent a single group. However, not all of the families overlap with the periodic table groups. The transition metals consist of all the elements from group 3 to group 12.

Molecules

As mentioned earlier, molecules are two or more atoms tightly bound together by chemical bonds. Molecules are a neutral, non-ionic species that can exist in all three physical states of matter. Molecules are in constant motion. Some move rapidly and some slowly, depending on the state of the material. For example, molecules in solid materials move slowly, while they move faster in liquids. They move even more rapidly in gases. Applying heat also increases this movement.

When a **physical change** occurs, such as a change from solid to liquid or from liquid to gas, molecules of a material are not changed by the process and remain intact. When a **chemical change** occurs, molecules are altered. These altered molecules do not have the same properties as those of the original.

One of the best and most commonly used examples of a physical change is water. At normal atmospheric pressure and temperature above 32°F (0°C), water is a liquid. When the temperature of water falls below 32°F (0°C) and the pressure remains the same, water changes state and becomes ice. At temperatures above the boiling point of 212°F (100°C), water changes state to a gas called steam.

A chemical change occurs when cement, water, sand, and stone are mixed in the correct proportion. When the cement heats up, undergoing a chemical change, it drives off the water. The new material is concrete.

Structure of Molecules

A **molecule** is the smallest unit of an element or compound that keeps the chemical traits of the original material. Atoms are smaller and joined together using a form of electricity to bond. **FIGURE 2-6** shows a piece of wood that is subdivided into a cell, molecules, and the cell wall. The cell wall contains the molecules and compounds that make up the wood. Note the complexity and interworking relationships needed to produce wood.

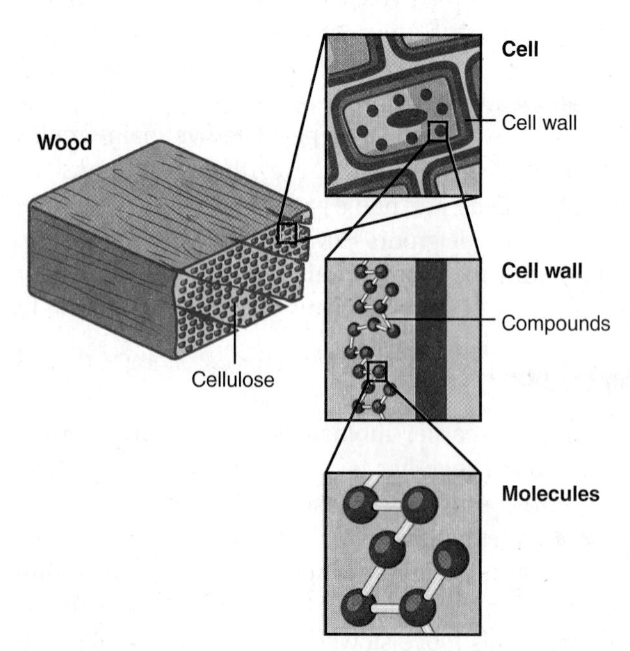

FIGURE 2-6 Structure of a molecule showing wood subdivided into the cell, molecules, and compounds.

Compounds

In chemistry, a **compound** is a substance formed from two or more elements joined in a fixed ratio. In general, the ratio must be fixed due to its **physical property**. This refers to the ball or chain configuration of the compound. This is a trait that is constant to the compound. The defining characteristic of a compound is its chemical formula. Formulas describe the number of atoms in a substance. For example, with H_2O, there are two hydrogen atoms and one oxygen atom. This is a fixed ratio of 2:1. It cannot be changed. Because a compound is always made of the same atoms in the same ratio, it always has the same properties.

A compound consists of molecules that are combined chemically. They are homogeneous and have a definite composition regardless of origin, location, size, or shape. The elements in a compound cannot be pulled apart by any physical means. Compounds are more abundant than elements. It is difficult to say how many exist because new ones are created each day.

An example of a compound is any material composed mainly of cellulose. Cellulose consists of carbon, hydrogen, and oxygen with smaller amounts of nitrogen and other elements. Cellulose is generally agreed to be 6 carbon atoms, 10 hydrogen atoms, and 5 oxygen atoms, represented by $C_6H_{10}O_5$. This combination can vary slightly depending on the amount of oil and the density of the material.

All compounds will break up into smaller compounds or individual atoms when subjected to heat. The temperature to which the material must be heated for it to decompose is called the **decomposition temperature**. These smaller compounds cannot be put back together to re-form the original compound. For example, when wood is burned, you cannot collect the ash, charcoal, and vapors released during combustion to return it to its original form before combustion. **Combustion** is a chemical reaction that creates heat and light.

Mixtures

Most natural forms of matter are mixtures of pure substances. A **mixture** is a combination of substances held together by physical rather than chemical means. Another way to think of it is as a solution of two or more different types of molecules. Soil and rock, plants, animals, coal, oil, air, and cooking gas are all mixtures. Mixtures differ from compounds because ingredients of a mixture retain their own properties, meaning that the substance has not been changed to form the mixture.

Thus, mixtures can be separated from other ingredients by physical means. An example of a mixture is a fruit salad. The combination of apples, oranges, and walnuts with dressing forms one salad. While they are combined, each of the three components retains its own properties. Another example is saltwater. When heat is applied to saltwater, the water will reach its boiling point and turn to steam. This steam can be captured and cooled to form water in its liquid state. Once the water has been boiled away, salt crystals remain. You can add the salt crystal to the water, and it will re-form the saltwater mixture.

Chemical Names

Each chemical element has a unique name as an identifier. Names complying with the International Union of Pure and Applied Chemists guidelines are those responders should use when looking for information about a product. The technical and proper shipping names can be found in the Department of Transportation (DOT) 49 CFR Section 172.101, Table of Hazardous Materials.

For international trade, the official names of the chemical elements are decided by the International Union of Pure and Applied Chemistry (IUPAC). IUPAC uses British spelling, and there have been a few changes. For example, *aluminium* and *caesium* replaced the U.S. spellings *aluminum* and *cesium*, and the U.S. spelling of *sulfur* was replaced with the British spelling *sulphur*. These kinds of changes take place on a continuing basis as the global economy accommodates more countries. Firefighters need to continue updating their knowledge and information. Chapter 10 will discuss more about finding chemical names and identifying their hazards.

Prefixes and Suffixes

Prefixes (syllables added to the beginning of a word) and suffixes (syllables added to the end of a word) are used to describe atoms that are added to a basic molecule. Suffixes often refer to groups of atoms that are added and behave as a group, not as individual atoms. Firefighters also need to know that prefixes and suffixes are identifiers that provide warnings. They indicate a higher concentration of oxidizers in the material. Examples are the uses of the prefix per- and suffixes -ate and -ite. Chemical names with these additions will cause the fire and oxidation process to increase at a more rapid rate. **TABLE 2-1** shows some common substances with suffixes and prefixes that firefighters may encounter. For example, ammonium perchlorate contains four additional atoms of oxygen. This makes the substance highly unstable.

Organic Chemicals

The words organic and inorganic, although not part of chemical names, are often used to describe the makeup of compounds. Organic compounds contain some form of hydrocarbons—that is, a combination of carbon and hydrogen molecules. The numbers of carbon and hydrogen atoms determine the properties of the substance. More importantly, the combined number determines how the substance will react under varying conditions. Organic compounds may have one or many carbons linked in the makeup of the molecule.

Carbon is a remarkable element. Carbon atoms have four electrons in the outer ring, making it able to form multiple bonds. Carbon has a great affinity to form chains of atoms that may be straight, branched, or formed into rings. Because of these properties, carbon has formed nearly 10 million different compounds.

TABLE 2-1 Prefix and Suffix—Warning Signs of Oxidizing Agents

Substance	Chemical	Description	Prefix	Suffix
Chlor**ine**	Cl	Chlorine alone		**-ine**
Sodium chlor**ide**	NaCl	Chlorine and sodium (Na)		**-ide**
Sodium chlor**ite**	$NaClO_2$	Chlorine with sodium and two oxygen (O)		**-ite**
Sodium chlor**ate**	$NaClO_3$	Chlorine with sodium and three oxygen (O)		**-ate**
Ammonium **per**chlor**ate**	NH_4ClO_4	Ammonia with perchloric acid (four atoms of hydrogen and four atoms of oxygen: a powerful explosive)	**per-**	**-ate**

Data from Hess, Fred C. 1996. *Chemistry Made Simple*. New York: Doubleday Publishing.

This exact number is unknown due to the daily additions of new compounds that contain carbon.

Organic peroxides are important because they can explode when involved in the combustion process. The reason these substances pose a threat is found in their chemical structure. An oxidizing agent is a substance that gains electrons in a reduction–oxidation reaction, or redox reaction, while the reducer loses an electron. **Redox** reactions involve the loss of electrons, which speeds up the oxidation process, and a gain of electrons, which slows the oxidation process. When these reactions occur, organic peroxides are chemically bonded with an oxidizing agent or additional oxygen. The oxidizing agent provides the oxygen, and the reducer breaks the material down, allowing it to become further oxidized. When an oxidizing agent is heated, a violent reaction may occur. For example, benzoyl peroxide in an undiluted form will ignite readily and burn rapidly, almost as rapidly as black powder. Heat will cause it to decompose rapidly, and if it is confined, it will explode. Decomposition can also start by a shock or heating by friction. This type of heating occurs by rubbing the surface of the material, which moves the molecules, creating heat.

Firefighters need to find facilities in their areas where these products are stored and used. They also need to use the hazardous materials warning label system to identify hazardous materials and then follow up with more detailed information on the specific material they may encounter (see Chapter 10).

Properties of Chemicals

Because all atoms of a certain element have the same structure, they also have the same properties. The structure and properties of atoms make their behavior predictable. This is a valuable tool for firefighters when they are working with hazardous chemicals.

Physical and chemical properties of elements and compounds determine their behavior. When assessing the properties of elements and compounds, it is important to note the temperature scale used. The **Fahrenheit scale** is used to describe the weather and ambient temperature during fire department operations. It is based on water freezing at 32° and boiling at 212°. It was established in 18th century by Daniel Fahrenheit, a German physicist. Other countries and most scientific references use the Celsius scale. The **Celsius scale**, also known as centigrade, is based on water freezing at 0° and boiling at 100°. It was founded by the Swedish physicist Andres Celsius. In this text, both scales will be used in examples and charts to help firefighters become accustomed to the Celsius scale (see Chapter 1 and Appendix A).

Boiling Point

The **boiling point (BP)** can be defined several ways. One explanation is the temperature at which the vapor pressure of the liquid exceeds the pressure of the atmosphere around it. This allows the liquid to convert to a vapor. Another way to explain the boiling point of a liquid chemical is the temperature at which the molecules in a liquid are heated, boil, and change to vapor. Once started, and as long as the heat is applied, this process will continue. **TABLE 2-2** provides a list of boiling points of specific liquids.

Vapor Pressure

Vapor pressure (VP) is the pressure placed on the inside of a closed container by the vapor in the space above the liquid. It takes energy to release the molecules through the surface of the liquid. This energy is in the form of heat. As the temperature of the liquid is raised, the molecules begin to move rapidly. Some are released at the surface. They will continue to be released more quickly as the temperature within the liquid increases.

Once the surface tension is broken, the molecules are released in the form of vapor. **Flash point** is the lowest temperature of a liquid at which vapors are produced and will ignite when an outside ignition source is present. Note that at the flash point, sustained combustion does not occur. Flash point varies depending on the pressure, the oxygen content, and the purity of the product.

TIP

The lower the flash point, the greater the fire hazard.

TABLE 2-2 Boiling Points

Substance	Temperature °C	Temperature °F
Benzene	80.1	176.18
Carbon dioxide	−78.5	−109
Chlorine	−34.1	−29
Hydrogen	−252.8	−423.17
Oxygen	−183.0	−297.3
Water	100.0	212

Another closely related concept is fire point. **Fire point** is the lowest temperature of a liquid sufficient to produce vapor that will ignite from an outside ignition source and sustain combustion. This is opposed to the instant flash of the flash point. Fire point is usually a few degrees higher than the rated flash point temperature. For example, the flash point of methanol is 54°F (12°C), while the fire point is usually 10°F to 30°F, or 5°C to 15°C higher than the flash point).

Two tests are commonly used to determine the flash point. The open cup test and the closed cup test are often used to find the lowest temperature that a volatile substance will turn into a flammable gas. In both tests, a source of ignition is used until there is a flash when the substance ignites.

The **open cup test** measures the release of the vapors in terms of the pressure being exerted at a specific temperature. The flammable vapors or molecules driven off the flammable liquid are not trapped by a lid. The temperature at which the vapors are released is higher than the temperature recorded using a closed cup test.

In the **closed cup test**, a lid is placed on the cup to keep the vapors in the cup. The sample is heated and stirred, and an ignition source is introduced at random times to find the flash point of the sample. A pressure reading (called a closed cup reading) is taken when the vapors are released from the liquid in large enough quantities to ignite. The devices used to perform this test are the Tag Closed Cup Test for liquids below 200°F (93°C) and the Pensky-Martens Closed Cup Apparatus for liquids above 200°F (93°C).

Open and closed cup ratings vary widely because of the pressure differences between the two test methods. In the closed cup test, a lid traps the molecules driven off by the heat of the product. This allows a lower temperature to be recorded as the flash point. The open cup test is affected by the outside temperatures, which require the higher ignition temperature. The closed cup test results are often used because they offer a larger safety factor.

Vapor Density

Vapor density (VD) is the mass of the given vapor in comparison to the same amount of air. Vapors with a vapor density greater than 1.0 will sink to the ground and may collect and displace breathing air. **Carbon dioxide** (CO_2) is 1.5 times as heavy as air. This means that a room filled with carbon dioxide will eventually have the lower portion blanketed with vapors at the floor level. The oxygen in the room will be replaced or forced upward. Firefighters will not be able to breathe if the entire room fills with carbon dioxide.

Another concern is when the gas is flammable. For example, butane gas is twice as heavy as air and has a flammability range of 1.9% to 8.5%. Flammable vapors heavier than air may pool at the lowest point, awaiting a source of ignition. The vapor density of butane coupled with its flammability range can present a serious explosion threat. This is a good example of how understanding chemical properties is relevant to assessing risks at an emergency incident scene.

Solubility

Solubility is the maximum amount of a substance that can be dissolved in another substance. Insoluble or slightly soluble materials will form a separate layer in another substance. They will either float or sink, depending on their specific gravity. This is called **insolubility**. Solubility is described with words and with numbers. Both indicate the percentage of the material that will dissolve.

Polar solvents are substances that allow firefighters' foam to be used on alcohol-based fires without breaking down the soap-based materials in the foaming agent. Alcohols and other polar solvents dissolve in water. This means that these materials may be diluted to a point that they will not burn.

In many cases, nonpolar solvents or hydrocarbon liquids that are not soluble in water will float on top of water. This presents firefighters with a serious fire hazard and safety problem. If they put out a fire in a hydrocarbon material floating on water, an ignition source can reignite the material. This is because hydrocarbons are highly flammable. The fire attack must be made from a place where firefighters cannot be surrounded with floating liquid that may reignite. For safety, firefighters should float a thick coating of foam over the top of the material. (See Chapter 4 for details on the use and application of foam.)

Specific Gravity (Water and Air)

Specific gravity is the density of the product divided by the density of water or air. Water and air are the standards that have been given the value of 1.0. Water is then measured against other liquids and air is compared against other gases. If the liquid being tested is heavier than water, it will sink to the bottom of the container. Firefighters fighting an oil tank fire must be aware that the water applied by hose lines may sink to the bottom of the tank and heat up as the fire warms the metal sides of the tank. This heat from the metal tank walls and the oil itself is transmitted to the water trapped below the oil at the bottom of the tank. Once the trapped water is

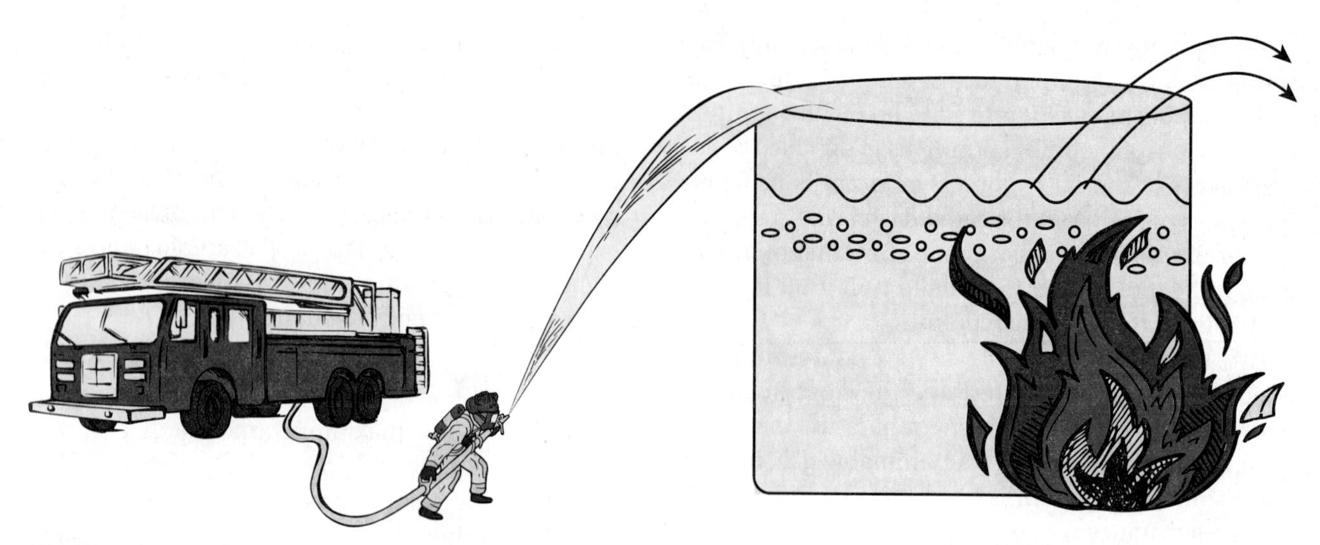

FIGURE 2-7 Boil over is the expulsion of the tank's contents by the expansion of water vapor that has been trapped under the oil and heated.

Courtesy of Lisa Zolfarelli, Graphic Designer; © Kate Macate/Shutterstock; © Artchi art/Shutterstock; © ArtMari/Shutterstock

heated to 212°F (100°C), it is converted to steam and expands, blowing the oil upward and out of the top. This is called a boil over and is a very dangerous safety problem for firefighters. **Boil over** is the expulsion of the tank's contents by the expansion of water vapor that has been trapped under the oil and heated by the burning oil and metal sides of the tank (**FIGURE 2-7**).

Specific gravity also allows firefighters to recognize the fuel hazard priority. Fuels that float on water will have a flammable property. Those that mix with water will be diluted, increasing their ignition temperatures. Fuels that sink will have a toxic property. This does not mean that these fuels do not share flammability and toxicity hazards. It helps to know which to be aware of for deciding the proper extinguishing agents.

The specific gravity of air is 1.0 at 70°F (21°C). Gases are measured against the air to determine if the gas is either heavier or lighter than air at the same temperature. This means that the gases that are heavier than air seek the lowest elevation in a floor area. For example, of the two **liquefied petroleum gases** (**LPGs**; a term given to butane and propane gases that have been pressurized and kept in a container in a liquid state), butane is 2.0 times heavier than air and propane is 1.5 times heavier. This means that firefighters can expect butane and propane vapors to seek the lowest areas and pool there. The fire code requires appliances with pilot lights or other sources of ignition to be above the floor area because these heavier-than-air gases pool in low spots.

On the other hand, natural gas (methane) is lighter than air. It has a specific gravity of 0.55 in relation to air, which is 1.0. This causes the vapors from methane to seek elevated locations in a confined room. Firefighters need to ventilate the higher elevations of a confined space first, if possible.

Knowing and understanding the specific gravity of the liquid helps firefighters to determine the best method to put out the fire. In the case of a gas, the best and safest means would be to move the gas to an open environment. **TABLE 2-3** shows specific gravities of selected liquids that first responders may encounter. **TABLE 2-4** shows examples and the vapor densities of gases that first responders may be encounter.

TABLE 2-3 Specific Gravities of Selected Liquids

Substance	Specific Gravity at 70°F/21°C	Heavier/Lighter Than Water
Acetic acid	1.05	Heavier than water
Allyl chloride	0.94	Lighter than water
Chlorobenzene	1.11	Heavier than water
Heptane	0.68	Lighter than water
Hydrochloric acid	1.19	Heavier than water
Nitric acid	1.50	1.5 times heavier than water
Sulfuric acid	1.84	1.75 times heavier than water

TABLE 2-4 Common Gases Encountered by First Responders—Heavier/Lighter Than Air

Substance	Vapor Density (Air = 1.00)	Heavier/Lighter Than Air
Acetylene	0.899	Lighter than air
Ammonia	0.589	Lighter than air
Carbon dioxide	1.52	1.5 times heavier than air
Chlorine	2.46	Almost 2.5 times heavier than air
Hydrogen	0.07	Lightest of all gases
Methane	0.553	Lighter than air
Nitrogen	0.969	Lighter than air
Oxygen	1.11	Heavier than air
Propane	1.52	1.5 times heavier than air
Sulfur dioxide	2.22	2 times heavier than air

The **auto-ignition temperature** is the temperature at which a material will ignite in the absence of any outside source of heat. We also refer to this temperature as the spontaneous ignition temperature, meaning that the material can be heated to a point where it self-ignites.

Explosive/Flammability Limits and Range

The **explosive/flammability range** is the range of concentrations of gases or materials (dusts) in the air that permit the material to burn. Flammability range is also the numerical difference between the flammable substance's lower and upper explosive limits in air. The lowest concentration of a substance (by percentage) in air that will burn is called the **lower explosive limit (LEL)**. The highest concentration of a substance (by percentage) in air that will burn is the **upper explosive limit (UEL)**.

The LEL creates an environment where the amount of flammable gas is less than the amount of oxygen in the air for the flammability range requirements. This produces a condition where combustion cannot occur. The environment would be described as "being too lean."

TABLE 2-5 Flammability Limits of Materials Encountered by Firefighters

Material	LOWER % to Air	UPPER % to Air	Result
Acetone	2.6	12.8	Explosion or fire
Butane	1.9	8.5	Explosion or fire
Kerosene	0.7	5	Explosion or fire
Methane	4.4	16.4	Explosion or fire
Gasoline	1.4	7.6	Explosion or fire
Carbon monoxide	12	75	Explosion or fire

The UEL produces an environment where the amount of flammable gas is greater than the amount of oxygen in the air for the flammability range requirements. Again, combustion will not occur. This environment would be described as "being too rich." **TABLE 2-5** shows a list of gases and their flammability limits.

Acetylene has a wide explosive range (2.5% to 82%) and a BTU output of 1,499 per cubic foot, while hydrogen gas has a narrower range (4% to 75%) with a lower BTU output of only 325 per cubic foot. Both present a serious fire and/or explosion hazard (NFPA 2022, Appendix A).

TIP

If the vapor concentration in air is less than the LEL or greater than the UEL, the material will not burn.

Hydrogen Ion Concentration

Hazardous materials can be divided into two groups: bases and acids. A **base** is a substance that will react with acids in aqueous solution to form salts, while releasing heat. An **acid** is a chemical that releases hydrogen ions when dissolved in water. It also has a pH less than 7. That means any product with a pH below 7 is an acid. The lower the pH, the stronger the acid; the higher the pH, the more corrosive the base (**FIGURE 2-8**). Materials with a pH of 2 or lower or of 12.5 or higher are classed as corrosive. These materials are specially marked because of their danger to people or the environment.

The pH of a chemical is a measure of acidity or alkalinity, where the number 7 is defined as the neutral

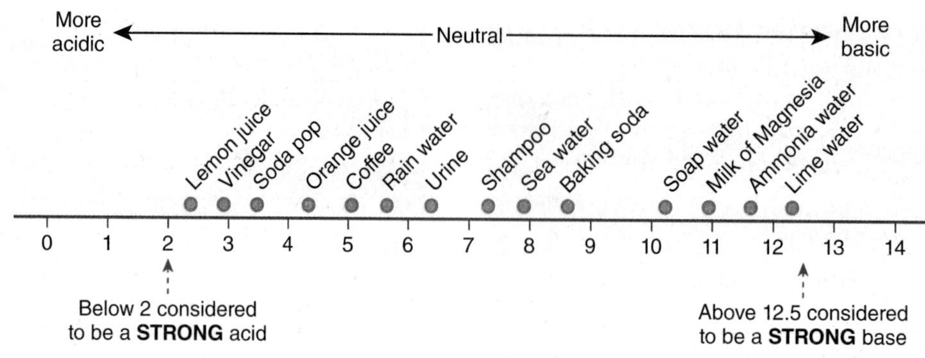

FIGURE 2-8 pH values of common substances.

point. **pH** is defined as a measure of a substance's ability to react as an acid (low pH) or as an alkali (high pH). The acidity or alkalinity of a substance is determined by the amount or concentration of hydrogen ions present.

Acidity is the amount of acid in a substance. **Alkalinity** is a measure of water's ability to neutralize or cancel out acids. This neutralization occurs when equal amounts of an acid and base are combined and reach a neutral pH balance of 7.

Appearance and Odor

Appearance and odor are properties that are important in the description of a material. The appearance and odor of a material include color, smell, physical state at normal temperature, and pressure. If all properties are carefully considered, one cannot find two chemicals with the same properties.

Matter

Matter is anything that occupies space and has mass, or something that occupies space and can be felt by one or more of the senses. It can be best described by its physical appearance or by its physical properties. These can be measured and observed, such as mass, size, or volume. Sometimes firefighters can detect matter by color or smell. **Mass** is basically a measure of the amount of matter objects contain. In informal usage, the words *weight* and *mass* often mean the same thing.

Chemical Properties of Matter

All materials are either inorganic or organic. **Inorganic** means that the matter is composed mainly of earth materials, such as rocks, soil, air, water, and minerals at or below the Earth's surface. Examples of minerals are quartz, sulfur, iron, and granite. Most inorganic minerals are not involved in the combustion process.

Organic means that the matter is in substances that were once living organisms. As mentioned earlier in the chapter, organic substances consist of carbon, hydrogen, and oxygen. Examples of organic substances are plastics, wood, gasoline, and oil. Matter that was a living organism at one time contains cells that were used to grow and feed the organism.

Physical States of Matter

There are three states of matter: solids, liquids, and gases. Matter can act and appear very differently depending on its state. Hazardous materials can also be solids, liquids, gases, or sludge, which may be part solid and part liquid. Some of the risk assessment of a hazardous material at an incident depends on the state of the material. A spilled or leaked liquid will spread and seek out the nearest low-lying areas, while a spilled solid will form a pile, usually in the vicinity of the spill. The shape of the spilled material greatly influences the extent and threat of the spill.

Solids

A solid is a state of matter with a definite volume and shape. The primary concerns with spilled solids on land are their flammability and reactivity with air. If solids spill into water, one must consider the traits of solubility and reactivity. A few solids will form vapors without first becoming liquid, which is called **sublimation**. Examples include naphthalene (moth balls), carbon dioxide (CO_2; dry ice), and paradichlorobenzene (a toilet bowl deodorizer). The toxicity and flammability of the vapors must be assessed if subliming solids are spilled.

Liquids

Liquid matter has a definite volume but not a definite shape. A liquid will take the shape of its container. Some liquids will turn into a gas when exposed to the atmosphere. Liquids can flow away from a leak, thereby extending the hazard area. The thickness of a liquid may be a factor in incident response. The more viscous (thick and sludgy) liquids tend to flow more slowly and to stick to surfaces, such as the ground, clothing, and equipment. Sludge that has thickened due to evaporation may be concentrated. It may be more dangerous than the original liquid.

Gases

There are some special considerations for emergency response to incidents that involve gases. Many gases are shipped and stored in a variety of containers, so it is common to encounter gases at incidents. Gases differ from liquids and solids in several ways. Their most unique characteristic, however, is that they have neither a definite shape nor volume. Their shape is determined by the amount of pressure placed on the gas and the shape of the container. In most cases, liquids and solids are considered to be incompressible. Gases, on the other hand, are very elastic, and the pressure exerted by the gas itself can measure this trait. This means gases can be compressed. They can retain their gaseous form or be compressed in a liquid form.

Flammable gases are of special concern to firefighters because they may be involved in emergency responses. A **flammable gas** is a compressed gas that can easily catch fire and continue to burn. As mentioned earlier in the chapter, flammability range is the numerical difference between a flammable substance's lower and upper explosive limits in air. An example would be acetone, which has a flammability range of 2.6% (LEL) to 12.8% (UEL).

Boyle's Law

According to **Boyle's law**, the more a gas is compressed, the more the gas becomes difficult to compress further. The law states that the pressure of a gas is inversely proportional to its volume at a given temperature. This means that if the pressure is doubled in a specific volume in a closed container, the volume of the gas is reduced by one-half. For example, consider a tire that has 2 ft^3 of air. If that air is released into a 4 ft^3 container, it will have one-half the pressure it had when it was in the 2 ft^3 tire.

Charles's Law

Dr. Jacque Charles, a French scientist, found that a gas would expand or contract in direct proportion to an increase or decrease in temperature. This is called **Charles's law**. He found that the molecules of a gas are in continual motion. They collide with each other and the walls of their container.

When these molecules are compressed, they bounce against the container walls with an even greater force. When the gas is heated, the movement of the molecules becomes faster, which increases the pressure. If the walls of the container are elastic like a rubber balloon, the pressure will expand the outer walls. Likewise, if the gas is confined so that it cannot expand, its pressure will increase or decrease in direct proportion to the temperature.

Compressed Gases

A **compressed gas** is any material that, when in a closed container, has an absolute pressure of more than 40 pounds per square inch (psi) at 70°F (21°C) or an absolute pressure exceeding 104 psi at 130°F (54°C), or both. **Absolute pressure** is the measurement of pressure exerted on a surface, including pressure from the atmosphere, measured in psi absolute. Pressurized gases become liquid when compressed and exist in a liquid-vapor relationship inside containers.

Cryogenics

The **cryogenics** are those materials with boiling points of no greater than –150°F (–101°C) that are transported, stored, and used as liquids. The super-cold temperatures allow a large volume of gas to be stored as a liquid in a much smaller container at lower pressures. The hazards of cryogenic gases depend on the nature of the gas, the large volume or ratio of vapor to liquid, and the extreme cold.

One characteristic of cryogenic materials is their unique storage vessels. While the pressures are low (under a few psi), keeping a constant temperature is a top priority. For this reason, they are stored in cylinders designed like thermos bottles. The containers have silver-coated linings to reflect heat and use the temperature of the materials to maintain the frigid environment. Because of the extreme cold, cryogens can cause severe damage to anything that contacts the liquid.

Cryogenic liquids and liquefied gases turn to vapor quickly when released from their containers. A liquid spill will boil into a much larger vapor cloud. These clouds can be extremely dangerous, especially if the vapors are flammable. Both cryogenic and liquefied gases can cause ice burns or severe frostbite that requires cold injury treatment. Clothing soaked with cryogenic materials must be removed immediately. This is extremely important if the vapors are oxidizers or flammable.

TABLE 2-6 Boiling Points and Expansion Ratios of Selected Cryogens

Substance	Boiling Point (°F)	Boiling Point (°C)	Expansion Ratio
Liquid argon	−302	−185.6	840:1
Liquid fluorine	−306	−187.8	980:1
Liquid helium	−452	−268.9	700:1
Liquid hydrogen	−423	−252.8	848:1
Liquid krypton	−243	−263.9	695:1
Liquid natural gas	−289	−178.3	635:1
Liquid neon	−411	−246.1	1445:1
Liquid nitrogen	−320	−195.6	694:1

Cryogenic gases include argon, fluorine, helium, hydrogen, krypton, liquid natural gas, liquid neon, liquid nitrogen, liquid oxygen, and liquid xenon. **TABLE 2-6** shows the boiling points of these substances and their expansion ratios. The **expansion ratio** is the volume of its liquid form compared to the volume of its gas form.

Changes in Physical State

Some chemical reactions that occur during a leak or a fire cause a change in physical state that can lead to problems. A leaking liquid may change to a gas with a tremendous increase in volume. As a vapor, it may pool in lower areas or ascend depending on its specific gravity. Containers heated in a fire may leak or explode due to an increased volume of the contained gas or liquid. The expansion ratio of a chemical is known for all chemicals. It is an important consideration in an incident involving the release of gases liquefied by pressure.

Combustible Dusts

Many materials in solid form do not combust. They can be made to combust or explode by converting them to dust. Under the right conditions, any organic dust can explode. In grain processing plants, dust explosions are common. The finer the dust is, the more likely it is to explode. Confectioner's sugar and cornstarch are both finely ground. As a result, they present the most severe explosion potential.

Many dust explosions occur in pairs. The first, smaller explosion occurs near the ignition source. It lifts and scatters loose dust into the atmosphere, which lights a second, more devastating explosion, as more fuel is made available for burning. Temperature and relative humidity (RH) can contribute to a dust explosion. RH is a factor in the combustion process because it supplies moisture to fuels. Thus, the lower the humidity, the drier the air, and the cooler the temperature, the greater the possibility of a dust explosion.

Many metals in the form of dust or shavings will also explode. Magnesium, aluminum, and titanium are the more common metals firefighters encounter. A well-maintained fire sprinkler system with a dust removal system and good housekeeping can decrease the potential for explosions.

Boiling Liquid/Expanding Vapor Explosion

A **boiling liquid/expanding vapor explosion (BLEVE)** is the result of the explosive release of the vessel pressure, portions of the tank, and the burning vapor cloud of gas with the accompanying radiant heat. A BLEVE occurs when a pressure tank has its container metal softened or weakened by heat or corrosion (**FIGURE 2-9**). When the expansion of vapors inside the container exceeds the pressure relief valve limit, the pressure causes the container to violently rupture into two or more pieces. The **pressure relief valve** is used on compressed gas cylinders to release pressure buildup within a cylinder. All cylinders can be expected to fail (or exceed the pressure limit of the relief valve) after being subjected to enough heating and expansion of the internal gas vapors.

BLEVEs most often occur when flames contact a tank shell above the liquid level. They can also occur when not enough water is used to keep the tank shell cool. In the direct attack mode, first responders should remember the following points to ensure their safety:

- In most cases, these fires should not be put out because it is safer for firefighters to control flames than to release gas vapors, which may result in an explosion.

FIGURE 2-9 Boiling liquid/expanding vapor explosion (BLEVE).

- The goal of firefighting operations is to cool the tank shell to a point that the liquid inside cools and reduces the vapors and pressure. The pressure should be kept below the pressure setting on the pressure relief valve. Generally, the pressure relief valve is preset to open at a given pressure and then reset. The pressure relief valve may be spring activated or a frangible disc that ruptures at a preset pressure rating. The disc does not reseal until after the pressure has been reduced. Even with a reset pressure relief valve, the continued application of water on the tank is important. This will prevent a re-venting because the material will no longer be escaping the container. In the case of a frangible disc, the remaining liquid will escape, and the product will be burned off. It is still important to continue to apply water to the tank.
- If control cannot be gained from hand lines cooling the vessel, then an unattended monitor should be put in place for the safety of the firefighters. A **monitor** is an industrial device used to deliver large amounts of water for firefighting purposes.
- First responders should never approach these tanks from either end. They should always be approached, and water applied, from the broad side of the tank and at a distance.

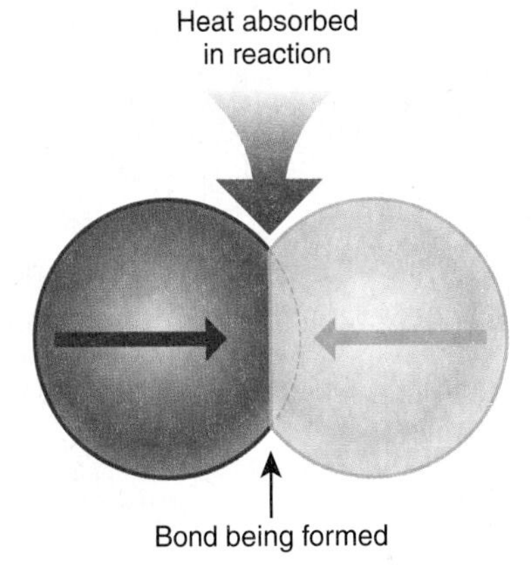

FIGURE 2-10 H_2O (the bond of water being formed).

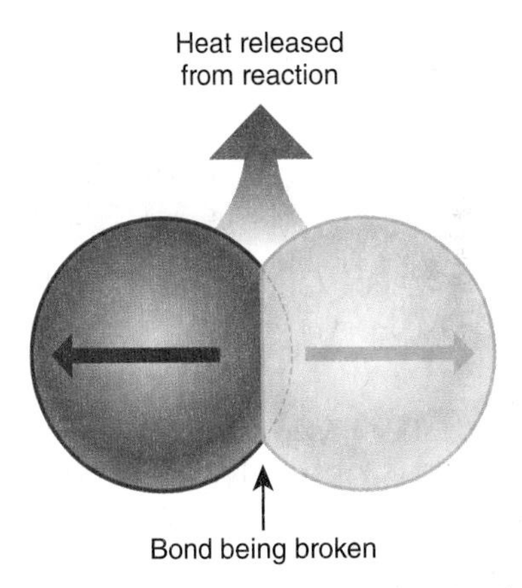

FIGURE 2-11 Exothermic reaction releasing heat.

Chemical Reactions

The stability of a material is the result of strong bonds between atoms. The bonds are formed when atoms exchange or share electrons. Covalent bonding allows atoms to resist changes in a normal environment or when exposed to shock or pressure, air, and water. This bonding and absorption of heat is known as an **endothermic** or heat-absorbing reaction. The releasing of the bond with the release of heat is known as an **exothermic** or heat-releasing reaction.

An endothermic reaction occurs when heat (energy) is absorbed from the reaction or the "joining" of the molecules. This occurs when the bond that holds the molecules together is set. In **FIGURE 2-10**, the bond between two molecules of water (H_2O) is being formed with heat being absorbed.

When a bond is broken, an exothermic reaction occurs, releasing heat (energy). Firefighters face this reaction in almost every fire situation where heat, light, and other products of combustion are released. In **FIGURE 2-11**, the bond between two molecules is being broken, thus releasing heat (energy).

It is vital for firefighters to understand these chemical reactions. They often create very serious hazards for emergency responders and the environment. Some materials that can cause problems are water-reactive or air-reactive materials. Some others are oxidizers, unstable or incompatible materials, and those that polymerize. **Polymerization** is a process of reacting (linking or rearranging) monomer molecules in a chemical reaction to create a polymer. Monomer molecules, or **monomers**, are usually gaseous or liquid small molecules. The reaction can be explosive or violent. This is referred to as runaway polymerization.

Water-Reactive Materials

Water-reactive materials react with water, often violently, to release heat, a flammable or toxic gas, or a combination of the two. If water is used to put out the fire, the presence of water-reactive materials can make the situation much more dangerous.

For example, firefighters enter a metal disposal yard and attack a small fire in cardboard boxes. They are unaware that the boxes contain magnesium filings.

FIGURE 2-12 Magnesium is a water-reactive substance.

They use the water-based extinguisher hanging on the wall in the storage area. Immediately there is an explosion as the magnesium reacts with the water. One firefighter gets seriously burned. **FIGURE 2-12** shows this reaction.

Air-Reactive Materials

Air-reactive materials, as their name implies, are reactive in the presence of air. Some water-reactive materials are also air reactive and will catch fire in air. Potassium metal is one example. Diborane and trimethylaluminum are organic metal compounds that are air reactive as well. **White phosphorus**, which is a toxic substance made from phosphate-containing rocks, must be stored under water or oil to prevent it from catching fire. This solid material is more dangerous than red phosphorus because of its ready oxidation and sudden ignition when exposed to air.

Oxidizers

Oxidizers, also called **oxidizing agents**, present special hazards because they react chemically with many combustible organic materials. Some examples are oils, greases, solvents, paper, cloth, and wood. All organic and inorganic peroxides are highly combustible, and some are highly reactive. The reaction is violent because the agents contain additional oxygen molecules. When these molecules combine with the fuel, they require only an ignition source to explode.

FIGURE 2-13 Federal building, Oklahoma City, Oklahoma.

Some inorganic peroxides such as sodium peroxide and potassium peroxide are very reactive and shock-sensitive. A **shock-sensitive** material will react violently when struck or compressed. Other commonly encountered oxidizers include ammonium nitrate, potassium permanganate, ammonium persulfate, and sodium nitrate. The 1995 bombing of a government building in Oklahoma City, Oklahoma, shows the strength of oxidizers. **FIGURE 2-13** shows the outcome of a terrorist attack in which ammonium nitrate, a powerful oxidizer, was mixed with fuel oil and used as an explosive, resulting in the death of 286 people.

Unstable Materials

When exposed to water, unstable air, shock, or pressure, materials decompose, polymerize, or become self-reactive. One group of materials that are chemically unstable consists of the monomers. As mentioned earlier, monomers are gaseous or liquid small molecules. They are the building blocks that form polymers, such as resins, plastics, and synthetic rubber materials. Some of these materials can polymerize spontaneously (runaway polymerization), causing a container to break.

One example of runaway polymerization is spray foam insulation in an aerosol container. Spray foam is used to seal walls and gaps around windows. When the foam is released from the container, it reacts with air to expand. Once started, the reaction is hard to stop. If the runaway reaction occurs in a container like a railroad car, the car could fail due to the amount of pressure being built up inside the container.

Incompatible Materials

A review of basic chemistry for first responders would not be complete without including the incompatibility of chemicals. Some chemicals, when mixed with other chemicals, can adversely affect human health and the environment in a variety of ways. These mixtures can cause:

- Generation of heat
- Violent reaction
- Formation of toxic fumes or gases
- Formation of flammable gas
- Fire or explosion
- Release of toxic substances if they burn or explode

Catalyst

Some reactions proceed more readily in the presence of a catalyst. A **catalyst** is a substance that speeds up a chemical reaction. It can also lower the temperature needed to start a reaction. In either case, it is not changed or used up by the reaction. An everyday example of a catalyst is platinum, which is used in the catalytic converter of a car. It causes the fuel to burn faster and cleaner. This reduces the output of carbon without consuming the platinum.

Toxic Combustion Products

A good example of chemical reactions and their hazard to firefighters is the toxic materials at a fire. The formation of these toxic products depends on what is burning and how much oxygen is present. For example, many home furniture and decor items are made with plastics. When exposed to fire, these items release toxic chemicals into the air. Wool and silk may give off hydrogen cyanide, a highly poisonous gas, when they burn. For these reasons, firefighters must wear respiratory protection or a self-contained breathing apparatus during firefighting activities. A **self-contained breathing apparatus (SCBA)** is a pressurized tank on a backpack that provides breathable air within a fully enclosed mask. Firefighters should also wear this gear after a fire during the overhaul phase. This is because gases are still present in the structure.

It is important to note that chemical properties and chemical reactions are not limited to hazardous materials incidents. This important concept is relevant in everyday firefighting activities. Fire is a chemical reaction. A solid understanding of chemistry is vital for firefighters in nearly every task they perform.

CASE STUDY

CONCLUSION

1. **In 1984, whom could the firefighters have consulted before entering the building?**

   The firefighters could have consulted with their dispatch, the local or national poison control center, and possibly government agencies, which were forming hazardous materials groups at the time. They should have considered the fire suppression system as well.

2. **What could the firefighters have done to protect themselves before entering the building?**

   Firefighters should have considered the traits of the material they were dealing with: ignition temperature, flammability range, level of personal protective equipment (PPE), and methods recommended to put out a fire in this material.

3. **Why is the relationship between the anhydrous ammonia and the conditions of the building of vital importance?**

   According to the Department of Transportation (DOT), anhydrous ammonia is not a hazardous material. This material should have been considered a flammable gas regardless of the DOT rating and handled accordingly.

WRAP-UP

SUMMARY

- Atoms are the basic building blocks of matter and the smallest units of an element. Atoms consist of a nucleus that contains protons and neutrons and an orbit that contains electrons.
- The bonding of atoms creates molecules and compounds.
- The periodic table is an arrangement of the elements.
- A molecule is the smallest part of a pure chemical substance that has all the properties of the material.
- A compound is a substance formed from two or more elements.
- A mixture is a combination of substances held together by physical rather than chemical means.
- All atoms of a certain element have the same structure, so they also have the same properties.
- Matter is anything that occupies space and has mass or something that occupies space and can be felt by one or more of the senses.
- The three states of matter are solids, liquids, and gas.
- The stability of a material is the result of strong bonds between atoms.
- This bonding and absorption of heat is known as an endothermic or heat-absorbing reaction. The releasing of the bond with the release of heat is known as an exothermic or heat-releasing reaction.
- A good example of chemical reactions and their hazard to firefighters is the toxic materials at a fire.

KEY TERMS

absolute pressure The measurement of pressure exerted on a surface, including pressure from the atmosphere, measured in pounds per square inch absolute.

acid A chemical that releases hydrogen ions when dissolved in water.

acidity The amount of acid in a substance.

alkalinity A measure of water's ability to neutralize or cancel out acids.

atom The smallest unit of an element that takes part in a chemical reaction; made up of a neutron, proton, and a cloud of orbiting electrons.

atomic number The number of protons in an element.

atomic weight The average number of protons and neutrons in atoms of a chemical element; sometimes referred to as atomic mass.

auto-ignition temperature The temperature at which a material will ignite in the absence of any outside source of heat; also referred to as the spontaneous ignition temperature.

base A substance that will react with acids in aqueous solution to form salts, while releasing heat.

boil over The expulsion of a tank's contents by the expansion of water vapor that has been trapped under the oil and heated by the burning oil and metal sides of the tank.

boiling liquid/expanding vapor explosion (BLEVE) The result of the explosive release of vessel pressure, portions of the tank, and the burning vapor cloud of gas with the accompanying radiant heat.

boiling point (BP) The temperature at which a liquid will convert to a gas at a vapor pressure equal to or greater than atmospheric pressure.

Boyle's law A theory that states the more a gas is compressed, the more the gas becomes difficult to compress further.

carbon dioxide A nonflammable liquid asphyxiant gas.

catalyst A substance that speeds up a chemical reaction but is not changed or used up by the reaction.

Celsius scale A temperature scale based on water freezing at 0° and boiling at 100°; also known as centigrade.

Charles's law A theory that states a gas will expand or contract in direct proportion to an increase or decrease in temperature.

chemical bond The attractive force that binds two atoms in a combination that is stable at room temperature, but becomes unstable at a high temperature.

chemical change Molecules of a material are changed by a process.

closed cup test A lid is placed on the cup to confine the vapors above the cup. In this test, a sample is heated and stirred, and an ignition source is introduced at random times to find the flash point of the sample.

combustion A chemical reaction that creates heat and light.

compound A substance formed from two or more elements joined with a fixed ratio.

compressed gas Any material that, when in a closed container, has an absolute pressure of more than 40 psi at 70°F (21°C) or an absolute pressure exceeding 104 psi at 130°F (54°C), or both.

covalent bond The sharing of electron pairs by combined atom(s).

cryogenic Material with boiling points of no greater than –150°F (–101°C) that are transported, stored, and used as liquids.

decomposition temperature The temperature to which the material must be heated for it to decompose.

electron A very light particle with a negative electrical charge, a number of which surround the nucleus of most atoms.

element The simplest form of matter.

endothermic The type of reaction in which heat (energy) is absorbed when the reaction takes place.

exothermic The type of reaction that will release or give off heat.

expansion ratio The volume of a substance's liquid form compared to the volume of its gas form.

explosive/flammability range The range of concentrations of gases or materials (dusts) in the air that permit the material to burn.

Fahrenheit scale A temperature scale based on water freezing at 32° and boiling at 212°.

families Groups of elements in the periodic table with like properties.

fire point The lowest temperature of a liquid sufficient to produce vapor that will ignite from an outside ignition source and sustain combustion.

flammable gas A gas that is flammable at atmospheric temperature and pressure within a mixture with air of 13% or less (by amount or volume) or that has a flammability range with air of more than 12%.

flash point The lowest temperature of a liquid at which vapors are produced and will ignite when an outside ignition source is present.

group A vertical column of elements in the periodic table, each with the same number of valence electrons, resulting in similar properties and reactivity.

halogenated compound Compound formed from any chemical reaction in which one or more halogenated atoms are combined with an existing compound.

inorganic A term that describes matter composed mainly of earth materials, such as rocks, soil, air, water, and minerals at or below the Earth's surface.

insolubility A term that describes materials that form a separate layer in another substance and will either float or sink, depending on their specific gravity.

ionic bond A bond formed between two charged atoms, or ions, to create a molecule.

liquefied petroleum gas (LPG) A term given to butane and propane gases that have been pressurized and contained in a tank.

lower explosive limit (LEL) The lowest concentration of a substance (by percentage) in air that will burn.

mass A measure of the amount of matter objects contain.

matter Anything that occupies space and has mass or something that occupies space and can be felt by one or more of the senses.

mixture A combination of substances held together by physical rather than chemical means.

molecule The smallest unit of an element or compound that keeps the chemical traits of the original material.

monitor An industrial device used to deliver large amounts of water for firefighting purposes.

monomers Gaseous or liquid small molecules that are the building blocks for polymers; also called monomer molecules.

negative ion Forces outside an atom cause it to gain an extra electron, giving it an overall negative charge; also known as an anion.

neutron A particle with nearly the same mass as the proton, but electrically neutral; it is part of the nucleus of all atoms except the most common isotope of hydrogen.

octet rule A rule that refers to the preference of atoms to have eight electrons in the valence shell.

KEY TERMS CONTINUED

open cup test A test that measures the release of the vapors in terms of the pressure being exerted at a specific temperature; the flammable vapors or molecules driven off the flammable liquid are not trapped by a lid.

organic A term that refers to matter in substances that were once living organisms.

oxidizer or oxidizing agent Substance that presents special hazards because it reacts chemically with many combustible organic materials, such as oils, greases, solvents, paper, cloth, and wood; halogens are also powerful oxidizers.

period A horizontal row of elements in the periodic table that identifies how many orbits revolve around the nucleus of the atom.

pH A measure of a substance's ability to react as an acid (low pH) or as an alkali (high pH).

physical change Molecules of a material are not changed by a process and the molecules remain intact.

physical property A trait that is constant to the compound; for example, a ball or chain configuration.

polar solvent A substance that allows firefighters' foam to be used on alcohol-based fires without breaking down the foaming agent.

polymerization The process of reacting (linking or rearranging) monomer molecules in a chemical reaction to create a polymer; the reaction can be explosive or violent.

positive ion An atom that is missing electrons; also known as a cation.

pressure relief valve Used on compressed gas cylinders to release pressure buildup within a cylinder.

proton A positively charged particle that is the nucleus of all atoms, including the most common isotope of hydrogen.

redox A chemical reaction that involves the loss of electrons, which speeds up the oxidation process, and a gain of electrons, which slows the oxidation process; shortened term for reduction–oxidation.

self-contained breathing apparatus (SCBA) A pressurized tank on a backpack that provides breathable air within a fully enclosed mask.

shock-sensitive A material that will react violently when struck or compressed.

solubility The maximum amount of the substance that can be dissolved in another substance.

specific gravity The density of the product divided by the density of water or air; water and air are the standards that have been given the value of 1.0.

sublimation The direct change from a solid to a gas without changing into a liquid.

upper explosive limit (UEL) The highest concentration of a substance (by percentage) in air that will burn.

valence electrons The electrons in the outermost shell of an atom that interact in the bonding process with other atoms.

vapor density (VD) The mass of the given vapor in comparison to the same amount of air.

vapor pressure (VP) The pressure placed on the inside of a closed container by the vapor in the space above the liquid.

white phosphorus A toxic substance made from phosphate-containing rocks that must be stored under water or oil to prevent it from catching fire.

REVIEW QUESTIONS

1. What is the difference between a physical change and a chemical change of a molecule?
2. What is the difference between flash point and fire point?
3. Define and give one example of each of the three states of matter.
4. Why is solubility critical when fighting a hydrocarbon fire?
5. What are the major concerns for firefighters when faced with incompatible materials?

DISCUSSION QUESTIONS

1. How can knowledge of the periodic table help firefighters in an emergency?
2. Vapor density has two definite hazards: a spill and release. How does this impact firefighting and what precautions should you take?
3. When considering specific gravity, there are additional hazards that can be associated with the three properties of a liquid when added to water. What are these properties and their hazards?
4. Given the physical states of matter and their properties, in which order is matter a greater danger to firefighters? Why?

APPLYING THE CONCEPTS

You are responding to the report of a natural gas leak inside a residence. The outside temperature is 98°F, and humidity is at 48%. On arrival, you find the residence is vacant and there are no windows open.

1. En route to the call, what should be your first consideration?
2. On scene, what would be your first mitigating task?
3. How would you best measure the outcome of your mitigation of the leak?

REFERENCES

Academic Library. "Shreveport LA September 17, 1984 Anhydrous Ammonia Fire Incident." n.d. Accessed January 3, 2023. https://ebrary.net/131214/health/shreveport_september_1984_anhydrous_ammonia_fire_incident.

BBC. "Group 0: The Noble Gases." n.d. Accessed March 14, 2023. https://www.bbc.co.uk/bitesize/guides/zy6cfcw/revision/2#:~:text=The%20atoms%20of%20noble%20gases,take%20part%20in%20chemical%20reactions.

Britannica Kids. "Combustion." n.d. Accessed January 3, 2023. https://kids.britannica.com/kids/article/combustion/399410#:~:text=Combustion%20is%20a%20chemical%20reaction,joins%20with%20carbon%20in%20wood.

CAMEO Chemicals. "Reactive Group Datasheet: Hydrocarbons, Aromatic." n.d. Accessed January 3, 2023. https://cameochemicals.noaa.gov/react/16.

Camlab Limited. "What Is the Difference Between Open and Closed Cup Flash Point?" November 23, 2015. Accessed January 3, 2023. https://www.energy-xprt.com/articles/what-is-the-difference-between-open-and-closed-cup-flash-point-622009.

Centers for Disease Control and Prevention (CDC)/National Institute for Occupational Safety and Health (NIOSH). "White Phosphorus: Systemic Agent." Last reviewed October 20, 2021. Accessed January 3, 2023. https://www.cdc.gov/niosh/ershdb/emergencyresponsecard_29750025.html#:~:text=DESCRIPTION%3A%20White%20phosphorus%20is%20a,a%20pesticide%20and%20in%20fireworks.

ChemicalSafetyFacts.org. "The Periodic Table of Elements Explained." n.d. Accessed January 3, 2023. https://www.chemicalsafetyfacts.org/chemistry-101/the-periodic-table-of-elements-explained/.

Corrosionpedia. "Total Alkalinity." n.d. Accessed January 3, 2023. https://www.corrosionpedia.com/definition/1100/total-alkalinity-ta.

Delmar Cengage Learning. *Firefighter's Handbook: Firefighting and Emergency Response*, 3rd ed. Clifton Park, NY: Delmar Cengage Learning, 2008.

Energy.gov. "DOE Explains...Catalysts." n.d. Accessed January 3, 2023. https://www.energy.gov/science/doe-explainscatalysts.

Faraday, Michael. *The Chemical History of a Candle.* Atlanta, GA: Cherokee Publishing Company, 1993.

FireSafetySearch. "Fire Monitors." n.d. Accessed January 3, 2023. https://www.firesafetysearch.com/fire-monitors/.

Friedman, Raymond. *Principles of Fire Protection Chemistry and Physics.* Quincy, MA: National Fire Protection Association, 1998.

Gantt, Paul. *Hazardous Materials: Regulations, Response & Site Operations.* Clifton Park, NY: Delmar Cengage Learning, 2009.

Hall, J. R., Jr. "Fire Risk Analysis: Model for Assessing Options for Flammable and Combustible Liquid Products in Storage and Retail Occupancies." *Fire Technology* 31 no. 4 (1995), 290–308.

Hawley, Chris. *Hazardous Materials Incidents*, 3rd ed. Clifton Park, NY: Delmar Cengage Learning, 2008.

Helmenstine, Anne Marie. "Element Families of the Periodic Table." ThoughtCo. Updated October 27, 2019. Accessed January 3, 2023. https://www.thoughtco.com/element-families-606670.

Helmenstine, Anne Marie. "Solubility Definition in Chemistry." ThoughtCo. Updated January 22, 2020. Accessed January 3, 2023. https://www.thoughtco.com/definition-of-solubility-604649.

Hess, Fred C. *Chemistry Made Simple.* New York: Doubleday Publishing, 1996.

Khan Academy. "Building the Periodic Table of the Elements." n.d. Accessed January 3, 2023. https://www.khanacademy.org/humanities/big-history-project/stars-and-elements/knowing-stars-elements/a/dmitri-mendeleev.

REFERENCES CONTINUED

Klem, Thomas J. "Explosion in Cold Storage Kills Fire Fighter." *Process Safety Progress* 5, no. 1 (1986), 27–30. Accessed February 28, 2023. https://aiche.onlinelibrary.wiley.com/doi/abs/10.1002/prsb.720050107.

LibreTexts Chemistry. "1.5: Octet Rule: Ionic and Covalent Bonding (Review)." 2019. Accessed January 3, 2023. https://chem.libretexts.org/Courses/Sacramento_City_College/SCC%3A_Chem_420_-_Organic_Chemistry_I/Text/01%3A_Introduction_and_Review/1.05%3A_Octet_Rule_-_Ionic_and_Covalent_Bonding_(Review)#:~:text=The%20octet%20rule%20refers%20to,the%20most%20stable%20s.

National Cancer Institute. "Acidity." n.d. Accessed January 3, 2023. https://www.cancer.gov/publications/dictionaries/cancer-terms/def/acidity.

National Fire Protection Association. *National Fire Protection Handbook*, 21st ed. Quincy, MA: NFPA, 2021.

Sawe, Benjamin Elisha. "How Many Elements Are There?" WorldAtlas.com. February 23, 2020. Accessed January 3, 2023. https://www.worldatlas.com/articles/how-many-elements-are-there.html.

Science Notes. "What Is a Solid? Definition and Examples in Science." 2021. Accessed January 3, 2023. https://sciencenotes.org/what-is-a-solid-definition-and-examples-in-science/.

U.S. Department of Transportation. 2022. *2020 Emergency Response Guidebook*. Accessed January 3, 2023. https://www.phmsa.dot.gov/sites/phmsa.dot.gov/files/2021-01/ERG2020-WEB.pdf.

Workplace Hazardous Materials Information Systems. *WHMIS Quick Facts: Flammable Gases*. 2006. Accessed February 28, 2023. https://www.canada.ca/content/dam/hc-sc/migration/hc-sc/ewh-semt/alt_formats/pdf/occup-travail/whmis-simdut/flam-gases-eng.pdf.

CHAPTER

3

Combustion Processes

LEARNING OBJECTIVES

After studying this chapter, you will be able to:

- Explain combustion and the fire models that apply to the combustion process.
- Describe the different ways fire is classified.
- Explain the five classes of fires, the combustible materials in each class, and how to put out fires in these materials.
- Describe the stages of fire as it progresses from the initial stage to its final stage.
- Explain heat flow path and the causes of flame over, flashover, and backdraft.
- Explain how to prevent and protect against flame over, flashover, and backdraft.
- Describe the various methods of heat transfer.

CASE STUDY

Residential Garage Fire

A homeowner has a garage attached to his 3,000-square-foot home. He is a welder by profession. He often camps and stores camp fuels and an oxygen/acetylene torch system in the garage. To prepare for a camping trip with his family, he added a 5-gallon butane tank and some gasoline for his generator to the stored fuels.

Before the trip, he had been doing some refinishing woodwork using a linseed oil stain. He left the stained rags on the floor of the garage in a cardboard box.

The temperature outside is 98°F. The inside temperature of the garage is 10 degrees warmer. There is limited ventilation.

1. Are conditions in line for a possible explosion or fire?
2. What can the homeowner do to limit or prevent the combustion of the rags?
3. What is the ignition source going to be in the above story?
4. What other factors would be a concern in a typical garage?

Introduction

This chapter considers the physical and chemical processes of fire combustion. We will examine how these processes relate to how the fire service confines, controls, and puts out uncontrolled fires. We will explore fires contained within rooms and structures. We will also track how a fire progresses through the various stages, including the events of flame over, flashover, and backdraft.

What Is Combustion?

Combustion is a chemical reaction between a fuel and oxygen that creates smoke, heat, and light. The process is usually related to the oxidation of a fuel by oxygen in the air. However, some fuels may contain oxygen, which becomes part of the oxidation process. One definition of combustion is a process involving chemistry, thermodynamics, fluid mechanics, and heat transfer. **Thermodynamics** is the relationship between heat and other forms of energy. **Fluid mechanics** is the study of fluids at rest or in motion.

Combustion is not the same as fire. **Fire** is a rapid oxidation process that involves heat, light, and smoke in varying intensities. Fire differs from combustion because of the control of the combustion event itself. Another way to think about combustion is as the same chemical reaction as the slow rusting of iron. It is also the same chemical reaction as an explosion of any flammable gas or substance. The only difference is the speed of the reaction.

In the past, the fire service model of fire was the fire triangle. The **fire triangle** is a three-sided model that represents the three elements needed for fire. The first element is an oxidizing agent, such as oxygen. Oxygen needs to be in the correct proportion (between 16% and 21% to air). The second element is a material or fuel that will burn. The third element is an ignition or heat energy source. The source must have enough heat or energy to raise the fuel to its ignition temperature. **FIGURE 3-1** shows the fire triangle.

Research led the fire community to decide that a fourth element, a chemical chain reaction, should be added to the fire triangle. This led to the fire tetrahedron (**FIGURE 3-2**). The **fire tetrahedron** is a four-sided model showing the heat, fuel, oxygen, and chemical

FIGURE 3-1 The fire triangle.

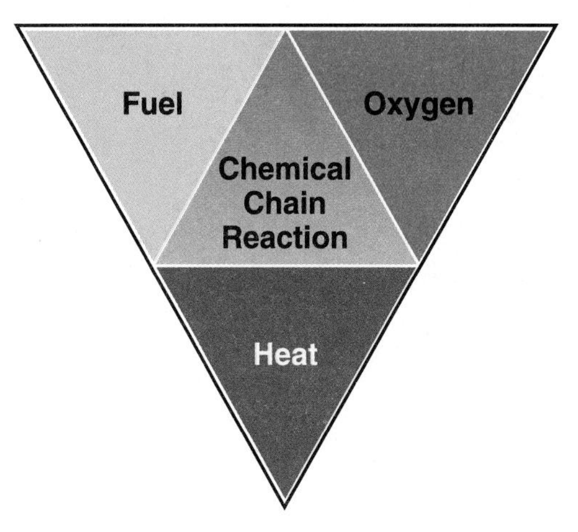

FIGURE 3-2 The fire tetrahedron.

reaction needed for combustion. The continual burning of a material is controlled in part by the heat of the flame. This heat is transferred back to the fuel. This causes the fuel to release free radicals that turn into vapor. A **free radical** is a fragment of a molecule that has at least one unpaired electron. This turning of free radicals into vapor can occur with or without chemical breakdown of the molecules. If chemical breakdown occurs, the process is called pyrolysis. **Pyrolysis** is the breaking down of a solid fuel with heat into gaseous parts. Once started, the burning continues until one of the following events occurs:

- The burning material is used up (fuel).
- There is not enough oxidizing agent (oxidizer).
- The heat is removed or prevented from reaching the combustible material (heat).
- The flames are curbed chemically or cooled to prevent further reaction (chemical chain reaction).

The removal of any side of the fire tetrahedron will put out the fire.

Spontaneous Combustion

One slightly different process of combustion is spontaneous combustion, or autoignition. This type of combustion does not require a separate ignition source. In **spontaneous combustion**, the material self-heats to its piloted ignition temperature before catching fire. The **piloted ignition temperature** is the temperature of a liquid fuel at which it will self-ignite when heated. At this point, the material will support flame spread. For example, coal is a porous solid material. Air can enter and spread through it. However, the heat produced within the coal is held in by the coal's insulating properties. The temperature of the coal rises because the heat is trapped. The coal eventually reaches its ignition temperature and catches fire.

Methods of Fire Classification

Fire has been classified in several ways: the stages of combustion, the type of substance burning, stages of fire, and fire events. These classifications are detailed in this section.

Stages of Combustion

One of the simplest methods of classification is the stages of the combustion process. The three stages of combustion are precombustion, flaming, and smoldering.

Precombustion

Precombustion is when fuels are heated to their ignition point. The heat causes the fuel to release vapors and particulates. **Particulates** are the unburned products of combustion one can see in smoke. The efficiency of the combustion process improves with added heat and oxygen. This decreases the size of the particulates. In some situations, the smoke (particulates) can be almost invisible to the human eye.

Precombustion continues in two ways. First, the heat raises the fuel to its ignition temperature. The vapors released then combust. Second, hot gases are released, which rise above the fuel. In doing so, entrainment gathers more oxygen from the surrounding environment. **Entrainment** is the physical process in which surrounding air or gases are drawn into a fire plume. A small portion (approximately 13%) of the heat energy is radiated back into the fuel. This in turn releases a stream of unburned vapors. These vapors keep the process in a continuous cycle. This cycle, if not broken, will continue until the fuel is used up (**FIGURE 3-3**).

Flaming Combustion

This is the phase that firefighters will most likely encounter in an emergency incident. **Flaming combustion** is an exothermic (heat-releasing) chemical reaction with flames between a substance and oxygen. This phase can often be identified by the presence of flames (**FIGURE 3-4**). Flaming combustion cannot occur unless a gas or vapor is burning.

There are two types of flaming combustion. The first category is premixed flaming. **Premixed flaming** occurs when a gaseous fuel mixes with air before catching fire. For example, one can find this combustion process in a number of appliances (water heaters, gas

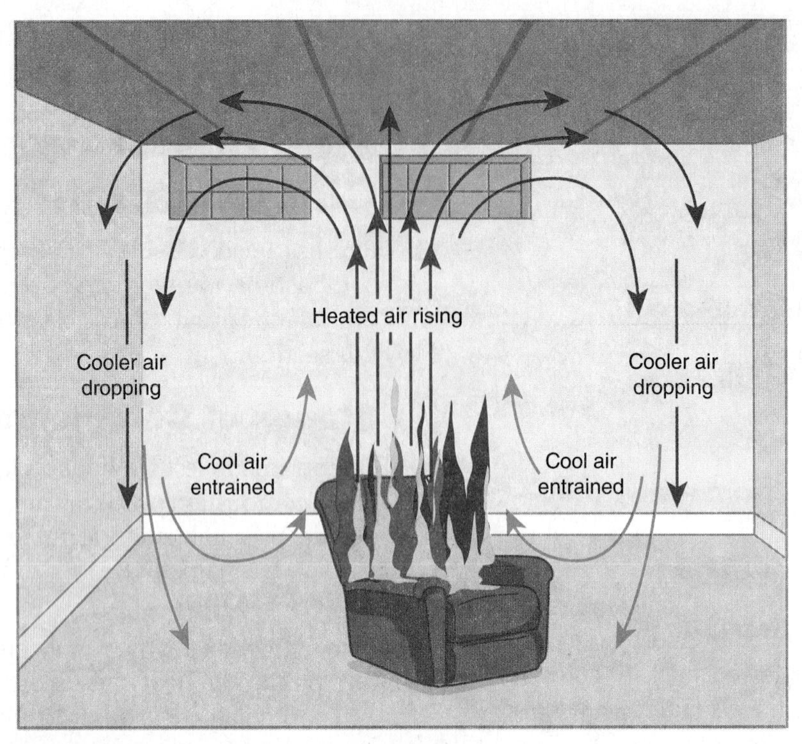

FIGURE 3-3 Fire plume with cooler air being entrained.

FIGURE 3-4 Flaming combustion is the phase that firefighters will most likely encounter.

furnaces, etc.). In these appliances, hydrocarbon fuel is premixed with air in a combustion chamber. A flame is created for heating purposes.

On most appliances, the combustion chamber can be adjusted. This ensures that the mixture will burn correctly. The ratios of oxygen and fuel are not an issue for the combustion process. The chemistry of the reaction still decides the total consumption of each. The proportions of fuel, oxygen, and end products are called the **stoichiometry of reaction**. The chamber of an appliance is designed to create a stoichiometric, or ideal, reaction. This **stoichiometric** reaction is a perfectly balanced reaction between the fuel, oxygen, and end products. The result is the almost complete consumption of all materials with little or no energy waste. In this example, the fuel's flame will be blue when burning, which means complete combustion, as seen when natural gas is burning. It should be noted that the color of the flames can range from blue to red, orange, yellow, and even green, depending on the type of fuel and the relative combustion process. Some examples are driftwood, which can produce a blue flame, or certain chemicals that produce multicolored flames.

Incorrect adjustment of the device may produce the deadly gas carbon monoxide. **Carbon monoxide (CO)** is the product of an incomplete combustion process. The unburned carbon causes the flames to appear yellow.

TIP

A flaming fire cannot occur unless a gas or vapor is burning. This is true even when the fuel supplying the flame is a solid material, such as wood, plastic, or coal. Firefighters who understand this can more effectively apply the extinguishing agent. They are aware of the base of the fire.

The second category is diffusive flaming. This is the most common type of flaming that firefighters will face. **Diffusive flaming** is characterized by yellow flames. In this case, the flame is part of the combustion process because it produces gaseous materials. Light and heat are emitted.

When these gaseous materials burn, they will produce yellow to orange flames. When a material such as alcohol burns, a blue flame will appear. Flame color is generally related to the oxygen mixture and the fuel itself. However, temperature does impact flame color in some cases. In a laboratory, flame colors from red to white have been seen. According to Fire Control Systems, a red flame could range in temperature from 1,112°F (600°C) to 1,832°F (1,000°C). The flame could turn orange between 1,832°F (1,000°C) and 2,192°F (1,200°C). Between 2,192°F (1,200°C) and 2,552°F (1,400°C), the flames become yellow. Hotter flames become blue-violet. It should be noted that these numbers may vary based on the reference source.

Smoldering Combustion

Smoldering combustion, also called glowing combustion, is identified by the absence of flame. It is also known for the presence of hot materials on the surface of the fuel. This is where oxygen diffuses into the fuel. The glowing means that the fire has a temperature over 1,000°F (538°C). Smoldering combustion has two phases: solid and gas. The solid material (fuel) is changed to gas, which allows the combustion process to continue at or near the surface. Three examples of smoldering combustion are shown in **FIGURE 3-5**.

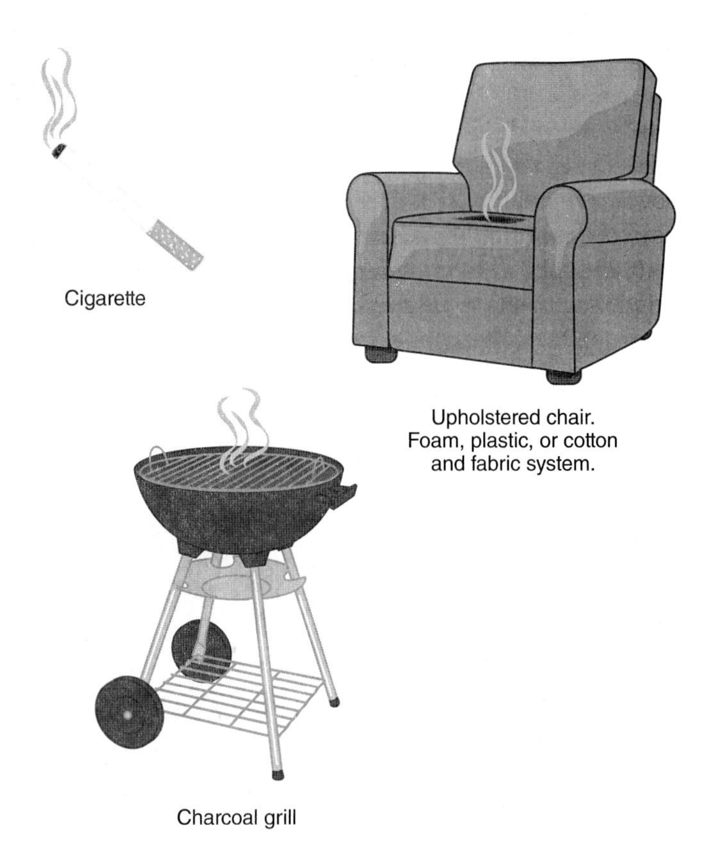

FIGURE 3-5 Three examples of smoldering combustion.

Smoldering combustion is a deadly process. The incomplete combustion creates very high levels of CO, which is a fire gas product that consists of a single atom of carbon and a single atom of oxygen. During smoldering combustion, more than 10% of the fuel is converted into CO. This colorless, tasteless, and odorless gas is present at every fire. The problem with CO for firefighters is that the body is attracted to it. In the lungs, CO competes with oxygen to bind with hemoglobin. Hemoglobin prefers CO to oxygen and accepts it over 200 times more readily. This prevents proper expulsion of CO from the body and prevents the intake of oxygen. CO is also flammable and has a very wide flammability range of 13% to 74%. Its ignition temperature is 1,128°F (609°C). Other toxic gases can be found in smoke as well (**TABLE 3-1**).

Fire Classification by Type of Substance Burning

Another method to classify fires is to identify them by the type of substance burning. In doing so, one can identify the most effective extinguishing agent for that class. **FIGURE 3-6** shows a list of fire classes. They are A, B, C, D, and K.

- **Class A fires** result from burning ordinary cellulosic materials. **Cellulosic materials** are those made by changing cellulose chemically. Some examples are wood, paper, and similar materials. Class A fires should burn with an ember or leave ash particles. Class A fires are typically put out with water.
- **Class B fires** are in flammable liquids and gases. Some examples are petroleum or petroleum-based products, such as gasoline, kerosene, and paint. In most cases, these fires are put out by smothering them to remove the oxygen supply. To do this, firefighters can use foam, CO_2 fire suppression equipment, or dry chemical extinguishers.

  A lithium-ion battery fire is another example of a Class B fire. These fires can be put out using a Class D dry powder or Class B foam. However, these fires are often left to burn out completely.
- **Class C fires** result from burning insulation materials in energized electrical equipment or wires. Water is not recommended to put out these fires. It will conduct electricity and may result in an electrical shock unless the electricity can be

TABLE 3-1 Toxic Fire Gases
Acrolein (C_3H_4O) is a strong respiratory irritant that is created when polyethylene is heated. It is also created when materials containing cellulose, such as wood and other natural materials, smolder. It is used in the production of medicines, weed killer, and tear gas.
Hydrogen chloride (HCl) is a colorless but pungent and irritating gas. It is given off during the breakdown with heat of materials that contain chlorine. Two examples are polyvinyl chloride (PVC) and other plastics.
Hydrogen cyanide (HCN) is a colorless gas with an almond odor. It is 20 times more toxic than CO. It can cause suffocation and can be absorbed through the skin. HCN is made in the combustion of natural materials, such as wool and silk. It is also made when polyurethane foam and other materials that contain urea burn. The bulk chemical is also used in electroplating businesses.
Carbon dioxide (CO_2) is a colorless, odorless, and nonflammable gas when produced in free-burning fires. Although it is nontoxic, CO_2 can suffocate a person by excluding the oxygen in a confined space. It can also increase a person's intake of toxic gases by speeding up their breathing rate.
Nitrogen oxides (NO_2 and NO) are two toxic and dangerous substances given off when pyroxylin plastics burn. Nitric oxide (NO) readily changes to nitrogen dioxide (NO_2) in the presence of oxygen and moisture. For this reason, NO_2 is the substance of most concern to firefighters. NO_2 irritates the lungs and can have a delayed effect on the whole body. The vapors and smoke from the nitrogen oxides have a reddish brown or copper color.
Phosgene ($COCl_2$) is a highly toxic, colorless gas with an odor of musty hay. It may be created when chemicals such as freon contact flame. Expect freon in fires at cold storage facilities. It may also be a factor in fires in heating, ventilating, and air conditioning systems. It is a strong lung irritant. The full extent of the damage is not evident until several hours after exposure.

FIGURE 3-6 Five classes of fire.

safely de-energized. A **clean agent suppression system**, one that uses nontoxic agents, might be the best solution. It will not leave a residue or damage electrical equipment.

CO_2 can be used in confined areas to extinguish a Class C fire. However, it may damage equipment. The CO_2 will freeze the water vapor in air, and this vapor will corrode certain metals used for electronics.

- **Class D fires** are fires in combustible metals. Some examples are titanium, magnesium, aluminum, and potassium. These fires often require the use of special agents to put out a fire in a specific metal. Firefighters inspecting locations with combustible metals should confirm that the proper type of extinguishing agent is near the metal being processed. Water can be used on some Class D fires. However, it must be applied only after it is deemed safe to use. An example of when to use water would be a car fire in which more than one combustible metal is present in small amounts.
- **Class K fires** are fires in liquid cooking materials, such as oils, grease, and vegetable and animal fats. These fires burn very hot. In certain cases, dry chemical extinguishing agents or water are not the best way to put out these fires. For example, to put out fires in kitchen ranges, hoods, ducts, and deep fat fryers, the extinguishing mechanism for both dry and wet chemicals is based on the process of saponification.

Saponification is chemically converting the fatty acid in cooking oil or grease to soap or foam. The soap or foam made by this process is readily broken down by exposure to heat. By adding air and water, the soap or foam forms bubbles of air in a water–soap membrane. The bubbles can be made lighter (more air) or wetter (more water) depending on the type of coverage needed by the fuel burning. The foam can be laid on top of a burning fuel to cool and smother it.

Fire Classification by Stages

The combustion process in enclosed compartments or rooms occurs in four stages. Some fires may display one or more of the three events described below in addition to the four stages. Firefighters will better understand the burning process if they can recognize the various stages and events in fire development. They will also be better able to efficiently and safely put out a fire.

The stages of fire are:

- Incipient/Ignition
- Growth
- Fully developed
- Decay

The fire events are:

- Flame over or rollover
- Flashover
- Backdraft

The first stage of fire development begins after ignition. The **incipient stage** (also called the ignition stage) is the point at which the four parts of the fire tetrahedron come together. The materials reach their ignition temperature, and the fire starts. The flames are initially small and contained within the materials. The oxygen supply is not reduced and is about 21%. A plume of hot gases rises from the flames. Flame temperatures can be as high as 1,000°F (538°C). The heated convection currents carry the products of combustion upward—in a house fire, for instance, to the upper areas of a room. The currents draw in more oxygen at the bottom of the flames. This sustains the combustion process. The spreading of the flames moves the fire to the second stage.

As the fire grows more intense, more fuel becomes involved. Additional heat energy is released. At this point, the fire enters the second stage. The second stage is called the growth stage. The **growth stage** is when the fire increases fuel consumption and heat generation. The convection currents carry the heated gases to the upper part of the room. On reaching the ceiling area, they spread horizontally until they reach the walls.

After reaching the walls, the heated gases drop into the lower portion of the room (**FIGURE 3-7**). These same gases radiate heat to the remaining parts and contents of the room. This action spreads and increases the heat intensity, which brings the materials in the room up to

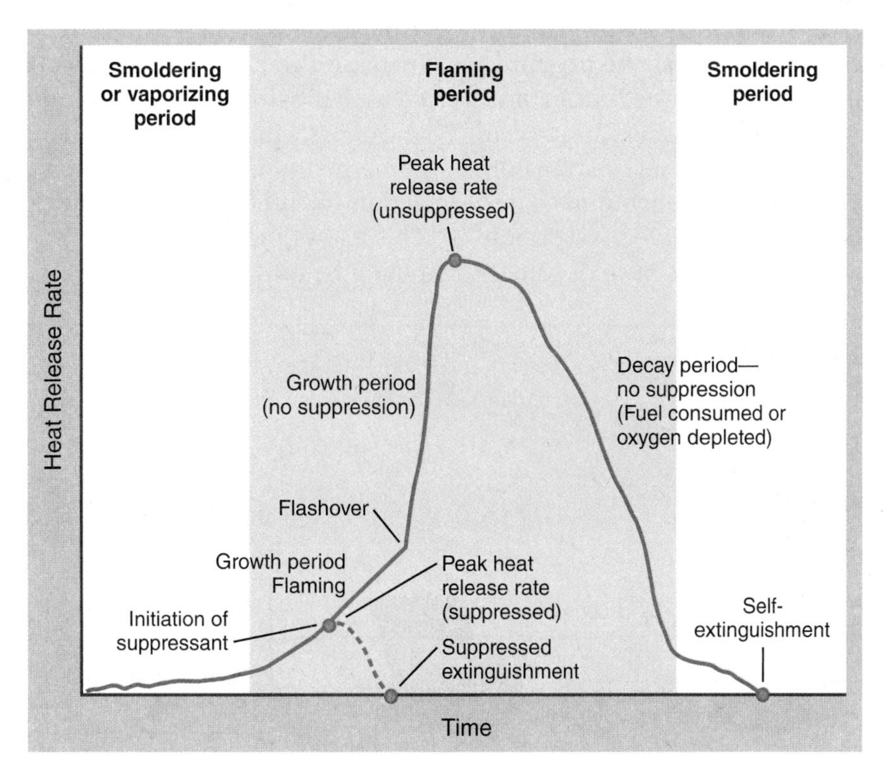

FIGURE 3-7 Temperatures associated with the stages of fire and the flashover event.

their ignition temperature. At this point, the fire is at the fully developed stage. In the **fully developed stage**, there is a maximum generation of heat and flames. The temperatures in the upper portions of the ceiling reach 1,800° to 2,200°F (982° to 1,204°C). The fuel in the room is eventually used up. Oxygen is reduced to a percentage between 9% and 12%.

The drop in oxygen slows the combustion process. However, the upper parts of the room continue to be packed with superheated fuel-rich particles. The floor area is still cool and has oxygen. A hot layer zone and cool layer zone develop. The upper layer of gases is hot. In some cases, it has a large concentration of unburned particles. This condition would exist due to lack of oxygen. If the heated layer of gases reaches the ignition temperature of the unburned heated particles, and there is enough oxygen, a flame over event will occur. A **flame over** is when flames travel through unburned gases in the upper portions of a confined area during the fire's development. If a flame over does not occur, and the heat intensity increases, the room contents in the lower cool zone begin to heat up toward their ignition temperature. Once they reach their ignition temperature, a flashover event occurs. A **flashover** is when all the contents of a room or enclosed compartment reach their ignition temperature at the same time. This results in an explosive fire. A flashover will occur during the transition between the growth stage and the fully developed stage.

The fire consumes the fuel and oxygen in the compartment. The air is still hot, but the rate at which the heat is being released begins to slow as the oxygen is consumed. Large quantities of unburned gases and particles are given off. At this point, because the oxygen has been used up and the compartment is still very hot, the introduction of oxygen into the environment results in a backdraft. A **backdraft** is a sudden reignition of room contents once the oxygen has been depleted. The introduction of oxygen then creates an immediate explosion. This introduction of oxygen can be from firefighters opening doors or windows to reach the fire.

If firefighters vent the fire under controlled conditions, the venting releases the confined fire gases. This will prevent a backdraft and the fire will continue at a much slower rate. The fire is then at the decay stage. In the **decay stage**, the remaining mass of glowing embers gives off heat for some time, but eventually all the fuel will be used up and combustion will cease.

Fire Events

Fires in compartments or containers, such as rooms and buildings, can become deadly. Conditions can change during the combustion process and lead to fire events. Let's examine these fire events in more detail.

Flame Over

As mentioned earlier, a flame over, also known as a rollover, occurs when flames travel through the unburned gases in the upper portions of a confined area (**FIGURE 3-8**). A flame over differs from a flashover because only the fire gases are involved. The surfaces of other fuels, such as furnishings and items within the confined space, are not involved. A flame over occurs during the growth stage as a hot gas layer forms at the ceiling or upper portion of the compartment.

If the mixture of the fuel and oxygen is within the explosive range, an explosion will follow. This will increase the pressure of the confined space. The increased pressure can break windows and destroy walls and doors. Firefighters trapped in this fire may be overcome by the intense rapid increase in heat. The best action for firefighters is to quickly drop to the floor. This allows the fire to blow over. They can then crawl out of the room.

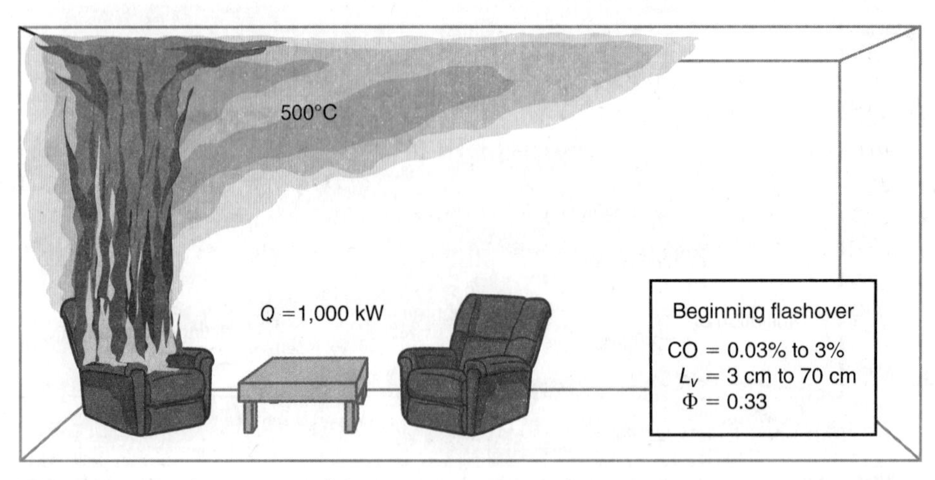

FIGURE 3-8 Flame over/rollover.

Flames may be seen in the heated gas layer as the gases reach their ignition temperature. Although the flames add to the total heat in the confined space, this situation is not a flashover. It can, however, signal that a flashover might happen soon.

Flame over may also be seen when unburned gases vent from a confined space to a nearby space during the fire's growth. In doing so, the gases mix with oxygen. If they are at their ignition temperature, flames often will become visible in the gas layer.

Flashover

A flashover occurs during the growth stage of a fire. Hot gases begin to build up at ceiling level. If the heat is enough to bring other materials in the room to their ignition temperature, all materials in the room will ignite into flaming combustion. This phenomenon is a flashover. It is a very dangerous event for firefighters and can be fatal. The temperatures during a flashover can reach 2,200°F (1,204°C) (**FIGURE 3-9**).

FIGURE 3-9 Flashover.

TIP

When a flashover is imminent, firefighters have two options. If water is available, they should immediately open the nozzle and cool the interior until no flames are seen. If this is not possible, they must get out of the environment immediately.

TIP

Firefighters need to know that they cannot survive a flashover because the entire confined space is spontaneously ignited. All materials have been brought up to their ignition temperature.

Backdraft

Combustion is a balancing act between the air, the fuel, and the rate at which the fuel is used up. In this process, if the ventilation is limited, the fire will progress slowly. The temperature will rise gradually. At the same time, the production of smoke and flammable fire gases will increase. This leads to a room or confined space filled with the products of incomplete combustion.

The fire could burn until the oxygen in the room is greatly reduced. If this happens, the fire will appear to breathe. The fire breathes as it draws in oxygen around cracks and small openings in rooms. As the drawn-in oxygen increases the combustion process, it only slightly raises the pressure in the room. This pushes out the smoke, thus relieving the pressure, and the cycle begins again. This "breathing" fire is awaiting a large influx of oxygen. Firefighters provide it by opening a window or door. Once this window or door is opened, the rush of oxygen into the room leads to a smoke explosion or a backdraft (**FIGURE 3-10**).

Backdraft can be prevented by proper ventilation techniques. A vented opening allows the hot, pressurized gases to be released outside the building. This reduces the temperature of the room and the amount of smoke. It also allows firefighters to direct the movement of fire. Most importantly, it relieves the pressure of the confined gases. See Chapter 6 for a discussion on special concerns in ventilation procedures.

Heat Flow Path

Heat flow path is the movement of heat, smoke, and fire gases from the high-pressure fire areas to all low-pressure areas both inside and outside of a building. There could be several heat flow paths in a building. The number depends on the building design and the available ventilation openings, such as windows and doors. Firefighters working in the heat flow path of a fire are at significant risk of injury or death.

There are several ways for firefighters to control heat flow paths. Outside the building, initial response units should consider the speed and direction of the wind in relation to the structure and then select the best access point to enter. They can also bounce a stream of water off the ceiling from outside to cool the fire gas layer of a room. Inside the building, fire crews can isolate the heat flow path by closing doors and using wind control

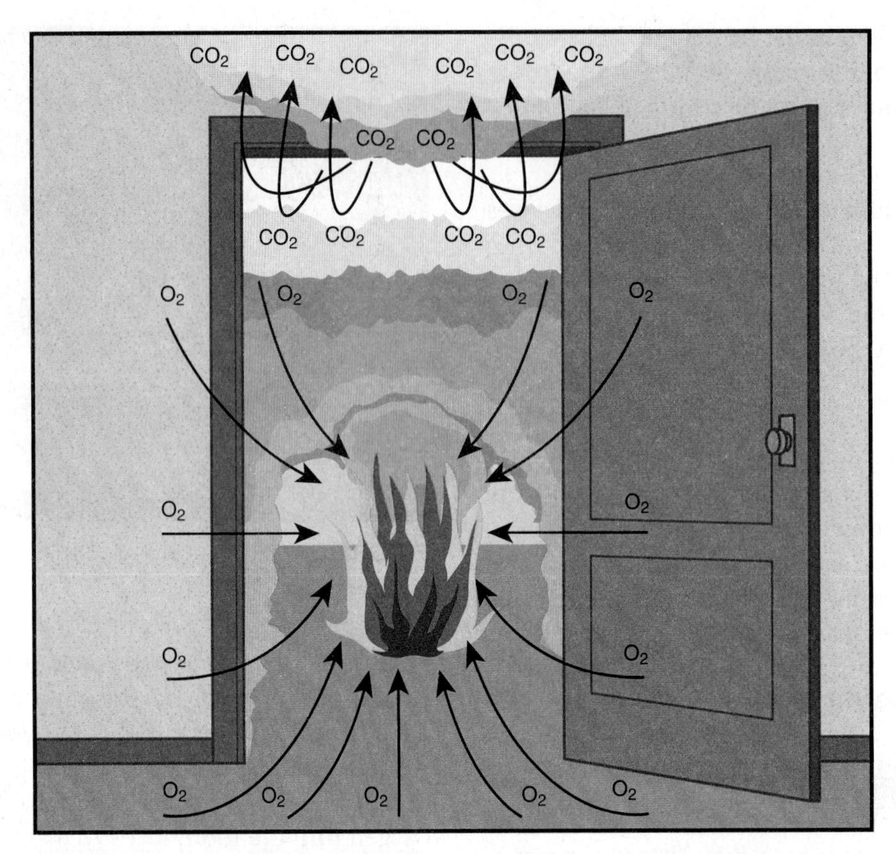

FIGURE 3-10 Backdraft conditions.

devices and smoke blockers to reduce the amount of air feeding the fire. Firefighters can also strategically ventilate to reverse the heat flow path. A change of floors for fire attack needs to be coordinated to allow for the protection of advancing hose lines.

These rules for engagement will help provide a safer environment for firefighters and can reduce line-of-duty deaths.

Building Construction and Fire Spread

The efficiency of firefighting declines quickly if the fire moves vertically through the building or bypasses horizontal construction barriers. For example, a pre–World War II high-rise, such as the Empire State Building in New York, included thick, concrete walls, floors, and ceilings. It had columns on 30-ft centers with windows and few or no vent shafts for air conditioning or other building uses. The concrete walls and columns would absorb a great deal of the heat. Thus, horizontal movement of the fire was blocked by the presence of thick concrete.

High-rise buildings since World War II are built with a truss construction and drywall in the interior. This provides fire resistance for the steel support of the building. Drywall cannot absorb as much heat as thick concrete. As a result, much of the heat from the fire is released back into the room. This quick buildup of heat spreads the heated fire gases into other portions of the building where they can endanger people trying to exit. An example of this type of construction was the original World Trade Center in New York.

Compartmentation, or the process of dividing large structures into compartments or units, is one option for high-rise building safety in the event of a fire. Compartments are built with fire-resistive materials rated at 30 minutes to 4 hours. They often include special protection for doors and other openings in the compartment. One of these areas could serve 20 floors. With such a compartment or fire floor, people from 20 floors above the fire floor could evacuate to this safe area. They could stay there until the fire was under control.

Compartmentation also restricts the fire to certain areas or floors of the building. It prevents the fire from reaching areas that are critical or contain hazardous materials. Common examples of these areas include plant rooms and server rooms.

Mass/Drying Time

Mass, or the quantity of material, impacts how long it will take a source of ignition to heat a material to its ignition temperature. The thicker or heavier the mass, the longer it takes to raise the temperature. This principle, when applied to the heating of water droplets, explains why smaller water droplets convert from water to steam more rapidly than larger water droplets. The smaller and less dense the water, the faster it rises to its conversion temperature. This is known as the law of latent heat of vaporization. **Latent heat of vaporization**, also known as the enthalpy of vaporization or evaporation, is the amount of energy (enthalpy) that must be added to a liquid (water) to change it into a gas (steam). Firefighters use the principle of latent heat of vaporization to calculate the conversion of water to steam. The steam absorbs the heat of the fire gases from the building and fire areas and moves it to the exterior of the building.

Weather Conditions

Hot, dry, windy conditions impact the burning traits of inside fires as well as outside fires. In fires inside rooms or confined areas, the hot, dry air works to dry out furniture, wall decor, and ceiling coverings. The added heat from the weather raises the potential fuel closer to its ignition temperature by driving the moisture out of the fuel.

The outside weather also affects the movement of air in high-rise buildings with what is known as the stack effect. The **stack effect** is the temperature difference between the inside and outside of a building that contributes to air movement inside the building. The interior of a high-rise building has air currents similar to ocean currents. They move in patterns driven by temperature differences. They can spread heated fire gases throughout a building by drawing air in from the outside through the lower floors. Think of how a chimney works to draw air in from the bottom and push it upward to the top.

Windy conditions outside the building can impact horizontal ventilation if windows are used to remove fire gases. Firefighters need to make sure that the opening of windows does not spread the fire into unburned areas of a building.

Relative Humidity

Moisture in the form of water vapor is always present in the air. This moisture is measured as relative humidity. **Relative humidity** is the ratio of the amount of moisture in a given volume to the amount that volume would contain if it were full. The amount of moisture in the air affects the amount of moisture in the fuel. This applies to both indoor and outdoor fuels. Dry air with a low moisture content will absorb water from fuels. Likewise, near the ocean or other water sources, dry fuel will absorb moisture from the air. The moisture content of the air and fuel is important in firefighting. Drier fuels will catch fire and burn much more rapidly, cleaner, hotter, and with less smoke and fewer unburned particles than wetter fuels. Warehouse fires with furniture stored for long periods are a good example of the drying effect of storage on the burning traits of fuels. After being stored for a long time, furniture tends to become dried out. This lowers its moisture content, and in turn, lowers its ignition temperature.

This same burning trait can be seen in vegetation fires. In those fires, the air layer near the ground influences both fire and fuel behavior. If the relative humidity is 30% or below, the fire will burn freely. According to the National Weather Service, the fire danger is critical when there is a combination of relative humidity of 15% or less and sustained winds of 25 mph or greater for at least 3 hours in a 12-hour period. Extreme fire behavior (i.e., the fire spreads rapidly and in unpredictable directions) will occur because the fuels are dried and preheated.

High humidity or a high percentage of moisture in the air also affects the movement of fire gases. Higher humidity means the air is heavier (more water moisture) and heated fire gases tend to move horizontally, spreading the fire outward.

Fire Flow

Fire flow is how much water is needed to put out a building fire. The National Fire Protection Association (NFPA) 1 Fire Code refers to the amount of water available for firefighters combined with the type of building construction. Fire flow is often decided during the pre-fire planning phase. For firefighters at an actual fire, the dimensions of the room or building are first estimated. That number is then divided by 100 to figure out how many gallons per minute are needed.

An example would be a room that is 24 feet wide by 30 feet long by 10 feet tall.

$$(24 \times 30)10 = 7{,}200 \text{ divided by } 100$$
$$= 72 \text{ gallons per minute}$$

This example is used to illustrate how much water was needed in the past. The increased use of plastics in most of today's product needs to be considered when calculating the proper amounts of water needed today.

Heat Measurement

The temperature of a material determines whether heat will transfer to or from other materials. Heat always flows from higher-temperature materials to lower-temperature materials. Temperature is measured in degrees using four scales. Two of these scales

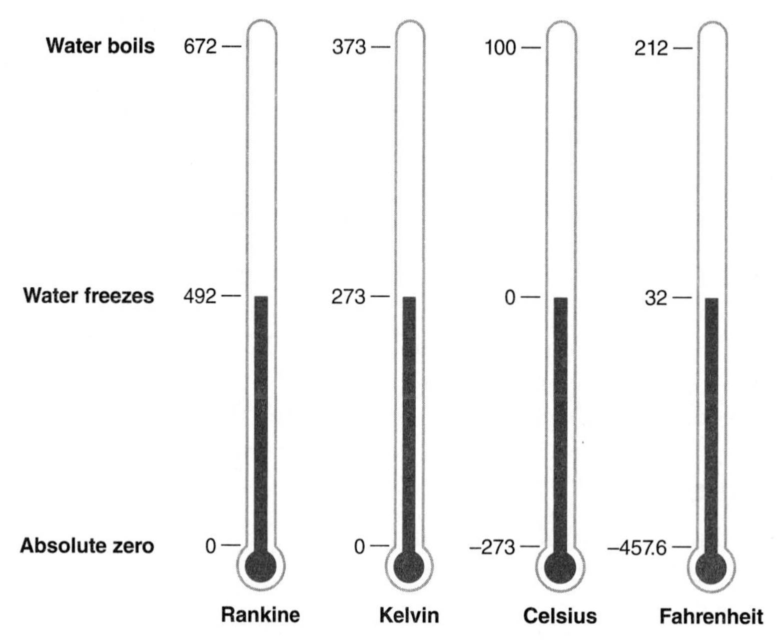

FIGURE 3-11 Relationship among temperature scales.

are mostly used in research laboratory settings. The four scales used are:

- Kelvin (mainly research)
- Rankine (mainly research)
- Celsius
- Fahrenheit

A comparison of these scales is found in **FIGURE 3-11**.

Heat Transfer

The transfer of heat is important in all phases of the combustion process from ignition through extinguishment. The presence of heat is responsible for continuing the combustion process. Heat releases the vapors and free radicals from the fuel. Fire extinguishment is a break in the transfer process.

Heat is transferred by one or more of four methods:

1. Conduction
2. Convection
3. Radiation
4. Direct flame impingement

Conduction

Conduction is the transfer of heat energy from the hot to the cold side of a medium. This takes place through collisions of molecules to molecules or atoms to atoms (**FIGURE 3-12**). Its impacts are most noticeable in solid materials. In solids, contact between molecules is at its highest and convection air currents do not occur.

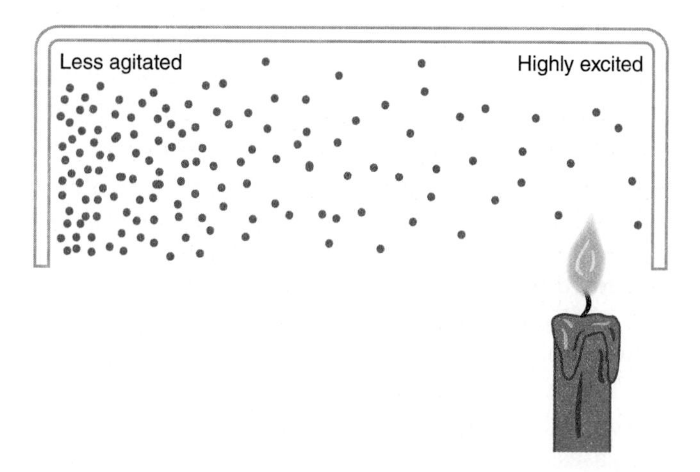

FIGURE 3-12 Conduction is the transfer of heat energy from the hot to the cold side of a medium.

One way to think of conduction is to visualize the heated material as moving molecules. When these molecules contact a colder material (where the molecules are not moving as fast), they transfer this movement. Thus, a transfer of energy occurs. For example, heat

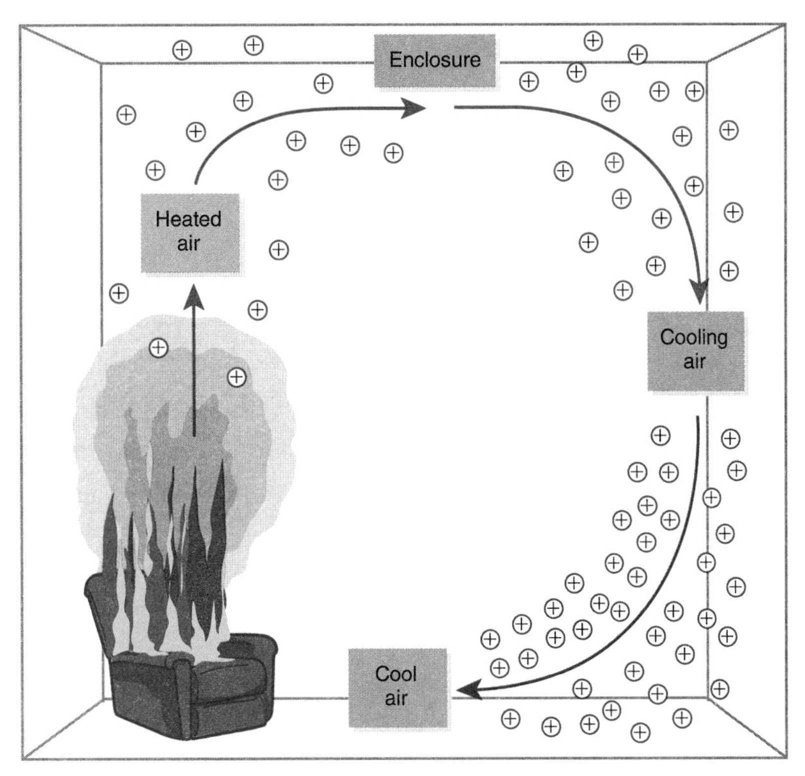

FIGURE 3-13 Convection is the transfer of heat by circulating air currents.

traveling by conduction through a rod will eventually raise the temperature at the other end of the rod. The amount of heat flowing through the rod is in proportion to the time, cross-sectional area, and difference in temperature between the ends. It is inversely proportional to the length of the rod.

Firefighters must be aware of thermal conductivity, or the rate at which heat is transferred through a material. It provides an estimate of the time that the material will be able to resist the effects of a fire.

Convection

Convection is the transfer of heat by circulating air currents. As heat rises, it draws cooler air in from below it, which is entrainment. This action moves the heated air upward and throughout a confined space, like a room. The heated air may then come in contact with other materials in the room, and heat is again transferred. This is a common example of conduction. The heated air will also carry smoke and fire gases up and throughout the building. This transfer of heat will lead to the fire reaching flame over stage, flashover stage, and its fully developed stage. Convection is the primary form of heat transfer in a building (**FIGURE 3-13**).

Radiation

Radiation is energy that travels across a space that does not need an intervening medium, such as a solid, gas, or fluid (**FIGURE 3-14**). It travels as electromagnetic waves and is very similar to light, radio waves, and x-rays. The radiation waves give off heat only when the waves can excite molecules. For this reason, firefighters can protect buildings from radiated heat by allowing water to run down the face of the building. The water cools the building, thus slowing the movement of the molecules. Radiation is the main form of heat transfer to exposures. An **exposure** is anything outside the immediate area of fire origin. This may be an outbuilding, adjacent house, fence, or vehicle, which will ignite when exposed to the radiant heat.

The difference between convection and radiation heat transfer can best be understood through the actions of a candle. A candle has all three forms of fuel: solid, liquid, and vapor. When the candlewick is lit, the wax nearest the wick melts and is drawn up into the wick. The wax turns to a vapor, and the proper air to fuel mixture is reached. A flame appears just above the wick. The flame does not touch the wick because the amount of flammable gas is greater than the amount of oxygen in the air for the flammability range requirements. The

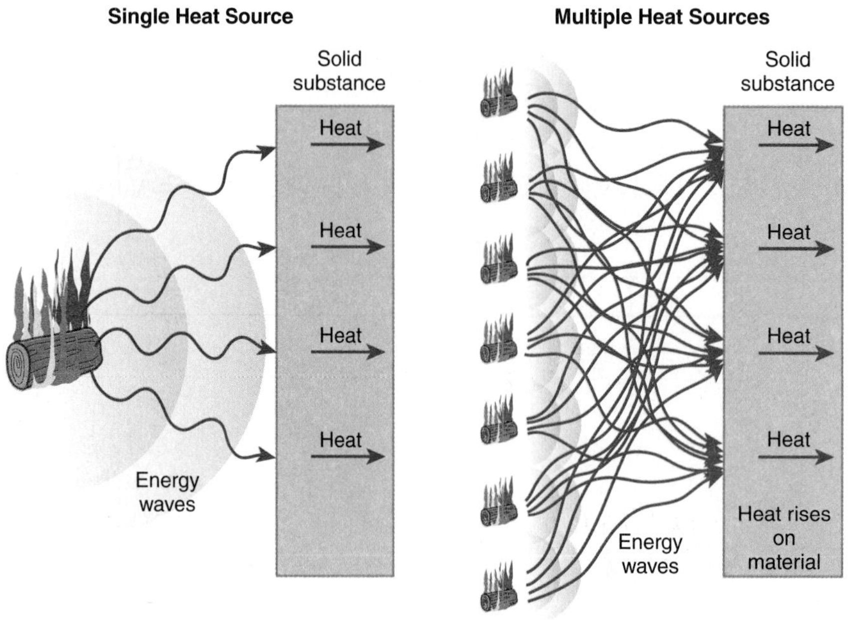

Single Heat Source

Solid substance

Heat

Heat

Heat

Heat

Energy waves

Multiple Heat Sources

Solid substance

Heat

Heat

Heat

Heat

Heat rises on material

Energy waves

FIGURE 3-14 Radiation is energy that travels across a space that does not need an intervening medium, such as a solid, gas, or fluid.

environment is "too rich." The heat produced from the burning wick creates a hot air current that rises and spreads out in a 360-degree pattern. The flame produces a radiant heat and light, which provide additional heat to the surrounding areas.

Direct Flame Impingement

At an actual fire, heat is also transferred by the flames directly impinging on materials. This raises the temperature of the materials to the point where they catch fire (**FIGURE 3-15**). Another way to visualize the action of direct flame impingement is to think of it as combining both convection and radiation. The heat is transferred from the plume of hot fire gases by convection. The plume gives off radiation in the forms of heat and light. This happens until the new fuel pyrolyzes and produces gaseous fuels. These fuels are then ignited by the flames.

FIGURE 3-15 Flames directly impinging on materials transfer the heat, raising their temperature to the point where combustion occurs.

CASE STUDY

CONCLUSION

1. Are conditions in line for a possible explosion or fire?

The conditions are in line for a possible explosion or fire. Warm heat and lack of circulating air could cause tanks to release gas into the air. The improperly stored furniture rags could also catch fire.

2. What can the homeowner do to limit or prevent the combustion of the rags?

The homeowner could prevent combustion by storing the oily rags in an airtight metal container. More circulating air in the garage area would also help to avoid a fire.

3. What is the ignition source going to be in the above story?

The ignition source would be the oily furniture stain (linseed oil) that had been improperly stored in the cardboard box.

4. What other factors would be a concern in a typical garage?

As personal property, the garage is not subject to an inspection by the fire department. Homeowners may store any type and amount of hazardous, flammable, or explosive materials on their property. Firefighters should factor in any such hazardous conditions when responding to an incident at a residence.

WRAP-UP

SUMMARY

- Combustion is a chemical reaction between a fuel and oxygen that creates smoke, heat, and light.
- The fire tetrahedron is a four-sided model showing the heat, fuel, oxygen, and chemical reaction needed for combustion.
- Fire is classified in several ways, including the stages of combustion, the type of substance burning, stages of fire, and fire events.
- There are five classes of fire that are organized by the substances that are burning. Using these five classes, firefighters can identify the most effective extinguishing agent for each.
  - Class A fires (in ordinary cellulosic materials)
  - Class B fires (in flammable liquids and gases)
  - Class C fires (in energized electrical equipment)
  - Class D fires (in certain metals)
  - Class K fires (in liquid cooking materials)
- Firefighters can also use the fire stages and events as a method to provide insight into the condition and extent of a fire. By using this information, they also can control the fire more safely.
- The physical and chemical properties of the fuels that feed the fires encountered by firefighters all affect how a fire will burn, how it will spread, and the quickness of its burning rate. This information is vital to firefighters because it enables them to determine how quickly and safely a fire can be controlled.
- Firefighters should understand how heat is transferred. Fire extinguishment is a break in the transfer process.
- Heat is transferred by one or more of four mechanisms:
  - Conduction
  - Convection
  - Radiation
  - Direct flame impingement

KEY TERMS

backdraft A sudden reignition of room contents once the oxygen has been used up; the introduction of oxygen then creates an immediate explosion.

carbon monoxide (CO) A fire gas product that consists of a single atom of carbon and a single atom of oxygen.

cellulosic material Material made by changing cellulose chemically.

Class A fire Fire involving ordinary cellulosic materials.

Class B fire Fire involving flammable liquids and gases.

Class C fire Fire involving energized electrical equipment or wires.

Class D fire Fire involving combustible metals.

Class K fire Fire involving liquid cooking materials.

clean agent suppression system Fire suppression system that uses nontoxic agents.

KEY TERMS CONTINUED

combustion A chemical reaction between a fuel and oxygen that creates smoke, heat, and light.

compartmentation The process of dividing large structures, such as high-rise buildings, into compartments or units to provide safe areas and prevent the spread of fire.

conduction The transfer of heat energy from the hot to the cold side of a medium through collisions of molecules to molecules or atoms to atoms.

convection The transfer of heat by circulating air currents.

decay stage The stage when the fire has used up all of the fuel and combustion will cease.

diffusive flaming A category of flaming combustion characterized by yellow flames that are part of the combustion process.

entrainment The gathered or captured cooler air that replaces the rising heated air surrounding the point of combustion.

exposure Anything outside the immediate area of fire origin.

fire A rapid oxidation process that involves heat, light, and smoke in varying intensities.

fire tetrahedron A four-sided model showing the heat, fuel, oxygen, and chemical reaction necessary for combustion.

fire triangle A fire requires the presence of heat, fuel, and oxygen, which is depicted in a model.

flame over The flames that travel through unburned gases in the upper portions of a confined area during the fire's development; also called rollover.

flaming combustion An exothermic or heat-releasing chemical reaction with flames between a substance and oxygen; can be broken into two categories: premixed flaming and diffusive flaming.

flashover When all the contents of a room or enclosed compartment reach their ignition temperature at the same time, resulting in an explosive fire.

fluid mechanics The study of fluids at rest or in motion.

free radical A fragment of a molecule that has at least one unpaired electron.

fully developed stage The stage of a fire where there is maximum generation of heat and flames; all fuel and oxygen are used up.

growth stage The stage where the fire increases fuel consumption and heat generation.

incipient stage The point at which the four parts of the fire tetrahedron come together and the materials reach their ignition temperature and fire starts; also called the ignition stage.

latent heat of vaporization The amount of energy (enthalpy) that must be added to a liquid (water) to change it into a gas (steam); also known as the enthalpy of vaporization or evaporation.

particulate The unburned product of combustion one can see in smoke.

piloted ignition temperature The temperature of a liquid fuel at which it will self-ignite when heated.

premixed flaming A category of flaming combustion in which a gaseous fuel mixes with air before catching fire.

pyrolysis The process of breaking down a solid fuel with heat into gaseous parts.

radiation Energy that travels across a space that does not need an intervening medium, such as a solid, gas, or fluid.

relative humidity The ratio of the amount of moisture in a given volume to the amount that volume would contain if it were full.

saponification The process of chemically converting the fatty acid in cooking oil or grease to soap or foam.

smoldering combustion The absence of flame and the presence of hot materials on the surface where oxygen diffuses into the fuel; also called glowing combustion.

spontaneous combustion An occurrence where a material self-heats to its piloted ignition temperature before catching fire.

stack effect The temperature difference between the inside and outside of a building that contributes to air movement inside the building.

stoichiometric An ideal burning situation or a condition where there is perfect balance between the fuel, oxygen, and end products that results in the almost complete consumption of all materials with little or no energy waste.

stoichiometry of reaction The proportions of fuel, oxygen, and end products.

thermodynamics The relationship between heat and other forms of energy.

REVIEW QUESTIONS

1. The fire tetrahedron model depicts combustion. How do the four sides of this model show the combustion process and how a fire could be put out?
2. What are the stages of fire and the events that may occur if conditions are right?
3. The term *stoichiometric* is used to describe a burning situation or condition. What does this term mean?
4. How have building construction materials and techniques changed to increase the amount of heat firefighters face in a confined space?
5. What are the four means of heat movement and how can firefighters stop or reduce some of this movement?

DISCUSSION QUESTIONS

1. What is the fire tetrahedron? What are its parts?
2. What is the purpose of classifying types of fires?
3. What are the four types of heat transfer and how can each spread fire through a building?
4. What is the main form of heat transfer in a building, and how does this contribute to fire spread?

APPLYING THE CONCEPTS

You have been asked for advice about storing flammable liquid and portable propane grill tanks at a remote vacation cabin. The cabin is type 5 A-frame construction with wood siding and a wood shake shingle roof. The area is heavily wooded and not easily accessible. This person tells you that there is already a 1,000-gallon propane tank on the property that is used for heating and cooking in the cabin.

1. What information and suggestions do you have for this person?
2. How would you describe where they store the flammable liquids?
3. Where would you recommend they keep the portable propane tanks?
4. Are there any additional precautionary instructions you might give regarding the safe space concept around their cabin?

REFERENCES

Angle, James, et al. *Firefighting Strategies and Tactics*, 2nd ed. Clifton Park, NY: Delmar Cengage Learning, 2008.

Bolton, Ian. "Editor's Pick 2014: Understanding Flow Paths." FIREFighting in Canada. April 24, 2014. Accessed April 6, 2023. https://www.firefightingincanada.com/understanding-flow-paths-18596/.

CLM Fireproofing. "Passive Fire Protection: Everything You Need to Know About Fire Compartmentation." n.d. Accessed January 14, 2023. https://clmfireproofing.com/what-is-compartmentation-in-fire-protection/.

CTIF. "Survive in the Flow-Path." March 29, 2018. Accessed April 6, 2023. https://ctif.org/news/survive-flow-path.

Delmar Cengage Learning. *Firefighter's Handbook: Firefighting and Emergency Response*, 3d ed. Clifton Park, NY: Delmar Cengage Learning, 2008.

Dickinson, Micah. "What Are the 5 Different Classes of Fires?" Vanguard Fire & Security Systems. June 10, 2021. Accessed January 14, 2023. https://vanguard-fire.com/what-are-the-5-different-classes-of-fires/.

Elgas New Zealand. "Blue Flame vs Yellow Gas Flame." n.d. Accessed January 14, 2023. https://www.elgas.co.nz/resources/elgas-blog/242-why-does-a-gas-flame-burn-blue-lpg-gas-natural-propane-methane/#:~:text=You%20get%20a%20blue%20gas,gas%20molecules%20in%20the%20flame.

Engineering ToolBox. "Water: Heat of Vaporization vs. Temperature." 2010. Accessed January 14, 2023. https://www.engineeringtoolbox.com/water-properties-d_1573.html.

Fire Consultancy Specialists. "The Fire Triangle vs the Fire Tetrahedron." July 7, 2022. Accessed January 14, 2023. https://fireconsultancyspecialists.co.uk/fire-triangle-vs-fire-tetrahedron-whats-the-difference/.

Fire Control Systems. "Flames Different Colors Explained." January 9, 2020. Accessed April 4, 2023. https://firecontrolsystems.biz/news/flames-different-colors-explained/.

Fire Protection Association. "What Is Fire Compartmentation?" November 16, 2020. Accessed January 14, 2023. https://www.thefpa.co.uk/news/fire-safety-advice-and-guidance/what-is-fire-compartmentation-.

REFERENCES CONTINUED

Firefighterinsider.com. "Is Carbon Monoxide Flammable? Should You Be Worried?" n.d. Accessed January 14, 2023. https://firefighterinsider.com/carbon-monoxide-flammable/.

Friedman, Raymond, and Gann, Richard. *Principles of Fire Behavior and Combustion.* Quincy, MA: National Fire Protection Association, 2002.

Frontline Wildfire Defense. "The Four Stages of Fire Combustion & Growth." n.d. Accessed January 14, 2023. https://www.frontlinewildfire.com/wildfire-news-and-resources/the-four-stages-of-wildfire-combustion/.

Hawley, Chris. *Hazardous Materials Incidents*, 3rd ed. Clifton Park, NY: Delmar Cengage Learning, 2008.

Helmenstine, Anne Marie. "Radiation Definition and Examples." ThoughtCo. Updated December 8, 2019. Accessed January 14, 2023. https://www.thoughtco.com/definition-of-radiation-and-examples-605579.

InterFire. "Term of the Week: Entrainment." n.d. Accessed January 14, 2023. https://www.interfire.org/termoftheweek.asp?term=1804.

Klinoff, Robert. *Introduction to Fire Protection and Emergency Services,* 6th ed. Burlington, MA: Jones & Bartlett Learning, 2019.

Marinucci, Richard A. *The Fire Chief's Handbook*, 7th ed. Tulsa, OK: Fire Engineering, 2015.

Monaghan, Jonathan. "Flow Paths and Fire Behavior." Firefighter CloseCalls.com. November 28, 2015. Accessed April 6, 2023. https://www.firefighterclosecalls.com/flow-paths-and-fire-behavior/.

Ottawa Fire Services. n.d. *Flame Types.* Accessed January 14, 2023. https://guides.firedynamicstraining.ca/g/structural-firefighting-fundamentals-of-fire-and-combustion/118139.

Pennsylvania Office of the State Fire Commissioner. "Fire Dynamics Terminology." n.d. Accessed April 6, 2023. https://www.osfc.pa.gov/AboutOSFC/Public%20Information/Documents/Fire_Dynamics_Terminology.pdf.

Quintiere, James G. *Principles of Fire Behavior*, 2nd ed. Boca Raton, FL: CRC Press, 2016.

Smith, Alan. "Weather Factors That Influence Fire Danger." OpenSnow. August 26, 2022. Accessed January 14, 2023. https://opensnow.com/news/post/weather-factors-that-influence-fire-danger.

Smokeybear.com. "Elements of Fire." 2021. Accessed January 14, 2023. https://smokeybear.com/en/about-wildland-fire/fire-science/elements-of-fire#:~:text=3-,Oxygen,%2C%20embers%2C%20etc.

Target Fire Protection. "What Is the Temperature of Fire?" n.d. Accessed March 15, 2023. https://www.target-fire.co.uk/resource-centre/what-is-the-temperature-of-fire/.

UCAR Center for Science Education. "Conduction." 2018. Accessed January 14, 2023. https://scied.ucar.edu/learning-zone/earth-system/conduction.

Wermac. "Lower and Upper Explosive Limits for Flammable Gases and Vapors." n.d. Accessed January 14, 2023. https://www.wermac.org/safety/safety_what_is_lel_and_uel.html.

CHAPTER

4

Extinguishing Agents

LEARNING OBJECTIVES

After studying this chapter, you will be able to:

- Describe the basic parts of the fire extinguishment process.
- Explain the various types of agents used to put out or control fires in each of the five classes.
- Describe in detail the variety of extinguishing agents.
- Explain how each of the extinguishing agents is applied.
- Describe the benefits of using fire extinguishing agents, such as compressed air foam and ultrafine water mist systems.

CASE STUDY

Putting Out a Class D Vehicle Fire

You are on an engine company and are dispatched to a vehicle fire. You pull a hose line and run off the tank water to fight this fire. As you apply water, you find that the intensity of the fire increases. The fire emits a sudden and extensive amount of bright light and smoke. Next, you try a Class B extinguisher. Your attempts to put out the fire have failed.

1. What is your explanation for the behavior of the fire?
2. Is water able to put out this fire?
3. What other extinguishing agents might be used on the fire?

Introduction

Fire is extinguished by removing elements needed for combustion. For years, the fire service believed that this was done by removing one or more of these elements: fuel, oxygen (air), and heat. A fire triangle or pyramid was used to depict this process. Today, we model the fire combustion process using a fire tetrahedron. This four-sided model includes the chemical chain reaction between the fuel, oxygen, and heat. The fire triangle and fire tetrahedron models are reviewed here because a better understanding of the fire combustion process helps us to be more effective at putting out fires.

Fire professionals use the fire model daily. Their task is to select the best method that will stop the fire quickly and safely. By doing so, they reduce the fire, smoke, and water damage.

Fire Extinguishment Theory

Heat can be described as the molecular motion of a material. The driving force is the effort to maintain equilibrium as it moves from warm bodies to cooler bodies. There is no such thing as cold. This state is relative to another body that is warmer. **Absolute zero** occurs when there is no movement of molecules. Heat energy above this point means that there is some movement of molecules.

Heat and temperature are not the same measurement. **Heat** is all the energy (both kinetic and potential) within the molecules. It is a "form of energy characterized by vibration of molecules and capable of initiating and supporting chemical changes and changes of state" (NFPA 2021). **Temperature** is a measure of the average molecular movement or the heat's degree of intensity. It is the "degree of sensible heat of a body as measured by a thermometer or similar instrument" (NFPA 2021). Therefore, heat is the sum of all the material's molecular energy. Heat is measured in calories or British thermal units (BTUs). A BTU is the heat needed to raise 1 pound of water 1 degree Fahrenheit. Firefighters often use the number of BTUs per pound of material. This helps them judge the relative heat generation during a fire. For example, 1 pound of burning wood will create about 8,000 BTUs. One pound of burning hydrocarbon products will create about 16,000 BTUs. This means twice as many BTUs need to be removed or absorbed when putting out a fire in a hydrocarbon product as would be needed if it were a wood or cellulose material fire. A **hydrocarbon product** is an organic compound, such as benzene or methane, that contains only carbon and hydrogen. **Cellulose material** is a complex carbohydrate of plant cell walls used to make paper or rayon.

Interruption of the Combustion Process

The combustion process can be interrupted by removing one or more of the parts of the fire tetrahedron. Firefighters choose an extinguishing agent based on the burning traits of the materials. The four ways to put out a fire include:

1. reducing the temperature/removing the heat;
2. removing the fuel;
3. reducing the oxygen; and
4. stopping the chemical chain reaction (stopping the release of vapors from the material).

FIGURE 4-1 Removing any of the sides of the fire tetrahedron will break the chain of combustion.

As shown in **FIGURE 4-1**, removing any of these parts breaks the chain of combustion.

Reducing the Temperature/ Removing the Heat

The most common method of putting out a fire is to use water. Water is cheap, plentiful, and can be easily stored and distributed in many ways. Enough water can cool the temperature of the fuel to below its ignition temperature. This prevents the fuel from releasing vapors that can continue to burn. Once the flammable material stops releasing combustible vapors, the combustion process stops. In structure fires, not enough cooling can lead to a rekindle. A **rekindle** is a fire thought to be out that reignites after the fire department has left. Solid and liquid fuels with high flash points can be put out by cooling. Flammable vapors, however, may continue to be released. If the fuel temperature is above the flash point, any ignition source with enough energy will cause a material to start burning again.

Removing the Fuel

Removing the fuel source extinguishes the fire. The fuel source may be removed by shutting down a gas valve or pipeline to stop the flow of liquid or gaseous fuel.

With wildland fires, firefighters remove vegetation in front of the fire, thus eliminating the fuel. This work can be done by bulldozers or hand crews cutting the fuel in front of the fire's path. Another fuel removal method used on wildfires is a backfire. A **backfire** is a fire set to burn the area between the control line and the fire's edge to remove fuel in advance of the fire, change the direction of the fire, and/or slow the fire's progress. The **control line** refers to all natural or created barriers used to stop a fire from spreading. It is where you begin the backfire. The control line ensures the backfire does not change direction.

Reducing the Oxygen

Reducing the amount of oxygen available for combustion reduces fire growth and may extinguish it. In its simplest form, this method is used to put out cooking fires when a cover is placed over a pan of burning food. Another example is when the oxygen content of a room can be reduced. One can flood the area with a gas, such as carbon dioxide, which will replace the oxygen (air) in the room. Because carbon dioxide does not support combustion, the fire will go out.

Oxygen can also be separated from the fuel. To do this, one can blanket the fuel with foam. The foam provides a protective film that, if unbroken, keeps the oxygen from the fuel. Neither of the last two methods works on burning fuels that are self-oxidizing materials. Some examples are peroxides, perchlorates, and nitrates. A **self-oxidizing material** has extra oxygen, which supports the process of combustion by making the fire stronger.

Stopping the Chemical Chain Reaction

Some extinguishing agents interrupt combustion, which is a self-sustaining series of chemical chain reactions needed to keep the fire burning. These agents react with the hydrogen atoms or the hydroxyl radicals. However, the exact mechanisms are not known. When applied, these powders also stop the combustion process by absorbing heat.

Extinguishment and Classification of Fires

One classification of fire is based on the type of fuel that is burned. For some classes, firefighters can use more than one type of fire extinguishing agent. Each of the five fire classes (A, B, C, D, and K) has its own requirements for fire extinguishment (**FIGURE 4-2**).

The Process and Agents of Extinguishment

Water can absorb more heat energy than all matter except for mercury. Water has the ability to draw the heat from the combustion process, cool the fuel to

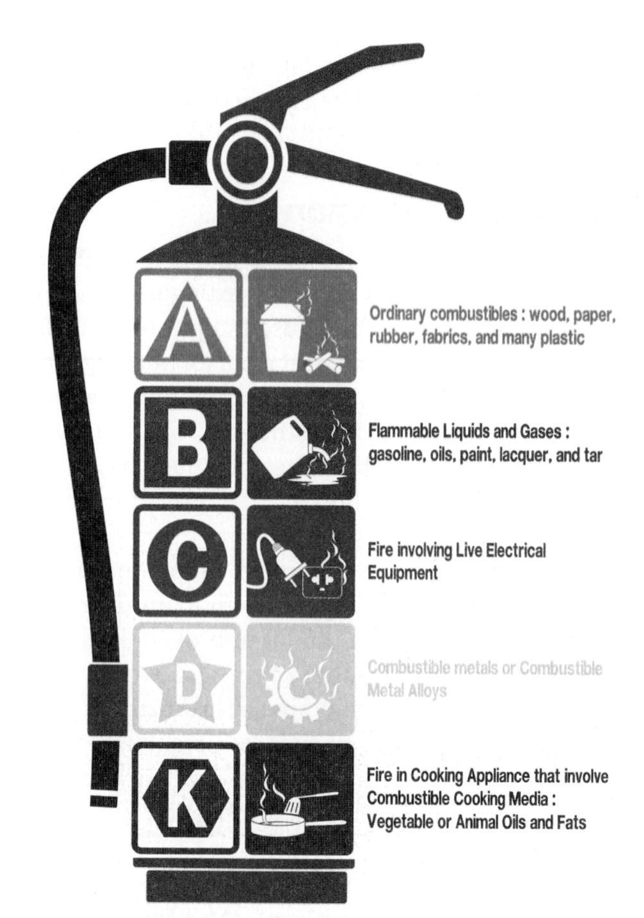

FIGURE 4-2 Fire classes.

temperatures below its vapor production point, and exclude the mixture with an oxidizing agent by smothering the fuel.

Water can expand and contract. The freezing point of water is 32°F (0°C). When heat is applied to water at this temperature, the water will contract instead of expanding. It will continue to contract until it reaches a temperature of 39°F (3.9°C). Above 39°F, water will expand as heat is applied until its boiling point of 212°F (100°C) is reached.

In reverse order, if water is cooled, it will contract until it reaches a temperature of 39°F. It will then expand until it freezes. Water sometimes seeps into cracks in rocks or concrete and freezes before it evaporates. It will cause further cracking as it becomes ice and expands with great force. This same force breaks sprinkler pipes because the water in fire sprinkler piping systems can freeze if they are not protected.

Of most importance to firefighters is that water can absorb a great deal of heat while converting from a liquid to a vapor (steam). This is called the law of latent heat of vaporization, as discussed in Chapter 3.

The latent heat of vaporization is the quantity of heat absorbed by a substance when it changes from a liquid to a vapor. At sea level, water boils at 212°F (100°C). When it reaches this temperature, it vaporizes or turns into steam. In doing so, the water absorbs heat. For every 1°F increase in the temperature in 1 pound of water, the water will absorb 1 BTU. So, when we raise the temperature of the water from 70°F to 212°F (21°C to 100°C), the water will absorb 142 BTUs. There are 8.345 pounds per gallon of water, so 142 BTUs × 8.345 pounds equals 1,184 BTUs absorbed from raising the gallon of water from 70°F to 212°F. Continued heating converts the liquid into steam and, in this process, another 970 BTUs per pound of water is absorbed during the conversion. Because there are 8.345 pounds in 1 gallon of water, another 8,094 BTUs (970 BTUs × 8.345 pounds) are absorbed.

Starting with water in a hose line at 70°F (21°C), when applied to fire, it will be brought up to 212°F (100°C) and then converted to steam. Once this conversion occurs, 1 gallon of water will absorb 1,183 BTUs + 8,094 BTUs = 9,277 BTUs.

In addition to absorbing heat, the high expansion ratio of water to steam can be used to purge areas of heated toxic fire gases. The steam also reduces the oxygen in a room.

The high expansion factor can be seen in **TABLE 4-1**. The table assumes a 90% efficiency of conversion. It shows that 50 gallons of water can expand and occupy a volume of space.

When Water Is Not Effective

Gasoline fires or other hydrocarbon liquids with flashpoints below 100°F (38°C) should not be put out with water. The use of water may move these flammable liquids away from the point of origin because they will float on top of the water, which could spread the fire. This can also be a problem when using water to put out a grease fire. The oils from the grease also float on top of water. They can then be transferred to other areas. In some cases, the fuel will produce too many BTUs for the water to be effective unless it is used in large amounts.

Water is not effective on most fires involving metallic dusts and shavings. It is also not useful for fires in pyrophoric metals. For example, a magnesium shavings fire can explode because the temperature at which magnesium burns is so high. It breaks the water down into its parts, oxygen and hydrogen.

TABLE 4-1 Water Expansion Properties

Temperature (°F)	Temperature (°C)	Cubic Feet of Steam	Volume of Space
212	100.0	10,000	8 ft × 25 ft × 50 ft
400	204.4	12,000	8 ft × 25 ft × 62.5 ft
800	426.7	17,500	8 ft × 25 ft × 100 ft
1000	537.8	20,000	8 ft × 25 ft × 100 ft

Additives to Improve Water Applications

Several chemicals can increase the effectiveness of water when combined with it. These chemicals can lower the temperature of water. This makes it thicker and reduces its surface tension to decrease friction loss. **Surface tension** is the tendency of molecules to be attracted to each other at the surface of a liquid. **Friction loss** is the pressure lost by fluids while they are moving through pipes, hose lines, or other limited spaces. Wet water and slippery water are additives often used by fire departments.

Wet Water

It is difficult for water to penetrate some kinds of fuel because water has a high surface tension. **Wet water** is an additive that reduces the surface tension of water. This allows water to go through burning materials more easily. This, in turn, greatly enhances the firefighters' ability to put out the fire. Firefighters use wet water when they need to reach the seat, or origin, of a fire. An example of this need would be a fire in tightly baled materials such as cotton or tightly packed cardboard boxes.

Wet water is an expensive product, and many agencies only use it as an additive to their high-pressure water extinguishers that are carried on the fire truck.

Slippery Water

Another additive that is often used by firefighters is slippery water. **Slippery water** uses polymers, which are plastic-like substances, that reduce friction loss in the hose. This increases the amount of water that can be moved through the hose.

In fire hoses, friction loss is caused by water rubbing against the rough interior as it moves through. This roughness creates turbulence in the flowing water, which prevents a smooth flow. As the flow increases, so does the friction loss. The pump must work harder to push the water through.

By using slippery water, it is possible to get as much water from a 1 1/2-inch (34-mm) line as it is from a 2 1/2-inch (64-mm) line using plain water. Firefighters should use caution, however, because the amount of nozzle reaction is not reduced. This means that flowing more water will increase the reaction backpressure, or nozzle reaction pressure. **Backpressure** is the pressure applied in the opposite direction of the water flowing from a nozzle.

In addition to wet water and slippery water, there is another product to improve water applications: viscous water.

Viscous Water

The low viscosity of water allows it to run off solid fuel surfaces quickly. This limits the ability of the water to smother a fire by forming a barrier on the surface. To improve this trait, one can add a thickening agent. **Viscous water** is a combination of thickening agents that is added to water so it will cling to the surface of a fuel.

If mixed correctly, it will:

- provide a coating over the fuel that is thicker than untreated water;
- project farther from a nozzle; and
- better resist wind and air currents, thus increasing the time it would take for the water to be removed by evaporation.

Viscous water also has negative aspects. It will not go through fuel as well as water. It creates a higher friction loss in the fire hose and piping and increases the size of water droplets.

Water Application Methods

The application of water can be performed in many ways. Water can be applied in a straight stream, or as a fog, spray, water mist, or foam. We will discuss these applications in the following section.

Straight Stream Applications

Using a smooth-bore nozzle should not affect the **thermal layer zones** (layering of air and fire gases based on their temperatures; **FIGURE 4-3**). It should also not affect the thermal balance of a room if the nozzle application is directed over the fire, opened quickly, and then shut down. This cools the temperature of the ceiling area over the fire. The room is then cool enough for firefighters to enter and make a direct application on the seat of the fire.

Straight stream applications use the conversion of water to steam, which is still an effective method to put out a fire. The steam absorbs the heat and blankets the combustion zone, thus depleting the oxygen.

Fog or Spray Applications

In the United States during the years 1900 to 1939, the fire service overlooked the use of spray streams or fog application on fires. The smooth-bore nozzle was the main application nozzle. It was used daily on structure fires, and the application methods were fine-tuned for the best results. The straight stream application is effective, but it requires firefighters to get close to the seat of the fire. Once they are near the seat of the fire, they can apply the water, reducing smoke and water damage to both the building and its contents.

Chief Lloyd Layman, assigned to training for the U.S. Navy, ran a series of tests on a fog nozzle that was perfected by the Germans. This nozzle divided the water stream into very small drops. This resulted in more water surface area. Increasing the surface area of the water droplets absorbs heat faster. The heat quickly converts the water droplets into a gas (steam). This conversion process results in more heat being absorbed from a room environment. This method of putting out a fire is very effective in some cases.

When the fog nozzle water pattern is used on a fire inside a building, the conversion from water to steam absorbs a great deal of heat. However, it can also create a thermal imbalance. A **thermal imbalance** is a condition that occurs through turbulent circulation of steam and smoke in the fire area.

Thermal imbalances occur between the hot smoke layer and the cool smoke layer. The hot smoke layer is near the ceiling or highest area. The cool smoke layer is in the lower portions of the room. There is a rapid increase in steam in the lower parts of the room, because water's ability to covert to steam is 1,700 : 1. Because steam occupies space, the firefighters would feel an increased pressure. They would find it hard to see, breathe, or even stand in the room. In **FIGURE 4-4**, firefighters have an imbalanced thermal layer driving the heated upper layer into the lower portions of the room.

The expansion of water to steam can purge an area of smoky and toxic gases. Although this is a factor in improving a dangerous environment, fires in ordinary

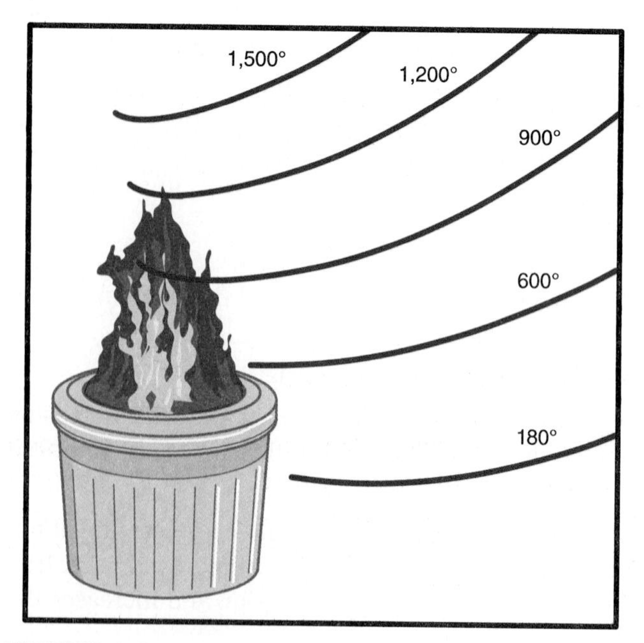

FIGURE 4-3 Temperature layering.

FIGURE 4-4 Thermal imbalance.

combustible materials are normally put out by the absorption of heat. They are not put out by the smothering effect created by the steam.

Both smooth-bore and fog nozzles have specific uses in firefighting procedures. It is up to the firefighter to size up the situation and select the best solution for the fire problem.

Water Mist Systems

High-pressure water mist systems are ideal for extinguishing fires and have replaced other systems in certain applications. They were originally used in maritime situations. The very fine mist reduced the amount of water needed to put out a fire. It also reduced the water damage. Modern advances now offer a highly efficient water mist sprinkler system. Tiny water droplets are created by high-pressure pumps. The combination of the tiny water droplets and the extremely high pressure penetrates deeply into the seat of the fire.

This system suppresses the fire on all three sides of the fire triangle. It cools the fuel to remove flammable vapors. It absorbs the heat by increasing the surface area of the water. It reduces oxygen in the room through the conversion to steam.

Firefighters can expect to see high-pressure water mist and clean agent systems anywhere water cannot be used for fire extinguishment. Some examples of where water cannot be used are areas with computers, servers, or electronic equipment.

Firefighters need to be aware that water mist has its hazards. First, 1 gallon of water converted to steam at 212°F (100°C) will increase 1,700 times in volume. This almost complete conversion of the water mist to steam drastically reduces visibility. Second, the conversion of water to steam will decrease the temperature in the room. This will upset the thermal balance, which is where firefighters have safe working space. An imbalance could allow the steam to interact with the firefighters' personal protective equipment (PPE) and burn the skin.

Foams

The application of foam on fires has a long history of success—as long as the correct type of foam is chosen for the class of fire and the situation. Originally, foam was made by mixing two separate powders with water in a generator. The generator combined the foam products and made carbon dioxide, which became trapped in the foam bubbles. The result was a thick blanket of foam with carbon dioxide bubbles. The bubbles worked to insulate the foam blanket as it spread over a burning liquid surface. This two-powder foam was eventually replaced with a single-powder foam.

TIP

Foam is made by a process of agitation with air injection or aspiration to mix three ingredients: water, foam concentrate, and air. Water is the main ingredient of foam. It ranges from 94% to 99%, depending on the ratio of water to foam concentrate.

In the 1940s, mechanical foam was introduced to the fire service. It was prepared from animal products and was commonly known as protein foam. It was mixed with air and had a very high expansion ratio. Protein foam was a free-flowing material and became popular because it was easy to use. However, it deteriorates rapidly. It needs to be used quickly after purchase due to its very short shelf life.

In the 1960s, surfactant or detergent foam became available (**FIGURE 4-5**). It was more compatible with dry chemicals that destroyed protein foam. These dry chemicals would cause the protein foams to break down and thus allow for a fire to rekindle. The surfactant foams have a much longer shelf life than protein foam. The surfactant or detergent foam could better go through a solid fuel source. It reduced the surface tension of water. As a result, it came to be known as wet water or a wetting agent.

Foam Classifications

Foams are classified as Class A, Class B, or special foams. **Class A foams** are used on a Class A fire (**FIGURE 4-6**). Class A fires are fires in combustible solids such as wood, fabrics, paper, or organic materials. **Class B foams** are used on a Class B fire (**FIGURE 4-7**). Class B fires include those on oil, greases, and other hydrocarbon products.

Foams are divided by their expansion ratio into three categories. The first is low-expansion foam. In low-expansion foam, the bubble expansion ratio is small (less than 20:1). The bubbles in low-expansion foam have a high percentage of water. The other two are medium- and high-expansion foams. In these foams, the expansion ratios are more than 20:1, and some are up to 1,000:1. Medium-expansion foam has an expansion ratio between 20 and 100, while high-expansion foam has an expansion ratio between 100 and 200.

FIGURE 4-5 Extinguishing foams.

© Jovanovic Dejan/Shutterstock

FIGURE 4-6 Class A foam applied to a structure.

Courtesy of the U.S. Department of the Interior.

FIGURE 4-7 Class B foam being applied to a vehicle.

© jackall211/Pixabay

At these expansion ratios, the bubble's water content is low. As a result, the bubble is light. The foam bubbles are made by mixing a foam concentrate with water and a foam solution. The foam solution is then shaken by a machine to form bubbles.

Compressed air foam systems (CAFSs), or foam-producing systems, are installed on a fire truck or special unit. These are becoming a popular option for fire departments. The amount of wetness or dryness of the foam can be programmed. Once the system has been set, the special unit regulates the amount of water, foam concentrate, and air pressure. It manufactures a uniform mixture of foam bubbles (**FIGURE 4-8**).

Foams are also described by their effectiveness on hydrocarbon fuels, fuels able to be mixed with water, or both. Low-expansion foams are applied to fires in flammable or combustible liquids in storage tanks. These tank systems discharge foam bubbles over the liquid surface. This provides a cooling, smothering blanket that covers the liquid surface and puts out the fire. This foam blanket can prevent the creation of vapors for some time. Low- and medium-expansion foams have a low viscosity.

FIGURE 4-8 Foam generator on a hose line.

Courtesy of Rosenbauer International AG.

They are used through the pump system on board the fire truck and can cover large areas quickly. High-expansion foams are used for enclosed spaces. One example is an aircraft hanger where quick action is needed.

In the early 1960s, research by the U.S. Navy led to the development of a film-forming foam called light water. It was so named because it leaves a thin film of water on top of the liquid fuel source. The very thin (0.001 inch) film prevents fuel vapor from forming. It works with all dry chemicals and has led to a dual-agent fire attack system. The Federal Aviation Administration (FAA) is encouraging the use of this system for airport fire protection. This type of foam is known as aqueous film-forming foam (AFFF). The combination of water and AFFF are what define it as a dual agent.

AFFF Concentrate

Aqueous film-forming foam (AFFF) is created by combining water and perfluorocarboxylic acid. Unlike most other foaming agents, it can be applied with simpler foaming devices, such as fire department spray nozzles and sprinklers. This foam can be adjusted to a final concentration of 1%, 3%, or 6% by volume with either freshwater or seawater. The strength of the film surface and its ability to flow on kerosene make it a good extinguishing agent for flammable liquid fires, such as those on jet aircraft fuel. However, this foam loses water content rather rapidly (drain time). It may also provide less burn back resistance compared to other protein-based foams. **Burn back resistance** is the ability to prevent any flame from breaking through the foam barrier. The **drain time** is the time needed for the water to drain away from the foam solution.

Application of Class A Foams

Class A foams are a mixture of water, foam concentrate, and air. The foam mixture can be made wetter or drier by adjusting the ratio of foam solution to water. The size of the air bubbles can be adjusted by the amount of air pressure. The foam can be made denser with less air and lighter with more air. The desired size of the bubbles also depends on the materials burning and if the foam (water) needs to go through the materials or cover the surface.

FIGURE 4-9 Making high-expansion foam.

Courtesy of Rosenbauer International AG.

Wet Foam

Wet foam has fast drain times and small bubbles. It has less expansion because it has less air. Wet foams are good for initial fire suppression and overhauling to search for hidden fire pockets once the fire is out. They also work to penetrate deep-seated fires.

Dry foam

Dry foam has a high expansion ratio and is the consistency of shaving cream. It has small bubbles. It is very fluffy and is mostly air. Dry foams have slow drain times and hold their shape for long periods. They are especially good for exposure protection, or protecting a structure from a nearby fire, because they are able to cling to vertical surfaces for long periods.

Fluid Foam

Fluid foam has the thickness of watery shaving cream. It has medium to smaller bubbles and moderate drain times. Fluid foam works well for direct attack and exposure protection. It is also useful for mop-up operations when firefighters put out residual fires to contain fire spread.

To apply the foam, high-pressure compressors first mix the foam solution with air. This process adds the needed bubble agitation to the finished foam product. **FIGURE 4-9** shows the process of making high-expansion foam.

Fire Extinguishing Chemicals and Other Agents

Several specially formulated dry chemicals, wet chemicals, carbon dioxide, halocarbon, and clean agents have been developed to fill in where water fails to do the job.

Dry Chemicals

Dry chemicals are often used as extinguishing agents. The principal chemicals used for dry chemical agents are sodium bicarbonate, potassium bicarbonate, potassium chloride, and urea-potassium bicarbonate. Using these agents with certain additives can improve how long they can be stored, how well they repel water, and their flow. Dry chemicals work very well on flammable liquid fires, such as grease, some oils, and gasoline (Class B fires).

Systems using dry chemicals are mainly used to protect flammable or combustible liquid storage locations. In these systems, the dry chemical agent can be released through fixed piping or supplying nozzles or hose lines. They can also be designed to protect hazardous operations, such as dip tanks.

Discharge nozzles can also be fixed to release the extinguishing agent on a burning surface, such as in paint booths. In these systems, the agent is delivered under high pressure (approximately 350 psi) by a gas expellant (**FIGURE 4-10**).

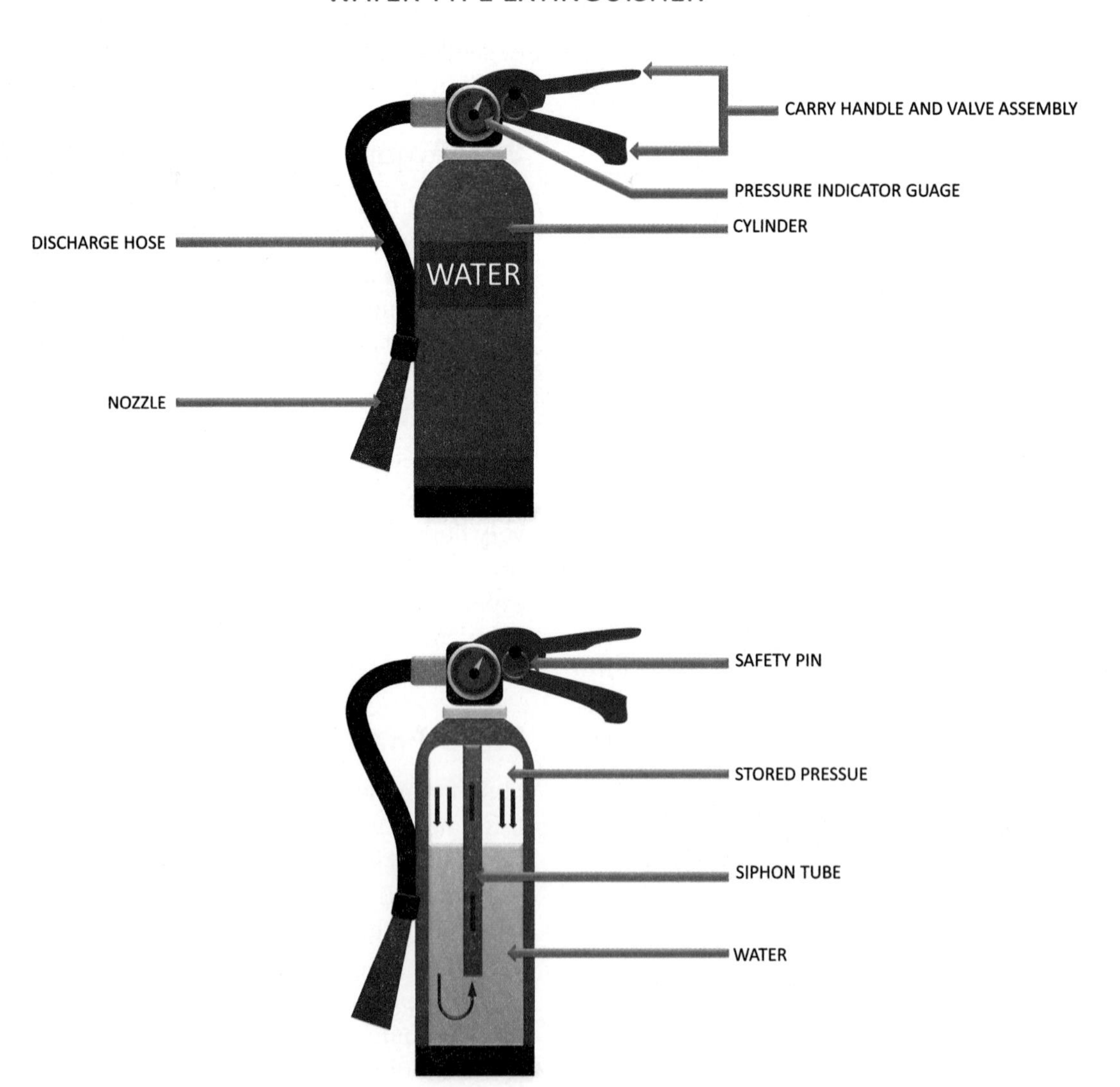

FIGURE 4-10 Stored pressure dry chemical extinguisher.

Application of Dry Chemicals

There are two common ways to store dry chemical expellant gas. It can be stored under pressure or in a separate container under atmospheric pressure. The container is usually joined to a pressurized cylinder (cartridge operated). If the dry chemical agent is stored in a separate container, a rupture disc is installed. It is added to the piping between the pressure tank and the agent storage tank. Activation of the system can be manual, automatic, or both. These systems usually have a manual release mechanism for the dry chemical storage cylinder. This mechanism is far enough from the hazard that the operator would not be in danger while turning it on.

Discharge nozzles are often attached to piping that extends down from the hood. The location of the nozzles is key because they need to be in the center above the cooking surface. Most cooking equipment has wheels so it can be moved for cleaning. The problem is that the system must be returned to the same location for it to be effective in the event of a fire (**FIGURE 4-11**).

Wet Chemicals

Wet chemical extinguishing systems have gained widespread use. Mainly used in restaurants, these systems use a combination of wet chemicals and a high-pressure network of nozzles to put out fires. The system's application provides complete coverage and a much quicker extinguishment than dry chemical systems. The chemicals are also easier to clean up, often needing only a wiping down of the area.

Wet chemical fire extinguishers use a few chemical combinations, such as water and potassium acetate, potassium carbonate, and potassium citrate. The chemicals turn flaming grease into a foamy, soap-like layer. This process, called saponification, prevents the grease from reigniting.

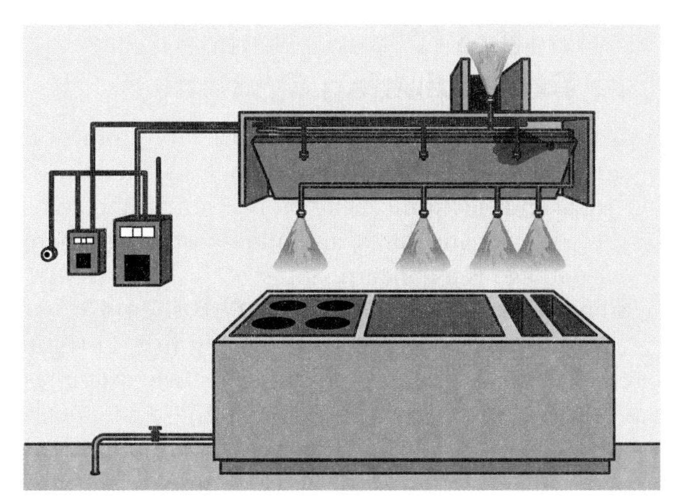

FIGURE 4-11 In dry chemical systems that protect kitchen equipment, discharge nozzles are often attached to piping that extends down from the hood.

Carbon Dioxide

Consumers have been able to buy carbon dioxide (CO_2) for years. It is known as "dry ice" and is used as a refrigerant. In its gaseous form, it is used for carbonating beverages. Carbon dioxide fire extinguishers have high-pressure cylinders or low-pressure tanks with CO_2 under pressure. These cylinders can be released through a special hose and nozzle or fixed piping with nozzles. Both systems work by flooding the area with CO_2, which is 1.5 times heavier than air. Carbon dioxide puts out fires by reducing the oxygen content below 15%, which smothers the fire. Such low oxygen levels also can be lethal to people. When CO_2 discharges, it creates a cooling effect. This can condense water vapor in the area, creating a fog that is hard to see through. **FIGURE 4-12** shows a portable CO_2 extinguisher.

TIP

Carbon dioxide is heavier than air. It will settle into low spaces, such as basements, pits, and recessed floor areas. A means to ventilate these areas and any low spaces near equipment must be considered.

Application of Carbon Dioxide

Carbon dioxide is a colorless and odorless gas. It does not conduct electricity and does not corrode most metals. Because of these traits, it is often used to protect electrical equipment. It is also effective on most flammable liquid fires. Total flooding systems are used when the protected hazard is enclosed in a room. Because total flooding systems put out a fire by smothering, the entire room must be enclosed for the system to be effective. Some of the systems have doors that close when the system is activated. For this reason, a firefighter must wear a self-contained breathing apparatus (SCBA) at all times when in the room.

For enclosed rooms that may be occupied, a predischarge warning alarm is required for total flooding systems. The alarm warns all people to leave before the carbon dioxide discharges.

Halocarbon Agents

The use of halocarbon agents has evolved over the years as we have learned more about chemical compounds.

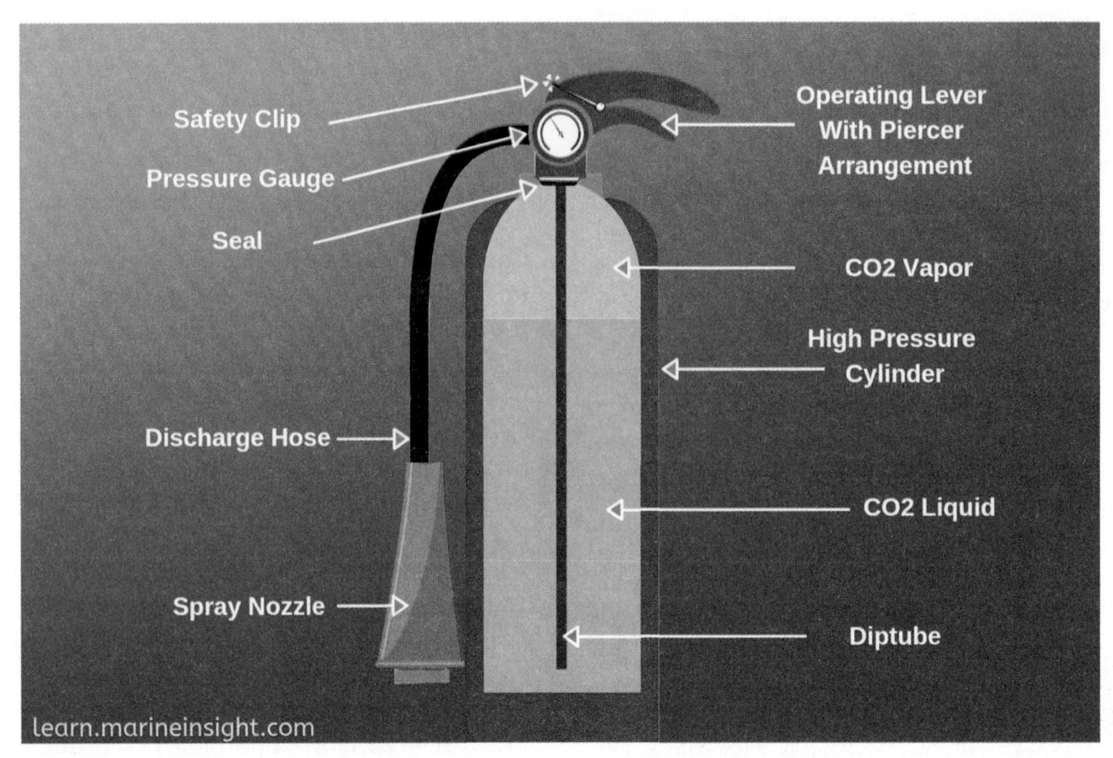

FIGURE 4-12 Portable CO_2 extinguisher.
Courtesy of Marine Insight.

Firefighters needed a more efficient and less damaging extinguishing agent than water. This was especially true for places with water-sensitive electrical equipment or water-reactive materials.

Some halocarbon agents put out fires but were deadly to humans when they were used in closed compartments. One example is carbon tetrachloride. After several deaths, the search began for an agent that could put out fires in electrical equipment but was not toxic.

After many years, a product known as Halon 1301 was developed. It was an effective agent and was nontoxic. Halon is a term used for halocarbon hydrocarbons. The fire service community embraced the use of Halon 1301. Many extinguishing systems were installed, especially where there could be flooding.

In the 1980s and early 1990s, research showed that the halogenated hydrocarbon agents destroy the ozone layer around the Earth. This loss of ozone allows the sun's harmful rays to enter the Earth's atmosphere. Due to this discovery, the halocarbon extinguishing agents were developed.

In 1994, production of certain halons used for putting out fires was banned in some countries. When they were banned, halocarbon systems were allowed to be phased out. When a system needed recharging, it was taken out of service. Certain halons are good fire extinguishing agents and are not toxic. However, they pose a serious threat to the environment. Their use should be limited to where release for a fire is necessary. The halon can then be collected and returned to the confining equipment. Once used, the system must be replaced with a system that does not impact the environment. The two banned halons are Halon 1211, a liquid form, and Halon 1301, a gaseous form. Both Halon 1211 and Halon 1301 were used to protect electronic equipment because they left no residue. As a result of the banning, environmentally clean agents were developed.

Alternative (Clean) Halocarbon Fire Extinguishing Systems

A clean fire extinguishing system is used for protection against many types of fires. It could be used for a fire in electrical and electronic facilities, flammable and combustible liquids and gases, and other properties where water damage is a concern. Some of the systems firefighters are likely to see are listed in **FIGURE 4-13**.

These clean agents do not work on fires in some chemicals or mixtures of chemicals. Two examples are cellulose nitrate and gunpowder. These chemicals erode quickly without air. These agents cannot be used on fires in reactive metals. Some examples are lithium, sodium, potassium, magnesium, and plutonium as well as metal hydrides.

Clean agents also do not work on fires involving chemicals that break down with heat. Some examples are certain organic peroxides and hydrazine. As these

- IG-55 (nitrogen and argon)
- IG-541 (nitrogen, argon, and carbon dioxide)
- IG-01 (argon)
- HFC-236a (hexafluoropropane)
- HFC-23 (trifluoromethane)
- HFC-227ea (heptafluoromethane)
- HFC 125 (pentafluoropropane)
- HCFC Blend A, HCFC-124 (chlorotetrafluoroethane)
- FC-3-1-10 (perfluorobutane)

FIGURE 4-13 Alternative clean halocarbon extinguishing systems.

liquefied agents move through the piping, they can create friction against the walls and build an electrical charge. A static spark may appear on discharge. This creates an electric arc that could explode. Grounding both the sending and receiving vessels lessens the energy and prevents an electrical charge from developing.

These systems are designed mainly for areas where people are not allowed. For places with people, a time delay is vital to allow them to leave before discharge of the extinguishing agent. All of the systems are meant to discharge into an enclosure that can maintain the gas concentration. Monthly inspection of the enclosure is needed. Inspectors must ensure that there are no openings that have not been properly sealed. Before the agent is discharged, air-handling systems need to be shut down. Doors and dampers must also be closed.

Clean Agent Application Methods

Tripping the detectors or a manual release starts an alarm in the control panel. The control panel can also shut down power and run the building fire alarm system. It can also perform other operations. One example is operating the control head to release the extinguishing agent.

Piping and nozzles are attached to the discharge port. Nozzles distribute the extinguishing agent at a controlled rate, smoothly and evenly in fan-shaped patterns. The agent is discharged at a right angle to the nozzle in 180-degree or 360-degree patterns. Nozzles should be covered with blow-off caps to prevent foreign material from getting into the pipes or nozzles.

Special Extinguishment Situations

When they are burning, many metals and some chemicals react with water and CO_2 extinguishing agents. These metals and chemicals react strongly when these agents are used. For this reason, firefighters need specialized agents. The fire service should know the traits of these specialized agents as well as how to apply them. Remember that specialized agents used on metals or chemicals do not contain any water or moisture.

Inert Gas Extinguishment

The use of inert gases is another form of clean extinguishing agent. Inert gases such nitrogen, argon, IG-541, or IG-55 can be used. These systems work by smothering the fire and reducing the amount of oxygen in the room and keeping it low.

These systems consist of a bank of pressurized cylinders. The cylinders hold the gas. The distribution system releases the gas through nozzles, and an alarm system sounds a warning that the system will activate. This warning alerts anyone in the area to leave.

These systems are most often used where expensive materials are being stored or electrical equipment is in operation. They can also be used in paint and powder coating rooms.

Combustible Metal Fires

Water cannot be used to put out fires in combustible metals. Several burning metals react strongly when in contact with water. On some of these fires, the water is separated into hydrogen and oxygen. This boosts the combustion process.

Some molten metal fires are very difficult to put out. Violent steam explosions can occur if water comes in contact with a pool of molten metal. Molten metal pools will also hold heat for long periods. In so doing, they may extend the fire.

Over the years, many specialized fire extinguishing materials have been made for specific metals. Firefighters must know the businesses using and storing these metals. They should check with the Fire Prevention Bureau to make sure that the correct extinguishing agent is on the premises. **TABLE 4-2** lists some of these agents.

Chemical Fires

Water also reacts with some inorganic chemicals. For example, calcium carbide reacts with water to form acetylene. Acetylene is a highly flammable gas with a wide explosive range (12% to 74%). Firefighters may see calcium carbide in high school or college chemistry labs. They could also find it at chemical plants where acetylene gas used for welding is produced. Other alkali metals, such as lithium hydride and sodium hydride, react with water. They form hydrogen gas, another highly flammable gas with a wider (than acetylene) explosive

TABLE 4-2 Extinguishing Materials for Some Metal Fires

Extinguishing Material	Main Ingredients	Metal Can Be Applied On
A liquid that can be applied		
TBM	Trimethoxyboroxine	Mg, Zi, Ti, Na, K
Gases that can be applied		
Helium	BF_3	Any metal
Argon	Ar	Any metal
Nitrogen	N_2	Any metal
Other materials that can be applied		
Met-L-X	$NaCl + (PO_4)_2$	Na
Foundry flux	Mixed chlorides + fluorides	Mg
Lith-X	Graphite + additives	Li, Mg, Zi, Ti, Al
Dry sand	SiO_2	Various metals
Sodium chloride	NaCl	Na, K
Soda ash	Na_2CO_3	Na, K
Lithium chloride	LiCl	Li
Zirconium silicate	$ZrSiO_4$	Li

range of 4% to 81%. Both lithium hydride and sodium hydride can be found at plants where porcelain, ceramic castings, and some greases are made.

The peroxides of sodium, potassium, barium, and strontium react strongly with water. They release huge amounts of heat. Some cyanide salts, such as potassium cyanide, react to water with acid. They form the highly toxic hydrogen cyanide. Firefighters can find these deadly gases at plants that tan leather and those that dye wool.

Another problem with using water on fires in toxic chemicals, such as bug killers, is the runoff of tainted water. It may impact wells and springs underground. In each case of a chemical fire, firefighters must make sure to have experts with them.

Lithium-Ion Battery Fires

Lithium-ion batteries are now used to power everything from your cell phone to your motor vehicle. These batteries pose an extreme fire danger. They can catch fire after short circuiting, overheating, or physical damage. Exposure to long periods of moisture can also lead to spontaneous combustion.

A lithium-ion battery fire is classified as a Class B flammable liquid fire. This is because lithium-ion batteries use liquid electrolytes to create a pathway that conducts electricity. As such, we fight lithium-ion battery fires in much the same way that we fight flammable liquid fires. It is essential to bring the battery below is ignition temperature, 932°F (500°C). This can be done by using a Class D dry powder or Class B foam. Unfortunately, there is no definite timeline for these cells to combust. This type of fire is one to let burn out completely, consuming the entire battery.

Pressurized Gas Fires

These fires are hard to put out because the gas is under pressure. A damaged tank, fitting, or valve will provide a nonstop supply of fuel. Even if there is no fire, people must leave the area. Sometimes firefighters can shut off the gas supply or plug the tank. These actions need to be taken under the protection of hose lines.

At other times, the leaking fuel is burning, or the tank is being impinged by fire from another burning object. In these cases, firefighters need to make sure

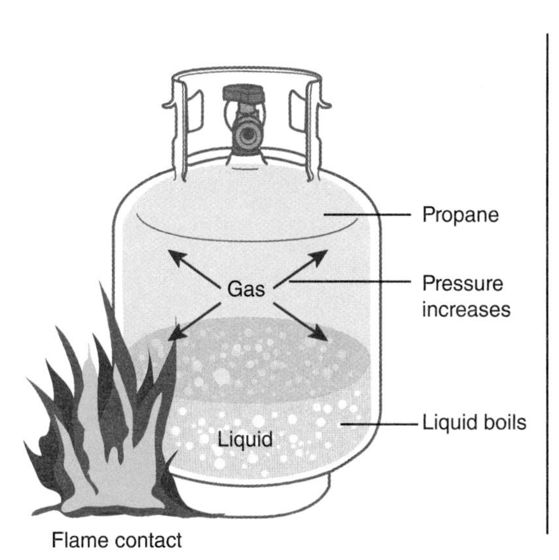

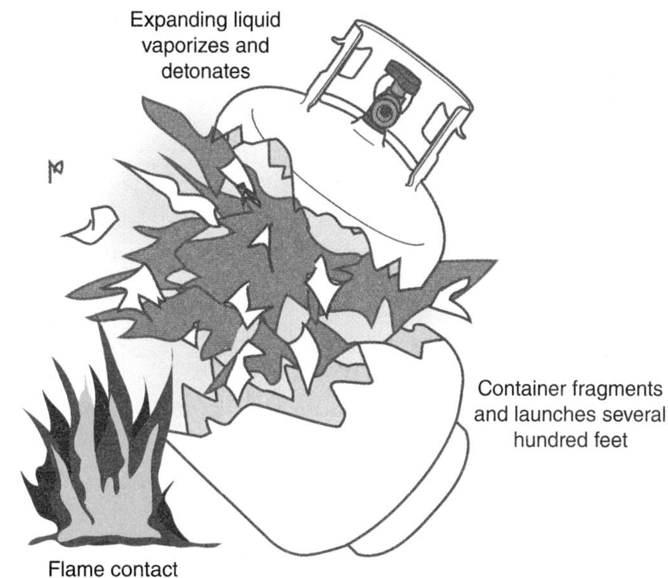

FIGURE 4-14 Diagrams of BLEVE events.

that once the gas supply is shut off, heated objects are cooled. Other sources of ignition must be put out to prevent reignition. If the damaged tank cannot be plugged or the valve repaired, firefighters must keep the tank cool. They must cool it until the pressure relief valve can safely release the gas pressure.

Sometimes it is not possible to cool the tank where a fire is touching it. If this happens, unmanned master streams should be put in place. All fire crews should leave the area. A rapid increase in tank pressure may cause a boiling liquid/expanding vapor explosion (BLEVE). BLEVEs are dangerous. The events leading up to a BLEVE are shown in **FIGURE 4-14**.

TIP

Firefighters should not attack a tank fire from either end. This is where the tank is welded together. These are the two weakest points.

TIP

Firefighters should craft prefire plans for areas where flammable liquid tanks are stored and where these tanks travel.

CASE STUDY

CONCLUSION

1. **What is your explanation for the behavior of the fire?**

   Based on the increasing intensity of the fire and the sudden appearance of bright light and smoke, it seems like there is a water-reactive material burning. The water has caused a chemical reaction.

2. **Is water able to put out this fire?**

   Water may put out the fire if enough is available to first cool down the fuel. It will also remove the heat generated to the surrounding materials, which will combust.

3. **What other extinguishing agents might be used on the fire?**

   Because this is a Class D combustible metals fire, it would be best to use a Class D dry chemical extinguisher.

WRAP-UP

SUMMARY

- Fires can be extinguished by cooling, removing the fuel, reducing the oxygen, or stopping the chemical chain reaction by stopping the release of fuel from the material.
- Water is still an inexpensive and available choice for the majority of fires encountered by firefighters.
- Water can also be made a more effective extinguishing agent by making it slippery, reducing its surface tension, and making it thicker.
- Water can be applied in straight stream applications, fog or spray applications, water mist systems, and foams.
- Some agents—dry chemicals, wet chemicals, carbon dioxide, halocarbons, and clean agents—are only effective on certain materials and under certain situations.
- In the 1980s and early 1990s, research showed that the halocarbon agents destroy the ozone layer around the Earth. This led to environmentally clean agents being developed.

KEY TERMS

absolute zero The temperature at which there is no movement of the molecules.

aqueous film-forming foam (AFFF) A foam created by combining water and perfluorocarboxylic acid that is used on flammable liquid fires, such as those in fuel.

backfire A fire set to burn the area between the control line and the fire's edge to remove fuel in advance of the fire, to change the direction of the fire, and/or to slow the fire's progress.

backpressure The pressure applied in the opposite direction of the water flowing from a nozzle; also called nozzle reaction pressure.

burn back resistance The ability to prevent any flame from breaking through the foam barrier.

cellulose material A complex carbohydrate of plant cell walls used to make paper or rayon.

Class A foam Foam used on a Class A fire.

Class B foam Foam used on a Class B fire.

compressed air foam system (CAFS) Foam-producing systems that are installed on the fire truck or special unit.

control line All natural and created barriers used to stop a fire from spreading; this is where you begin the backfire.

drain time The time needed for the water to drain away from the foam solution.

friction loss The pressure lost by fluids while they are moving through pipes, hose lines, or other limited spaces.

heat All of the energy, both kinetic and potential, within the molecules.

hydrocarbon product An organic compound, such as benzene or methane, that contains only carbon and hydrogen.

rekindle A fire thought to be out that reignites after the fire department has left.

self-oxidizing material Material that has extra oxygen, which supports the process of combustion by making the fire stronger.

slippery water Water that has polymers, plasticlike substances, to not only reduce friction loss in the hose but also increase the amount of water that can be moved through a hose line.

surface tension The tendency of molecules to be attracted to each other at the surface of a liquid.

temperature A measure of the average molecular movement or the heat's degree of intensity.

thermal imbalance A condition that occurs through turbulent circulation of steam and smoke in the fire area that makes it hard for firefighters to see and breathe.

thermal layer zone Layering of air and fire gases based on their temperatures.

viscous water A combination of thickening agents that is added to water so it will cling to the surface of a fuel.

wet water An additive that reduces the surface tension of water.

REVIEW QUESTIONS

1. Saponification is a process that converts a burning substance to a soap. Once the substance is a soap, how does it put out a fire?
2. What would happen if one tried to put out a fire in magnesium shavings with water?
3. How do water mist systems put out fires? What are the problems with and benefits of their use?

DISCUSSION QUESTIONS

1. Why is water the most often used extinguishing agent and what is its greatest value in putting out a fire?
2. What are the basics properties that distinguish Class A foam from Class B foam?
3. What is the best extinguishing agent for a Class D, metals fire?
4. How does high-expansion foam efficiently extinguish fire in a basement?
5. What is the best way to prevent a BLEVE? Why is this the best way?

APPLYING THE CONCEPTS

You are dispatched to a structure fire at a private residence. On arrival, you find the front door open and the owner of the home standing on the front lawn holding her child.

She tells you that she was preparing to fix her breakfast and went to change the child. When she came back, she saw that the stove top was on fire.

1. What are your immediate concerns?
2. What type of fire is it?
3. Which agent would you use to put out this fire?

REFERENCES

Bevelacqua, Armando S. *Hazardous Materials Chemistry,* 3rd ed. Burlington, MA: Jones & Bartlett Learning, 2018.

Bushie, Lance. "Steam: What You Need to Know." Canadian Firefighter. September 17, 2018. Accessed April 7, 2023. https://www.cdnfirefighter.com/steam-what-you-need-to-know-42913/.

Delmar Cengage Learning. *Firefighter's Handbook: Firefighting and Emergency Response,* 3rd ed. Clifton Park, NY: Delmar Cengage Learning, 2008.

Dickinson, Micah. "What Is a Clean Agent?" Vanguard. March 11, 2020. Accessed January 19, 2023. https://vanguard-fire.com/what-is-a-clean-agent/.

Fire Cap Plus. "Plain Water vs. Fire Cap Plus." n.d. Accessed January 19, 2023. https://www.firecapplus.com/how-it-works/.

Gagnon, Robert. *Design of Special Hazard & Fire Alarm Systems,* 2nd ed. Clifton Park, NY: Delmar Cengage Learning, 2008.

Gagnon, Robert. *Fire Protection Systems and Equipment.* Clifton Park, NY: Delmar Cengage Learning, 2003.

Kinetix. "Types of Clean Agent Fire Suppression Systems and Their Applications." n.d. Accessed January 19, 2023. https://kinetixfire.com/types-of-clean-agent-fire-suppression-systems-and-their-applications/.

Klinoff, Robert. *Introduction to Fire Protection and Emergency Services,* 5th ed. Burlington, MA: Jones & Bartlett Learning, 2016.

Lowe, Joseph D. *Wildland Firefighting Practices.* Albany, NY: Delmar Cengage Learning, 2001.

Mein, Suzi. "The Search for an Effective Halon Replacement." Firetrace International. June 19, 2019. Accessed January 19, 2023. https://www.firetrace.com/fire-protection-blog/the-search-for-an-effective-halon-replacement.

National Fire Protection Association (NFPA). "NFPA Glossary of Terms." National Fire Protection Association. 2021. https://www.nfpa.org/-/media/Files/Codes%20and%20standards/Glossary%20of%20terms/glossary_of_terms_2021.ashx.

Quintiere, James G. *Principles of Fire Behavior,* 2nd ed. Albany, NY: CRC Press, 2016.

REFERENCES CONTINUED

Staughton, John. "What Is Surface Tension?" Science ABC. Last updated February 9, 2022. Accessed January 19, 2023. https://www.scienceabc.com/pure-sciences/what-is-surface-tension-definition-causes-examples.html.

Sprinkler Age. "Exposure Protection: What Is It?" October 23, 2019. Accessed January 19, 2023. https://www.sprinklerage.com/exposure-protection/.

State of Alaska Department of Environmental Conservation. "Aqueous Film Forming Foam." n.d. Accessed January 19, 2023. https://dec.alaska.gov/spar/csp/pfas/firefighting-foam/.

Sturtevant, Thomas B. *Introduction to Fire Pump Operations*, 2nd ed. Clifton Park, NY: Delmar Cengage Learning, 2005.

U.S. Department of Agriculture. "What Is a Control Line in a Wildfire?" 2019. Accessed January 19, 2023. https://ask.usda.gov/s/article/What-is-a-control-line-in-a-wildfire.

CHAPTER 5

Foundations of Firefighting Strategies and Tactics

LEARNING OBJECTIVES

After studying this chapter, you will be able to:

- Describe the process of creating firefighting strategies and tactics involved in finding, limiting, and putting out fires in buildings and special fire situations.
- Discuss how to choose the proper fire operating mode: offensive, transitional, defensive, or nonattack.
- Define the term *size-up* and explain the steps and factors involved in sizing up an incident.
- Apply fire behavior traits to the special occupancies discussed.

CASE STUDY

Hackensack Ford Dealership Fire

On July 1, 1988, firefighters rushed into a building to fight a fire. They thought it was no different from any other fire. There was no one in the building. Thirty-five minutes later, five firefighters had died. Three died when the 60-ton bow truss roof fell on them. The other two died later. They ran out of air when they were cut off from escape and trapped in a utility room. Incident commanders did not hear the mayday alarms, which delayed the rescue. Only supervisors had radios at the time. All of the fire crew was on the same radio frequency.

1. Which mode of operation would have been best for this incident?
2. What information about the building would have helped the incident commander in decision making?
3. What process may have improved radio communications?
4. When observing an incident, what is a critical step in the decision-making model?

Introduction

This chapter reviews the most common structure fires and fire behavior that firefighters will face. We begin with a description of the decision-making process used by firefighters. They use this process to decide the strategies and tactics to use in emergency incidents. This chapter also looks at how to apply fire behavior tactics to specific occupancies. This includes special fire situations and difficult or dangerous areas.

Development of Strategy and Tactics

The overall mission of the fire service is always to save lives and protect property. To do this, firefighters need to address the three parts of an incident:

1. find the fire;
2. limit the fire; and
3. put out the fire.

These help to prioritize the activities in the plan. Once these parts have been considered, the incident commander (IC) sizes up the incident. A **size-up** is an ongoing review by firefighters to identify the problems at an incident. The size-up process is used to pick the best overall strategy to solve the problems at an incident. The three parts can be handled quickly and safely when the correct strategy and tactics are in place. A **strategy** is a general plan to meet incident goals. The **tactics** are the actions needed to complete the strategy or plan. **FIGURE 5-1** shows arriving units finding, limiting, and putting out a fire.

FIGURE 5-1 On fire incidents, firefighters must find, limit, and put out the fire.

To develop an overall strategy for emergency incidents, firefighters use the decision-making model. The **decision-making model** is a simple five-step process used to solve problems (**FIGURE 5-2**).

The first step is to quickly identify the problem. Once the heart of the problem has been found, the second step is to list possible solutions. Third, choose the best solution(s) to solve the problem based on the situation. Once the best solution has been chosen, the fourth action is to implement the solution. After the strategy is put into place, the fifth step is to reexamine the solution to see if it is working. If not, reexamine the problem and the solutions, choose the best solution, and start over.

Remember that the strategy is based on the use of the decision-making process. If used in every response, this process becomes natural.

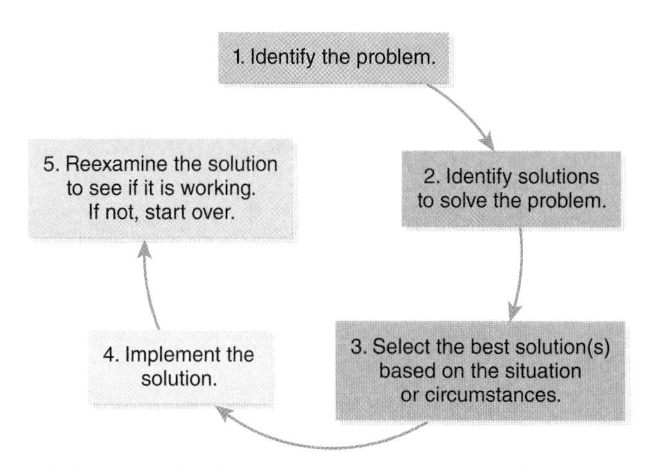

FIGURE 5-2 The five steps of the decision-making model.

Attack Modes

A crucial part of the size-up process is choosing the best mode of attack for a fire. Life safety concerns for people inside the building and firefighters determine the mode. Concerns for the environment factor into the decision of which mode to use as well. The attack modes are called the offensive mode, transitional mode, defensive mode, and nonattack mode. The nonattack mode is sometimes also called the passive approach.

Offensive Attack Mode

Offensive attack mode refers to firefighting that makes a direct attack on a fire. For structural firefighting, this usually means fire suppression or rescue (or both) within the building or buildings.

In many situations, the most effective way to put out a fire is the offensive or direct attack mode. Firefighters using the direct attack mode enter the building with a solid hose stream to deliver water directly to the fire base. The water cools the fuel below its ignition temperature. The water is applied in short bursts to conserve it. A fire in a building with people in it may call for the offensive attack mode.

There is a rule of thumb used by fire officers. If the fire is still progressing after 20 minutes of an offensive attack, then change to a defensive or transitional mode.

Transitional Mode

Transitional mode is the process of shifting from the offensive attack mode to the defensive attack mode, or vice versa. This phase is passed through as operations shift from one mode to another. This mode redirects the strategy due to changing events. It can be dangerous because hose lines protecting firefighters' positions move. The lines can be accidentally shut off, leaving firefighters exposed to fire, heat, and smoke.

Defensive Attack Mode

Defensive attack mode is a fire tactic used outside a building to put out a fire. The purpose of this mode is to limit the fire to the inside of a structure. During a fire attack, this mode is used when the temperature is rising and there are signs of a nearing flashover. From outside the building, firefighters send a short burst of water to the ceiling, which cools the heated gases at the top of the room. This action can delay flashover long enough for firefighters to apply water using the direct attack method to the seat, or origin, of the fire. It can also allow firefighters to make a safe exit from the building. Water applied to the ceiling should only be enough to cool it down slightly. Too much cooling will affect the balance or the thermal layering in the room. See the discussion in Chapter 4 on the effects of an unbalanced thermal layer.

The defensive mode is also used when a building is enveloped in fire and is threatening nearby structures. In this mode, firefighters try to protect exposed properties.

Nonattack Mode

The **nonattack mode** is a fire tactic used when a fire attack is too dangerous, or suppression activities are prevented. When this happens, the IC can choose this mode and let the fire burn out without an attack. The nonattack mode is often used in wildland fires where access to the fire is blocked by land. This mode might also be chosen for wildland fires due to a lack of available resources and because these fires pose little danger to people or structures. Wildland fires are addressed in Chapter 8.

Size-Up at the Incident Scene

Initially, the first officer of the first unit or company is responsible for the size-up. This person will be the first IC. This duty is passed on to higher-ranking officers as they arrive and assume command.

Before entering the building, the first-arriving units pull up past the building to see the front and sides. They then make a 360-degree tour of the scene. This gives them a complete picture of what the outside of the structure looks like. It can also help firefighters find the exact location of the fire.

Often, the first-arriving units assign letters to the sides of the building, such as **A** for the front, **B** for the left side, **C** for the rear, and **D** for the right side. This allows firefighters to identify the sides of the property even if it is built on a slope and additional floors are below street level.

To improve the size-up process, the fire service adopted a couple of acronyms. The first one, RECEO-VS, became known in 1953 with the publication of Lloyd Layman's book, *Fire Fighting Tactics.* **RECEO-VS** is used to devise strategies and tactics on the fire ground. It is also used for deciding incident command priorities. Each letter stands for one of the seven parts used to create a strategy. The first five letters should be done in order. The last two may be used at any point to support the first five. RECEO-VS stands for:

Rescue
Exposures
Confinement
Extinguishment
Overhaul
Ventilation
Salvage

SLICE-RS is a newer acronym used to develop strategies and tactics on the fire ground. It was created for the first-arriving officer to help in decision making. Like RECEO-VS, the first five letters should be used in order. The last two can be done at any time. SLICE-RS stands for:

Size up
Locate the fire
Identify the flow path
Cool from safe location
Extinguish
Rescue
Salvage

Each of the letters provides firefighters with a mental checklist (**FIGURE 5-3**).

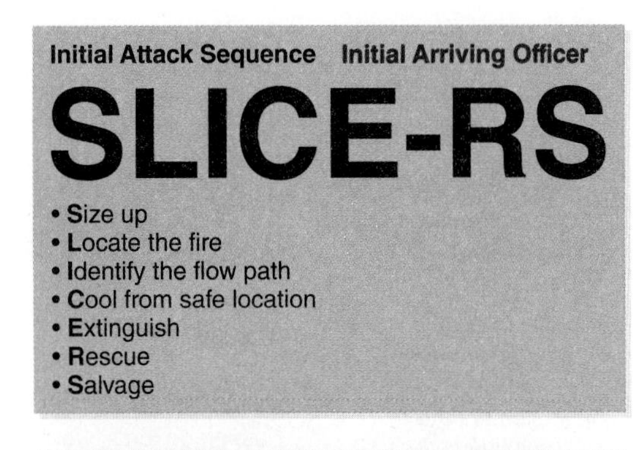

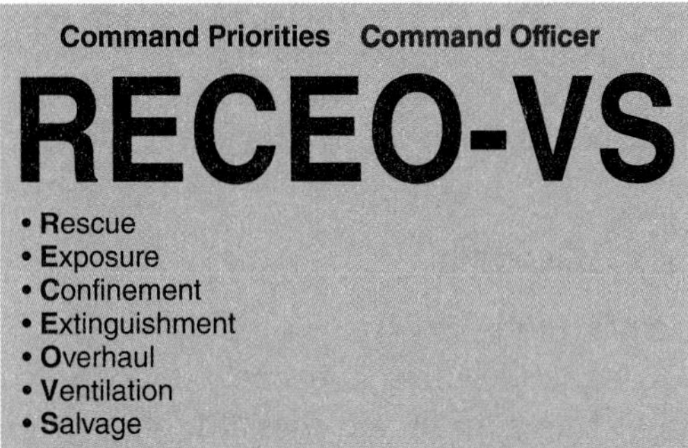

FIGURE 5-3 Two acronyms that can help a firefighter with scene size-up.

Construction

The International Building Code (IBC) classifies buildings into five different types of construction. Each type has its own set of parameters. The main reason for classifying types of construction is to uphold a certain level of safety for people inside the building in the event of a fire. The second reason is to preserve the property.

The National Fire Protection Association (NFPA) uses the IBC's construction classifications. Local building codes use more subdivisions. However, for prefire planning and strategy, the five IBC/NFPA divisions (Types I, II, III, IV, and V) will serve as a sample for this text.

The type of construction determines the fire-resistance rating of the structure. It includes the building parts and materials and the ability of those things to resist fire. Type I follows the most rigid requirements. That means that building materials and methods offer the highest level of protection against fire. Type V has the least rigid requirements.

Fire ratings are measured in terms of the time a material can be exposed to fire before it collapses. The time can be from 20 minutes to 4 hours. Fire ratings are the results of fire tests on materials. They show the ability of a door, floor, or wall to block the spread of fire. They can also show how much protection the structural members, such as beams and columns, provide for safe passage. There are also tests on the fireproofing added to building construction.

Many factors affect a building's ability to withstand fire. Differences can be caused by variations in materials, workmanship, installation, construction method, ventilation, and other factors.

Type I or Fire-Resistive Construction

Type I construction is called fire-resistive construction (**FIGURE 5-4**). The columns, beams, floors, walls, and roofs are made of materials listed as noncombustible according to the certification tests in NFPA standard 251, *Standard Methods of Fire Tests of Building*

FIGURE 5-4 Type I or fire-resistive construction.
© richardjohnson/Shutterstock

FIGURE 5-5 Type II or noncombustible construction.
© Dennis Tokarzewski/Shutterstock

Construction and Materials. This standard has been removed from the NFPA Standards and is here only for historical reference. The American Society for Testing and Materials (ASTM) E-119 is the new standard and honors this testing process.

In fire-resistive construction, the structural members are the main parts that support the framework of a building. These are made from materials that will not burn or that have a specified fire resistance. Type I is for buildings with more than one floor. The resistance is 4 hours for the columns and beams, 3 hours for the floor, and 2 hours for the roof materials.

There are some important size-up considerations for Type I construction. The intensity of the fire, the strength of the fire-resistive materials, and the number of hidden void spaces where fire can travel undetected are a few. A **void space** is an area in the structure's construction that has no outside or inside entrance and is not used for anything. Another factor n is the number of open areas where fire can rapidly weaken the structure until it collapses. These buildings are usually very well constructed and stable. If a collapse occurs, it is usually limited to one area of the building.

Type II or Noncombustible Construction

Type II construction is referred to as noncombustible construction (**FIGURE 5-5**). It can be either protected or unprotected. In unprotected construction, the major parts are noncombustible but do not have fire-resistant materials added for protection. The best example is unprotected steel (noncombustible) found as a roof support or unprotected steel beams. When exposed to fire, steel stretches and bends. It begins to lose its strength at about 1,000°F (538°C). In Type II construction, protected steel provides a degree of protection that is less resistive to fire than Type I. This is referred to as protected noncombustible construction.

TIP

Firefighters need to understand that calling a structure noncombustible does not mean it will not collapse. Noncombustible buildings have many combustible features, such as roofs, balconies, and overhangs. The contents of these buildings, too, especially plastic furnishings, add to the fire load. The **fire load** is the total amount of fuel that might be involved in a fire. This is measured by the amount of heat given off when that fuel burns. This, in turn, increases the amount of heat affecting structural parts.

Type III or Outside Protected/ Ordinary Construction

The **ordinary construction** type of building has outside walls made of masonry materials (**FIGURE 5-6**). The inside walls and materials can be partially or wholly combustible. The number of combustibles per square foot is limited by the occupancy or how the building is used. However, the masonry outside of these buildings can be deceptive. The buildings can give an impression of safety, while the insides can contain large numbers of burnable materials.

Type IV or Heavy Timber/Mill Construction

Type IV or **heavy timber construction** buildings are masonry. However, the inside columns and beams are solid or heavy wood. They are built without hidden spaces (**FIGURE 5-7**). The hidden spaces can allow for

FIGURE 5-6 Type III or ordinary construction.

FIGURE 5-7 Type IV or heavy timber/mill construction.

the fire to progress unnoticed for a time. They can also make finding the origin of smoke and fire difficult.

Because of the original design or through alteration, heavy timber buildings have many dangerous departures from ideal mill construction. Fire in a Type IV construction is often called slow burning, which helps when firefighters attack the fire from inside the building. However, once an outside defensive attack is underway, a better description might be a slow-burning fire producing high amounts of BTUs—that is, a very hot fire.

Type V or Wood Frame Construction

In Type V construction, all main structural members can be made from burnable materials (**FIGURE 5-8**). The chief support for the building is from wood components that have little or no protection from fire. The hidden voids and channels formed inside the walls between the wooden wall studs can present a major problem in a fire, as fire can easily spread to the upper parts of the building.

FIGURE 5-8 In Type V construction, all main structural members can be made from burnable materials.

This type of construction in a large apartment complex can also lead to a conflagration. Buildings under construction are in various stages of completion. Some buildings would have walls, windows, and roofing in place. Other buildings would have only the framing in place, which would allow for the rapid spread of fire through the complex.

Modern wood-frame buildings may be considered platform frame buildings (different building construction methods, including platform construction, will be discussed later in this chapter). However, such buildings still contain void areas, such as soffits, utility chases, and truss floors and ceilings. These are joined by electrical and plumbing areas. **Soffits** are false spaces under stairways and projecting roof eaves. They could also be false spaces above cabinets in kitchen or bathrooms. **Utility chases** are channels used for electrical, telephone, and plumbing lines and pipes for services in a building (**FIGURE 5-9**). These areas provide paths for fire and heated gases, which creates a dangerous condition for firefighters. Today's Type V construction includes homes and businesses made with lightweight steel members and recycled materials.

TIP

Firefighters need to be concerned with how sturdy these buildings are because the structural strength of the wood declines very quickly when it burns.

FIGURE 5-9 Utility chase.

© Douglas Sacha/Moment/Getty Images

Occupancy or Use

Occupancy is the building code term that provides standards to match a building's use and those who will use it with features to address fire hazards and life safety concerns. This category helps to figure out the building height and size limits, features, occupant safety requirements, exits, fire protection systems, and inside finishes.

Apparatus and Staffing

It is important to know the resources responding to an initial alarm. This could include the number of fire trucks, type of pumping, and ladder ability. By knowing the limits of their resources, fire crews can use standard operating procedures to ask for more resources. **Standard operating procedures (SOP)** or standard operational guidelines (SOG) are specific instructions on how to complete a task or assignment. Individual departments can choose either term to identify their operating directives. SOP/SOG can be used to pick the type and number of resources needed.

Life Hazards

The priorities of the fire department at the fire scene are to protect life, limit the fire, and protect property. To save lives, it is sometimes necessary to search for victims in the first few moments after arrival on the scene. This process is referred to as the primary search.

The search team should wear full protective clothing, including a self-contained breathing apparatus (SCBA) with a personal alert safety system. A **personal alert safety system (PASS)** is a small device that is sensitive to motion. Firefighters wear it with SCBA when they enter an area that is considered immediately dangerous to life and health (IDLH). The PASS device senses when the wearer is not moving. If motion stops for longer than 30 seconds, the system sounds an alarm to help other firefighters find the motionless firefighter. Older PASS devices needed to be turned on, and some wearers failed to do so. Newer devices turn on when the SCBA air bottle valve is activated.

It is important for a primary search to be carried out in an ordered way. The search pattern can vary, but it needs to cover the entire structure. This includes areas behind doors and in closets and other places where a person may try to find safety from the heat and hot gases. See Chapter 6 for a discussion of overhaul and salvage work.

Terrain

The terrain or the ground level of a building is important to firefighters. The reason is because the building can be built on land with different grade levels (**FIGURE 5-10**). The front side of the building may be only two stories from the ground level, whereas the rear could be as high as four or more stories.

In some situations, buildings are set back from the street. In other cases, buildings may have fences, trees, or electrical wires in front of the building. These things can make it difficult and time consuming for emergency

FIGURE 5-10 Variation of building heights on a sloped hillside.

© Marek Lambert/Shutterstock

trucks to access the building. Careful prefire planning can help to reduce the time needed to deal with these problems.

Water Supply

Water supply must be considered in terms of whether it can be delivered in enough gallons per minute to stifle the number of BTUs being given off by the fire. The water supply appraisal should be part of prefire planning. For example, prefire planning may include learning the locations and sizes of water mains and hydrants. This may also include identifying the availability of water from lakes, ponds, or rivers. Another consideration for water is the location of water lines of other areas or districts. Connecting valves could redirect water during an emergency.

Backup Sprinkler Systems

The presence or absence of backup fire protection systems is important in the size-up process. Sprinklers have proven to be effective in large structure fires. From 2015 to 2019, U.S. fire departments responded to about 51,000 structure fires per year in which sprinklers were present. The sprinklers worked in 92% of these fires and controlled the fire in 96% of the incidents. The most common reason these systems failed was because they were shut off before the fire. Overall, sprinkler systems were effective in 88% of the fires large enough to activate the systems (NFPA, 2021).

Three types of water sprinkler systems are available:

1. Wet pipe
2. Dry pipe
3. Deluge system

Wet pipe is a system that has water in the sprinkler piping at all times. Each sprinkler head is activated by heat (**FIGURE 5-11**). **Dry pipe** is a system that has air in the sprinkler piping until it is activated. Again, each sprinkler head is activated by heat (**FIGURE 5-12**).

A **deluge system** is a dry pipe system that protects areas that are being consumed by a fast-spreading fire. In this system, all sprinkler heads are activated at the same time (**FIGURE 5-13**). All of its sprinkler heads open, and the piping has air in it. When the system is on, water flows to all heads, providing complete coverage. The system has a deluge valve that opens when triggered by a separate fire detection system.

Some buildings are equipped with standpipes, most of which are for fire department use. **Standpipes** are a manual firefighting system with piping and hose connections inside the building (**FIGURE 5-14**). Firefighters can use these systems rather than pull hose lines into the building. Standpipes are divided into three categories. A Class I system is used for manual firefighting with a 2 1/2-inch (64-mm) hose connection and a Class II system is used for manual firefighting with a 1 1/2-inch (38-mm) hose connection. This would be where trained occupants would begin a fire attack. The Class II systems are being phased out. A Class III system is used for manual firefighting with a combination of 2 1/2-inch (64-mm) and 1 1/2-inch (38-mm) connections.

Street Conditions

Street conditions, such as narrow streets, traffic, double-parked cars, and construction work, can severely impact fire operations. All of these factors impact the ability of the arriving units to attack the fire quickly. In some cases, it may not be possible to get the fire truck close to the actual building on fire. Close coordination with the public works and traffic departments can ease such problems.

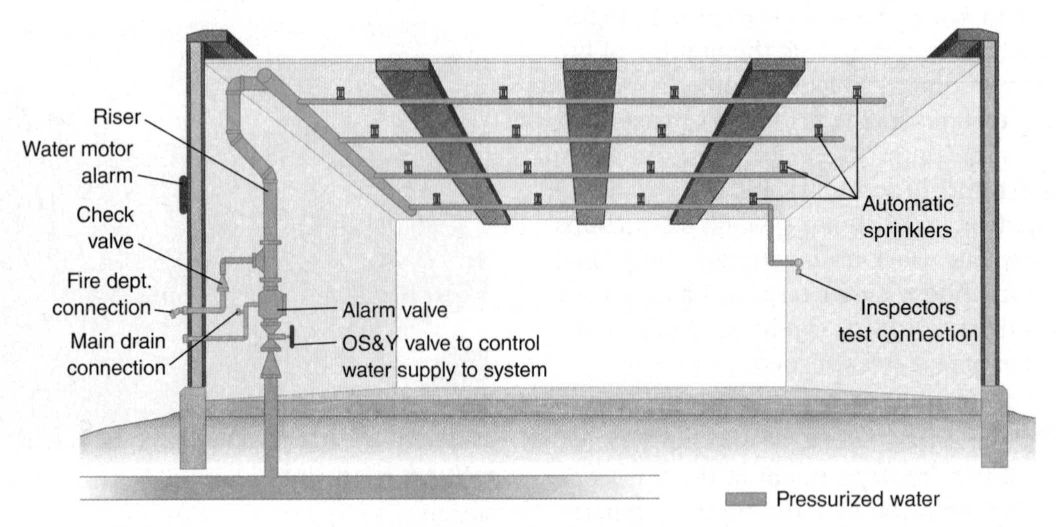

FIGURE 5-11 A wet pipe has water in the system at all times.

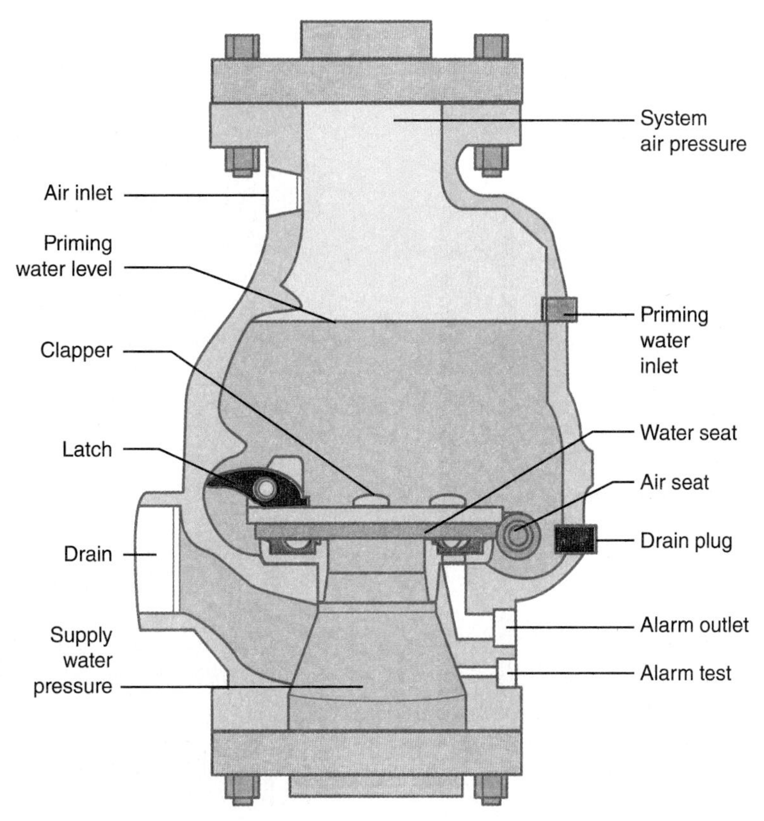

FIGURE 5-12 A dry pipe system contains water only when the system is activated.

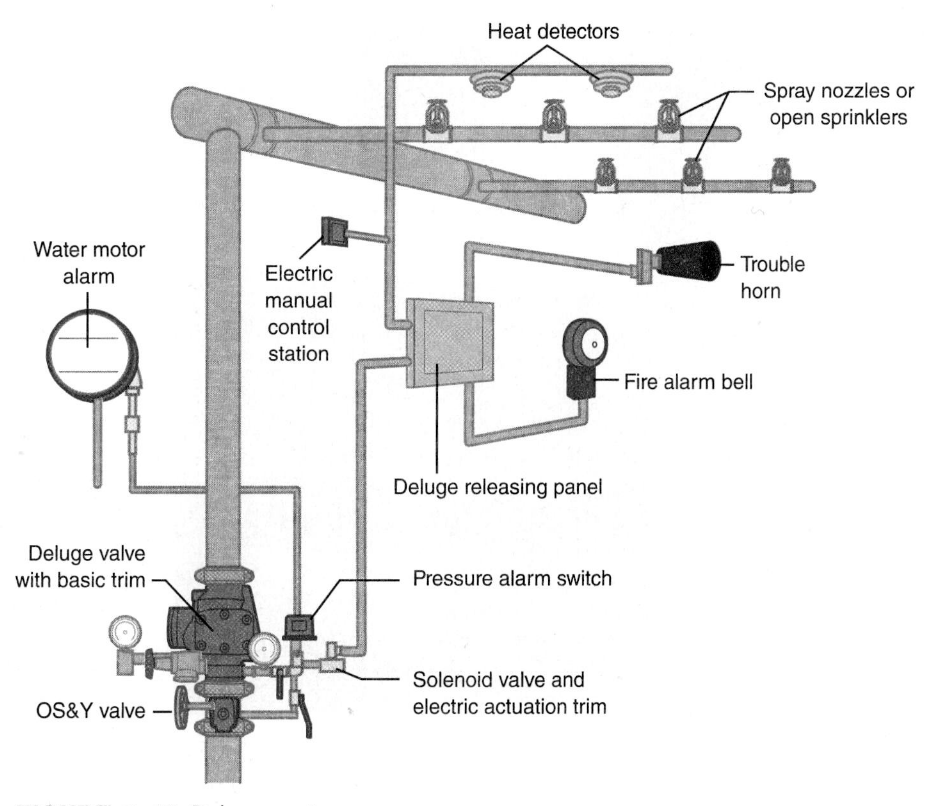

FIGURE 5-13 Deluge system.

A.

B.

C.

FIGURE 5-14 Examples of **(A)** a Class I standpipe system, **(B)** a Class II standpipe system, and **(C)** a Class III standpipe system.

A: Courtesy of Larsen's Manufacturing Co.; **B:** Reproduced from Mahoney, Shawn. 2021. "Standpipe System Designs and Calculations." NFPA Today. National Fire Protection Association. https://www.nfpa.org/News-and-Research/Publications-and-media/Blogs-Landing-Page/NFPA-Today/Blog-Posts/2021/11/19/Standpipe-System-Design-and-Calculations; **C:** Courtesy of the National Fire Protection Association.

TABLE 5-1 Subfreezing Temperatures That Can Slow Emergency Operations

	Temperature, °F												
Wind Speed, MPH	45	40	35	30	25	20	15	10	5	0	−5	−10	−15
5	43	37	32	27	22	16	11	6	0	−5	−10	−15	−21
10	34	28	22	16	10	3	−3	−9	−15	−22	−27	−34	−40
15	29	23	16	9	2	−5	−11	−18	−25	−31	−38	−45	−51
20	26	19	12	4	−3	−10	−17	−24	−31	−39	−46	−53	−60
25	23	16	8	1	−7	−15	−22	−29	−36	−44	−51	−59	−66
30	21	13	6	−2	−10	−18	−25	−33	−41	−49	−56	−64	−71
35	20	12	4	−4	−12	−20	−27	−35	−43	−52	−58	−67	−75
40	19	11	3	−5	−13	−21	−29	−37	−45	−53	−60	−69	−76
45	18	10	2	−6	−14	−22	−30	−38	−46	−54	−62	−70	−78
	A						B						C

	Wind Chill Temperature, °F (°C)	Danger
A	Above 25°F (−3.9°C)	Little danger for a properly clothed person
B	25°F (−3.9°C) to −75°F (−59.4°C)	Increasing danger, flesh may freeze
C	Below −75°F (−59.4°C)	Great danger, flesh may freeze in 30 seconds

Weather

Temperatures that drop below freezing can slow emergency response (**TABLE 5-1**). Ice that collects on hose lines, ladders, and clothing makes movement slow, exhausting, and unsafe. Also, severe cold causes mechanical failures, such as hydrant freezing and sluggish or downed fire truck operation.

High temperatures and high or low relative humidity can decrease the amount of work firefighters can do. Protective clothing holds in body heat and thus raises body temperature. This higher body temperature increases the body's fluid needs, so firefighters must drink more fluids. It can also lead to heat exhaustion or heat stroke (**TABLE 5-2**).

The key to successful operations under extreme weather conditions is to make sure working crews have a rehabilitation system. **Rehabilitation systems** are a group of activities for the health and safety of responders at emergency incidents.

Rehabilitation systems have:

- Fluids
- Food
- Places to rest and recover
- Medical help

High temperatures also raise the temperature of fuels, such as wood, flammable liquids, and other substances. This rise in fuel temperature brings the fuel closer to its ignition temperature. This lowers the temperature needed for ignition.

High humidity, or the percentage of water in the air, can and does impact the burning process. An increase in moisture in the air affects the movement of the smoke and heated gases. The more humid air tends to be heavier than the less humid air. This causes the smoke and heated fire gases to move horizontally. Both vertical and horizontal ventilation may be needed to speed up the ventilation process.

Exposures

An **exposure** is a property threatened by radiant heat from a fire in another structure or an outside fire. Property within 40 feet is considered an exposure risk, but larger fires can endanger property much farther away.

TABLE 5-2 Heat Stress Index Developed by the U.S. Fire Administration

Relative Humidity										
		10%	20%	30%	40%	50%	60%	70%	80%	90%
Temperature, °F	104	98	104	110	120	132				
	102	97	101	108	117	125				
	100	95	99	105	110	120	132			
	98	93	97	101	106	110	125			
	96	91	95	98	104	108	120	128		
	94	89	93	95	100	105	111	122		
	92	87	90	92	96	100	106	115	122	
	90	85	88	90	92	96	100	106	114	122
	88	82	86	87	89	93	95	100	106	115
	86	80	84	85	87	90	92	96	100	109
	84	78	81	83	85	86	89	91	95	99
	82	77	79	80	81	84	86	89	91	95
	80	75	77	78	79	81	83	85	86	89
	78	72	75	77	78	79	80	81	83	85
	76	70	72	75	76	77	77	77	78	79
	74	68	70	73	74	75	75	75	76	77

NOTE: Add 10°F when protective clothing is worn and add 108°F when in direct sunlight.

Humiture, °F[a]	Humiture, °C	Danger Category	Injury Threat
Below 60°	Below 15.6°	None	Little or no danger under normal circumstances
80°–90°	26.7°–32.2°	Caution	Fatigue possible if exposure is prolonged and there is physical activity
90°–105°	32.2°–40.6°	Extreme Caution	Heat cramps and heat exhaustion possible if exposure is prolonged and there is physical activity
105°–130°	40.6°–54.4°	Danger	Heat cramps or exhaustion likely, heat stroke possible if exposure is prolonged and there is physical activity
Above 130°	Above 54.4°	Extreme Danger	Heat stroke imminent!

[a]Humiture is the combination of humidity and temperature.

Reproduced from U.S. Fire Administration. "Command and Control of Fire Department Operations at Target Hazards," 2nd Edition, 10th Printing. August 2017. https://nfa.usfa.fema.gov/ax/sm/sm_n0314.pdf.

FIGURE 5-15 Large water appliance (monitor) in use.

Courtesy of Rosenbauer International AG.

Water does not soak up radiant heat well because the heat will pass directly through the water spray. Using water for exposure protection against radiating heat transfer is best done by running water down the side of the exposed building. This way, the water transfers the built-up heat on the building surface by conduction as it flows down the surface of the building.

Water spray can also protect against the motion of heated air currents (convection) and direct flame impingement. Firefighters use large water appliances to reduce or redirect air currents and to cool building surfaces (**FIGURE 5-15**).

Area and Height

The area and height of a building reveal the maximum potential fire area. Firefighters must check the entire border of the building before sending resources inside. This is because the total size and area can be misleading when looking at it from outside.

Firefighters should also be aware of the collapse zone. The **collapse zone** is the border around the structure. It is usually 1.5 times the height of the building where debris would fall during a collapse. Fire crews are not allowed to work in this zone. If a structural collapse is expected, crews use unmanned master streams to replace hose lines.

> **TIP**
>
> The height of a building is a factor in how far the debris from the walls will travel during a collapse. Establish a collapse zone and have a preplanned warning system in place.

Location and Extent of Fire

The lower the fire is in the building, the more serious the threat of fire is to the building. A low building fire exposes most of the building to the upward movement of the fire and heated fire gases.

A second concern is a fire below grade, such as a basement fire, subway fire, or one below deck on a ship. **Below grade** means the floor lower than ground level. These fires are hotter and more complex than the same fire above ground level. There are few ways to ventilate them horizontally. The stairs in the lower basement area provide a vent channel for the hot fire gases to travel upward. This dangerous situation requires all safety precautions to be in place before the attack.

Time

The time of the incident gives firefighters insight into whether the fire could be deadly. The life hazard in homes where people sleep is far greater at night than during the daytime. The time of day also affects the time required for a fire truck to arrive. Morning and evening traffic peaks can double the response time. In certain buildings, the number of people varies with the time of day or the day of the week. For example, an incident at a school at 10:00 AM on a weekday may pose a serious life problem, while for an incident in the same location at 2:00 AM on Sunday or on a holiday, the life threat is not as great.

Special Concerns

There are other special concerns during size-up. The first concern is the need for a personnel accountability system. A **personnel accountability system (PAS)** is a tracking system to follow the entry and exit of crew into the working area during an incident. This system ensures all members can be found at any time.

The second concern is to have a rapid intervention team in place. A **rapid intervention team** or **rapid intervention crew (RIT/RIC)** is a group of people whose purpose is to make a rapid response to reports of firefighters who become trapped or confused in the building.

The third concern is when the situation worsens rapidly. In this case, a personnel accountability report becomes vital. A **personnel accountability report (PAR)** is a roll call taken of all firefighters working on the fire ground. When needed, all crew must report their status via a special radio frequency or another wireless communication platform. Fire departments or the IC can request a PAR. They can be ordered at set time intervals or be requested when firefighters change from offensive attack mode to defensive attack mode. This ensures that all crew has left the building.

Fire Behavior in Specific Occupancies

Firefighters recognize fire behavior patterns and issues for specific occupancies. Different procedures must be followed for different types of buildings. It is important to understand the differences in building methods and types when assessing how to approach a fire.

Building Construction Methods and Occupancy Types

In a building made using the **platform construction method**, the floors are built separately from the outer walls. This means that the ceiling and floor serve as a fire block. In most cases, the fire block stops the upward motion of hot fire gases between floors. However, utility chases and poke-through construction (where contractors or owners poke holes in walls to add or rerun electrical, phone, or plumbing lines) provide channels for the movement of hot fire gases through the building.

The **balloon frame construction method** is an outdated method in which the wood studs run from the foundation to the roof. The floors are nailed to the studs. The wall space provides a channel for hot fire gases to spread vertically. This framework does not stop the fire between floors. As a consequence, a fire started in the basement can quickly spread up the vertical channels of the walls to the attic. The channels can also allow a fire to reenter the building on the upper floors.

Single-Story Family Homes of the Past

The types of fuels found in most homes today have dramatically changed from those of the 1940s and 1950s. A typical home during those years had walls over wood or wire mesh. The wood or wire mesh was painted or covered with heavy paper or cloth. Bare wood, rugs, or wool carpets blanketed the floors. Some floors were covered with linoleum. The furniture varied from bare wood to furniture overlaid with cotton, wool, or leather. Latex rubber was used only in some cushions and pillows. The mattresses were filled with either cotton stuffing or feathers. The rooms were poorly insulated. They had single-paned windows that allowed air in. The fire load was low, and the rooms were well ventilated.

Today's Homes

The fire load and fire behavior problems in homes have changed significantly over the last two decades. First, buildings are insulated better. The double-paned glass better insulates the inside from both heat and the cold. This insulation, however, holds in the heat from an inside fire. Also, homeowners tend to use more plastics or products made from hydrocarbons. Some examples are plastic furniture, decorations, carpets, and window coverings. These add to the fire load.

The combination of the better insulation and the increased fire load has made inside firefighting hotter. It has also decreased the time to flashover, making fighting fires more dangerous. Firefighters should expect wide variations in the intensity of fires in these buildings.

Fires in one-story homes vary from those that can be put out with a garden hose to those that involve the whole building. The fire service responds to more single-room fires because they are more common than a fully involved home fire. Single-room fires can usually be put out with a 1 3/4-inch (45-mm) hose line. That hose line can be backed up with a larger line to protect against a fast-spreading fire where more water may be needed. The fire crews bring the line into the home from the uninvolved area to confine the fire to the area of origin.

A fully involved one-story home of average size can usually be attacked with two 1 3/4-inch (45-mm) lines. The main objective with a totally involved building is to keep exposures from becoming involved. In some areas of the United States, homes are built close together. If wood is used as siding, then the exposure problem can be severe.

On a large home fire, firefighters have the option to start the attack using 2 1/2-inch (64-mm) lines. Some larger fires in single-family houses, those in which greater than 50% of the structure is involved in flames, can be put out with only one of these larger lines. However, the use of large-diameter attack lines for interior attacks makes it difficult to advance the hose line and move during the suppression activity. Once the fire is out, these larger lines can be reduced to 1 3/4-inch (45-mm) lines. These smaller lines are easier to handle and can reduce water damage. The ease of these lines also allows firefighters to be freed for work in other areas.

Multiple-Family Homes

Living arrangements in multiple-family homes vary from city to city, as does building construction. This results in different fire problems. In some cities, homes are three-story buildings built with wood frame construction. A separate family lives on each floor.

In contrast, other cities have three-apartment buildings, each with three stories and one family per floor. Most of them were built with brick wall construction, weak masonry work, and open attic areas. These

FIGURE 5-16 Open attic and truss construction.

© trongnguyen/Adobe Stock

apartments require firefighters to always look for signs of a possible building or wall collapse. **FIGURE 5-16** shows the common open attic with truss floors often found in these buildings.

Commercial Fires

Most fires in commercial stores happen on the ground-floor store or in the basement. If the property does not have fire sprinklers, fire can travel rapidly because large amounts of combustible goods are stored there. Always check for the horizontal spread of fire because separations between storefronts may not be fire resistant. Also, if the Sheetrock material was installed at the time of construction to the walls, it may now have openings or holes that allow fire and smoke to travel. These openings are often created by tenant upgrades for electrical, digital, or plumbing reasons. Often they are not properly sealed to prevent the spread of heat or fire.

The subdivision of the basement into areas or rooms makes fighting fires more difficult. Because of the extreme heat and smoke, advancing a line into the basement in most cases is not possible.

Strip Mall Fires

A type of commercial building that presents difficulty for firefighters is the strip mall. The name comes from the fact that the building was built with the idea that it would be quickly sold for profit. As a result, contractors follow the lowest construction requirements, making the structure as cheaply as possible. Fire protection of these buildings also barely meets the building code requirements. Strip malls are usually built as a row of stores, one story high. In some cases, two-story buildings are built. The construction is usually of poor quality and uses lightweight materials.

A few strip malls have fire-resistant walls, but usually not between every store. When these walls are present, they are often lacking fire protection. In some strip malls, there is a common basement under the row of stores.

Because of the construction type, a fire in one store of a strip mall has a good chance of spreading to others. One reason is because the fire can go into the cockloft area. A **cockloft** is a void space, about 3 feet deep, between the ceiling area and the underside of the roof. Once there, the fire will move horizontally with little or no resistance.

Ventilation at the roof is crucial. A large hole should be cut to draw and release the heated fire gases out and delay the horizontal travel. The hole should be cut directly over the fire. If possible, it should be about 80 square feet. There should be at least a small hole in the roof before fire crews open the ceilings. This will prevent a backdraft condition. The objective is to get in front of the fire to be successful.

Hotel Fires

Experience with fires in older hotels tells us that these buildings were not designed with fire safety in mind. Many were built with vertical openings that allowed fire to move to the top floors where tenants sleep.

Older hotels were built with open stairways and rooms off halls, often with transoms over the doorways. A **transom** is a small window in an older building, such as a hotel, at the top of the ceiling, usually over the room entrance. These windows allowed heated air to flow into the room because most rooms did not have heating units. The windows also allowed fire and hot gases in hallways into rooms. These older windows are being replaced with central air and heating units, which require the room and hallway to be enclosed.

Central heating, ventilation, and air conditioning systems present a different problem. **Heating, ventilation, and air conditioning (HVAC) systems** are central systems used to heat and cool large buildings. These units can move hot fire gases throughout a building if the unit is not protected with fire dampers. **Fire dampers** are items used to prevent transmission of flame where air ducts go through fire barriers. They use a fire-resistant shutter that works with a fusible link inside the duct to stop the movement of the gases. Some newer HVAC systems automatically shut down the fan in the event of fire. This prevents the air and smoke in the duct from being spread throughout the building. If the system is not automatic, firefighters must shut down the system to prevent fire spread. Some HVAC systems are reversible, meaning the fans can be set to remove smoke from the building. During prefire planning, firefighters should locate the HVAC system(s) and any operational instructions.

Hotels, apartments, and high-rise buildings without eyebrow dormers have serious fire behavior problems. An **eyebrow dormer** is an extension over the top of openings, such as windows, doors, and balconies. These eyebrow dormers may be made of concrete or Styrofoam with a stucco shell. Eyebrow dormers prevent fire and smoke from moving to the upper floor(s) in a multistory building. This lapping of fire and heated fire gases from one floor to an upper floor on the outside of the building, sometimes through windows, is called **auto-exposure**. The concrete extension prevents auto-exposure, which restricts fire movement.

Industrial Occupancies

Industrial buildings vary because manufacturing processes differ widely. This has resulted in a mix of hazards and concerns for firefighters. Many hazardous materials and processes can be found in these buildings, which require preplanning to identify. Hazardous materials and operations can be found in both old and new industrial buildings.

One of the common types of older industrial buildings is the tenant factory or loft building. This non-fire-resistant building is often four to six stories high with brick-and-wood-joist construction. The building fronts vary from 20 to 60 feet (6.1 to 18.3 m) in width and from 40 to 200 feet (12.2 to 61 m) in depth. In many cases, the front is between two streets. There is usually one factory per floor, but in some cases the buildings will have two tenants on each floor. Manufacturers of similar products, such as clothing, tend to be in one building. However, it is common to find several unrelated products being made in the same building.

In these types of buildings, the chance of a fast-spreading fire is high. This is because of the large numbers of combustible goods and products or manufacturing processes that use heat or flame.

Some older factory buildings have open stairs, elevator shafts, and unprotected light and ventilation shafts. The skylights are often aligned to allow light into the work area. If the skylight is broken or open, the fire will vent and turn into a rapidly spreading fire that burns with great strength. Large hose lines with ample water are needed to stop the fire from spreading and to protect exposures.

Newer manufacturing buildings use concrete tilt-up construction. Some are built with concrete panels, which are either raised or poured at the job site. The walls are raised and held together at the roof level with premade wood trusses. They are then covered with a wood sheeting material.

The trusses provide open spaces where fire gases can collect. The truss parts are joined with lightweight metal connectors, which are not protected with a material that is fire-resistant. Fire quickly weakens the metal truss connections or wooden trusses, which can lead to a roof collapse.

Churches

Services vary among religious groups. However, churches and other buildings where services are held are similar in construction. So, when we refer to church fires, we include any building used regularly for religious services that fits the occupancy description.

In church fires, the construction feature that usually leads to the church's destruction is a large hanging ceiling or cockloft. Once the fire reaches this area, it will burn unchecked because the area cannot be reached. The ceiling is too high to be opened from below, the walls are too thick to be pierced, and the roof is too unsafe to work from.

If the church has a steeple, chances are the fire will reach there and result in its collapse. The height of the steeple may require aerial streams. When placing the fire truck, firefighters should consider the distance the steeple may fall. Always protect crew and equipment by leaving enough space for collapse.

Many churches are attached to a rectory or living quarters by an unprotected hall. The hall could allow fire to travel unless it is prevented by large hose streams. The roofs are steep with wide-open attic areas, which also promote rapid fire spread.

The life safety record for church fires is good. This could be because they are not often filled. Most church fires occur when no one is there. Although the fire load from contents may not be high, church fires can still be very difficult and dangerous.

Schools

Most states require schools to practice fire drills and follow construction standards for fire safety. There has been an improvement in fire-resistant construction in schools. However, we still face fatal school fires. Firefighters should be aware that older schools are still in use.

Firefighters should ventilate at school fires. This will direct the smoke and heat away from halls and stairways. Hose lines should be placed to protect these vital escape routes.

Schools may use modular classrooms. These are limited in height and need to have at least 2-hour fire-resistant construction. They also have special requirements for the materials used for walls, floors, and ceilings. They are arranged in groups of two to four units with two exits per classroom. These structures are not expensive and have a good fire record.

TIP

Problems facing an IC at a school fire are the victims trapped in the building and the chaos. It is difficult for the first-arriving companies to find out the size of the fire and the rescue problems at hand. The IC should put an ordered search plan in place immediately. A prefire plan can help firefighters prepare and drills can greatly improve the actions of the school students during a fire or emergency.

Basement Fires

A fire in a basement can quickly become fully developed. It can present a very hot, smoky fire situation that is hard to see through (see Chapter 3 on fire stages). In many cases, the basement is stuffed with old furniture and discarded items, which increase the fire load.

Basements are usually large. Some are not sectioned off, which leaves a wide-open area of combustibles. There is increased risk to firefighters due to a lack of ventilation and limited entry and exit routes. There are also meters, utility panels, and hanging wires.

Because of the extreme heat and smoke, advancing a line into the basement may not be possible. First-arriving officers should do a size-up. The size-up should include the use of a thermal imager. This will help to locate the fire. A survivability profile and risk assessment should be done to decide if inside operations are needed. If needed, a fire attack must be done through a hole in the basement ceiling. From the floor above the fire, a cellar, or distributor, nozzle is placed in the basement ceiling. This special nozzle disperses the water in a circular pattern at various room levels.

In some cases, an apartment basement is where the utilities enter the building. In these situations, a gas could catch fire. The fire could melt the piping connections at the gas meter. There could also be an explosion from a leak. For these reasons, the gas service should be shut off as soon as possible.

When the electrical service enters through the basement, the electricity should be shut off. This can be done by cutting the electrical lines at the drip loop or meter head outside the building. A **drip loop** or meter head is the loop formed by the electrical supply lines (**FIGURE 5-17**). A circuit breaker or shut-off is found below the meter head at the circuit breaker box.

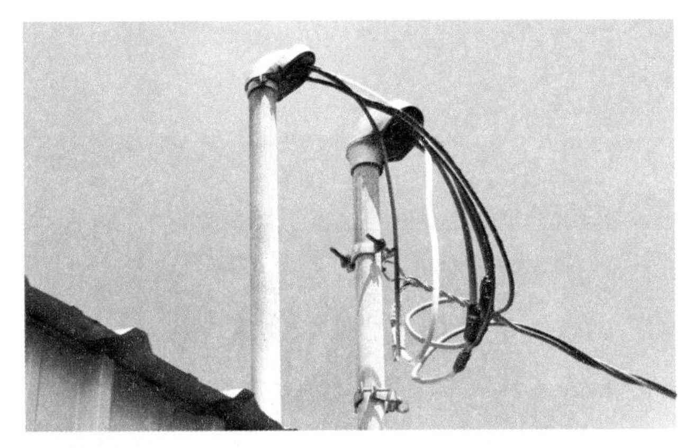

FIGURE 5-17 Drip loop or meter head.

TIP

In firefighting, the rule is "heat rises," so it is important to think about the vertical extension of fire. This is especially true of basement fires. Heat rises to the underside of the first floor. The heat has access to the floor joists, flooring, walls, stairways, dumbwaiters, and open elevator shafts. All of these features provide ways for fire to travel to the upper parts of the building.

Attic Fires

Attic fires share some of the same traits and hazards as basement fires. Attics are also filled with combustible items, insulation, and utilities. A firefighter's first job is to find the fire. The use of a thermal imaging camera can be considered. However, they are not always reliable for attic fires due to insulation in the attic floor.

For small fires with little ventilation, crews should begin to save what they can on the floor below the attic. Crews should roll up rugs, cover up furniture, lay down hall runners, and use debris bags. Any utilities running through the attic should be shut down.

From outside the building, an attic fire can be accessed through gables or the roof. Firefighters should be in full PPE, including SCBA. Such an attack would put firefighters in the way of smoke and hot gases.

Inside the building, attic fires should be attacked as fast as possible with a 1 3/4-inch (45-mm) line with a fog or spray nozzle. The lines should be advanced to the attic through a scuttle hole or attic stairway. If a scuttle

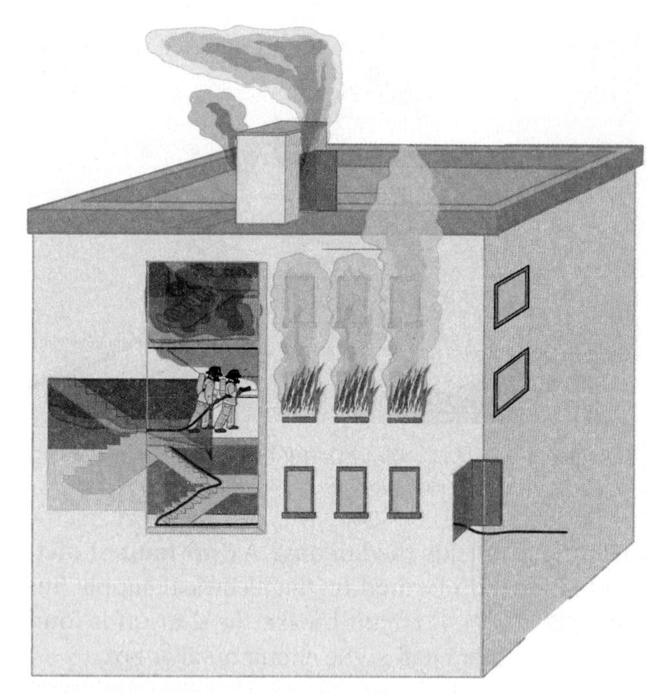

FIGURE 5-18 Ventilation of attic and roof area.

hole or attic stairway cannot be readily found, then the ceiling should be opened from below. The firefighter should stand between the area being pulled and the exit and pull away from the body. The key is to keep the fire small by keeping the ventilation limited. **FIGURE 5-18** shows the ventilation of an attic and roof area.

Flat Roofs

In a flat-roofed building, the attic space is about 3 feet deep and provides an open space between the underside of the roof and the top side of the ceiling. This open area allows fire gases to collect, thus heating the underside of the roof area. Several forces work to spread the fire. First, the heat from the fire gases preheats the burnable construction materials. This lowers the amount of heat needed for ignition. This also increases the speed of fire spread. Second, when these flammable fire gases gather and catch fire, they produce a very fast-spreading fire.

One flammable gas present at most fires is carbon monoxide. Carbon monoxide (CO) is an odorless, colorless, tasteless, flammable gas that has an explosive range from 12.5% to 74%. Its ignition temperature is 1,204°F (651°C). It is produced by incomplete combustion and is responsible for much of the fire spread.

When conducting topside ventilation on a flat roof, fire crews should always work with the wind at their backs or from the windward side. **Windward** is the direction from which the wind is blowing. This will keep the smoke and hot gases downwind. The opposite side is called the **leeward** side or the side facing the wind.

Peaked Roofs

The steep inclines of peaked roofs encourage the use of aerial ladders. An aerial ladder provides a stable, flat work surface. However, aerial platform access is not always available. In that case, the ventilation would need to be done using a roof ladder. The roof ladder is held in place by securing the hooks over the roof ridge. In this position, the roof ladder evenly distributes the weight of the firefighter over a larger area of the roof. This is important to remember because a sloped or peaked roof will support less weight than a flat roof.

A vent hole is cut as close to the ridge as possible without damaging the underlying rafters. The cut should be directly over the fire and, if possible, on the leeward side of the roof. Use a pike pole to make sure the vent hole goes through the ceiling into the area being vented.

Carport/Garage

A carport is a freestanding structure that is open on all sides. It is built with a flat roof only large enough to keep the elements off the stored vehicle. Such carports are open, and the vehicles are subject to vandalism and arson. There have been multiple-vehicle fires where a flammable liquid was used to set fire to more than one of these vehicles at the same time. This situation could allow for gasoline from exposed fuel tanks to catch fire.

This type of fire can be attacked using one 2 1/2-inch (64-mm) line or two 1 3/4-inch (45-mm) lines supported by a good water supply. A fast, direct attack with the hose nozzle should be enough to quickly control these fires. A carport is not much of a fire problem because it was built using a minimum amount of combustible materials.

Attached garages may be found next to or under one-story and two-story homes. A one-story home requires that a line be taken inside the house. This is to prevent the fire from spreading into this area. The two-story home requires a direct attack on the fire in the garage area and a line upstairs to stop the upward fire spread.

CASE STUDY

CONCLUSION

1. **Which mode of operation would have been best for this incident?**

   With no lives at risk, a defensive fire operating mode should have been used.

2. **What information about the building would have helped the incident commander in decision making?**

   Knowledge of the building and prefire planning would have greatly helped in firefighting. Fire crews could have found those areas that would pose problems. An example is the utility room where the firefighters were trapped.

3. **What process may have improved radio communications?**

   Better mayday policies might have helped. The proper mayday process would have allowed only the trapped firefighters and the IC to use a specific radio channel.

4. **When observing an incident, what is a critical step in the decision-making model?**

   When observing an incident, reassessing fire tactics is critical. This step provides feedback on whether the initial attack plan is working.

WRAP-UP

SUMMARY

- The overall mission of the fire service is always to save lives and to protect property.
- Firefighters responding to fire incidents need to:
  - Find the fire
  - Limit the fire
  - Put out the fire
- Once these have been considered, the incident commander sizes up the incident.
- There are five steps used in the decision-making process to determine the needed strategy and tactics of firefighting:
  - Identify the problem.
  - Identify solutions to solve the problem.
  - Select the best solutions(s) based on the situation or circumstances.
  - Implement the solution.
  - Reexamine the solution to see if it is working. If not, start over.
- These efforts are made while also deciding if the fire attack will be in offensive mode, transitional mode, defensive mode, or nonattack mode. Life safety concerns for people inside the building and firefighters determine the mode. Concerns for the environment factor into the decision of mode as well.
- Acronyms such as RECEO-VS and SLICE-RS identify specific factors that impact the strategy and tactics decision.
- Factors that influence a size-up include construction type, occupancy or use, apparatus and staffing, life hazards, terrain, water supply, backup sprinkler systems, street conditions, weather, exposures, area and height, location and extent of fire, time, and any special concerns.
- Fire behavior patterns and issues are different for different construction methods and occupancy types.
- Two types of construction are the platform construction method and the balloon frame construction method. Platform construction creates a fire block, whereas the balloon frame construction method allows hot gases to spread vertically to the upper regions of the house.

KEY TERMS

auto-exposure The lapping of fire from one floor to an upper floor on the outside of the building, sometimes through windows.

balloon frame construction method An outdated method in which the wood studs run from the foundation to the roof and the floors are nailed to the studs.

below grade A floor lower than ground level.

cockloft A void space, about 3 feet deep, between the ceiling area and the underside of the roof.

collapse zone The border around the structure that is usually 1.5 times the height of the building where debris would fall during a collapse.

decision-making model A five-step process used to solve problems.

defensive attack mode A fire tactic used outside a building to put out a fire.

deluge system A dry pipe system that protects areas that are being consumed by a fast-spreading fire; all sprinkler heads are activated at the same time.

drip loop A loop formed by the electrical supply lines; also called a meter head.

dry pipe A system that has air in the sprinkler piping until it is activated; each sprinkler head is activated by heat.

exposure A property threatened by radiant heat from a fire in another structure or an outside fire.

eyebrow dormer A concrete extension over the top of openings, such as windows, doors, and balconies.

fire damper Item used to prevent transmission of flame where air ducts go through fire barriers.

fire load The total amount of fuel that might be involved in a fire, measured by the amount of heat given off when that fuel burns; expressed in BTUs.

heating, ventilation, and air conditioning (HVAC) system A central system used to heat and cool large buildings.

heavy timber construction Masonry; however, the inside columns and beams are solid or heavy wood and are built without hidden spaces; also called Type IV or mill construction.

leeward Facing the direction toward which the wind is blowing.

nonattack mode A fire tactic used when a fire attack is too dangerous or suppression activities are prevented.

occupancy The building code term that provides standards to match a building's use and those who will use it with features to address fire hazards and life safety concerns.

offensive attack mode Firefighting that makes a direct attack on a fire.

ordinary construction A type of building construction in which the outside walls are made of masonry materials.

personal alert safety system (PASS) A small device that is sensitive to motion; worn with SCBA when firefighters enter an immediately dangerous to life and health (IDLH) area.

personnel accountability report (PAR) A roll call taken of all firefighters working on the fire ground.

personnel accountability system (PAS) A tracking system to follow the entry and exit of crew into the working area during an incident.

platform construction method A construction method in which the floors are built separately from the outer walls; this means that the ceiling and floor serve as a fire block.

rapid intervention team or rapid intervention crew (RIT/RIC) A group of people whose purpose is to make a rapid response to reports of firefighters who become trapped or confused in the building.

RECEO-VS An acronym used to develop strategies and tactics on the fire ground.

rehabilitation system A group of activities for the health and safety of responders at emergency incidents.

size-up An ongoing review by firefighters to identify the problems at an incident.

SLICE-RS An acronym used to develop strategies and tactics on the fire ground.

soffit A false space under stairways and projecting roof eaves or the false space above cabinets in kitchen or bathrooms.

standard operating procedure (SOP) Specific instructions on how to complete a task or assignment; also called standard operational guidelines (SOG).

standpipe A manual firefighting system with piping and hose connections inside buildings.

strategy A general plan to meet incident goals.

tactic An action needed to complete the strategy or plan.

transitional mode The process of shifting from the offensive attack mode to the defensive attack mode, or vice versa.

transom A small window in an older building, such as a hotel, at the top of the ceiling, usually over the room entrance.

utility chase A channel used for electrical, telephone, and plumbing lines and pipes for services in a building.

void space An area in the structure's construction that has no outside or inside entrance and is not used for anything.

wet pipe A system that has water in the sprinkler piping at all times, and each sprinkler head is activated by heat.

windward The direction from which the wind is blowing.

REVIEW QUESTIONS

1. What is the five-step decision-making process? How can one apply it to the size-up process?
2. Why does the fire service use acronyms to help create strategies and tactics?
3. How does the fire load of a building affect fire behavior and how firefighters put out the fire?
4. What fire behavior problems might a firefighter face at a church fire?
5. What actions would you take to vent a peaked roof? Why would you vent at the peak of the roof?

DISCUSSION QUESTIONS

1. What is the difference between a strategy and a tactic in firefighting?
2. In the acronym RECEO-VS, why are V and S not considered to be in chronological order?
3. How can sprinkler and standpipe systems help firefighters to put out a building fire?
4. How did the balloon construction method encourage rapid fire spread?

APPLYING THE CONCEPTS

In reference to the Hackensack Ford Dealership Fire and using lessons learned in Chapter 5:

1. What would be your first decision on arrival to this incident?
2. What additional resources should be considered?
3. Do the location and extent of the fire factor into your decision making?

REFERENCES

Ahrens, Marty. "US Experience with Sprinklers." National Fire Protection Association. October 2021. Accessed January 30, 2023. https://www.nfpa.org/-/media/files/news-and-research/fire-statistics-and-reports/suppression/ossprinklers.pdf.

American Society for Testing and Materials (ASTM). *Standard Test Methods for Fire Tests of Building Construction and Materials.* Report E-119-22. 2022. https://www.astm.org/e0119-22.html.

Angle, James S. *Occupational Safety and Health in the Emergency Services,* 4th ed. Burlington, MA: Jones & Bartlett Learning, 2015.

Angle, James S., Michael F. Gala, Jr., T. David Harlow, William B. Lombardo, and Craig M. Maciuba. 2020. *Firefighting Strategies and Tactics,* 4th ed. Burlington, MA: Jones & Bartlett Learning.

Avillo, Anthony. "The Fire Attack–Ventilation Connection: Street Considerations." Firefighter Close Calls. October 25, 2015. Accessed January 30, 2023. https://www.firefighterclosecalls.com/the-fire-attack-ventilation-connection-street-considerations/.

Brannigan, Francis L. *Building Construction for the Fire Service,* 4th ed. Quincy, MA: National Fire Protection Association, 2009.

Brennan, Christopher. "The Mission of the Fire Service Warrior." *Fire Engineering.* November 18, 2010. Accessed January 30, 2023. https://www.fireengineering.com/leadership/fire-service-warrior-mission/#gref.

Centers for Disease Control and Prevention. *Preventing Deaths and Injuries of Fire Fighters Working at Basement and Other Below-Grade Fires.* July 2018. Accessed January 30, 2023. https://www.cdc.gov/niosh/docs/wp-solutions/2018-154/pdfs/2018-154.pdf?id=10.26616/NIOSHPUB2018154.

Clark, William E. *Fire Fighting/Principles & Practices.* New York, NY: Dunn & Bradstreet, Technical Publishing–Fire Engineering, 1991.

REFERENCES CONTINUED

College of the Canyons. *Attic Fire Tactical Considerations* [Slide deck]. n.d. Accessed January 30, 2023. https://coc.instructure.com/courses/939/files/115173/download?wrap=1.

Diamentes, David. *Fire Prevention: Inspection & Code Enforcement*, 4th ed. Scotts Valley, CA: CreateSpace Independent Publishing Platform, 2015.

Dugan, Michael M. "Conducting a Proper Size-Up." *Firehouse Magazine.* August 1, 1998. Accessed March 17, 2023. https://www.firehouse.com/operations-training/article/10544552/conducting-a-proper-sizeup.

FirefighterNation.com. "Prioritizing Tactical Objectives on the Modern Fireground." December 1, 2013. Accessed January 30, 2023. https://www.firefighternation.com/firerescue/prioritizing-tactical-objectives-on-the-modern-fireground/#gref.

Foskett, Janelle. "FHWorld16: Strategies & Tactics for Residential Attic Fires." *Firehouse Magazine.* February 3, 2016. Accessed January 30, 2023. https://www.firehouse.com/home/news/12165977/fhworld16-strategies-tactics-for-residential-attic-fires.

Law Insider. "Interior Structural Firefighting Definition." n.d. Accessed January 30, 2023. https://www.lawinsider.com/dictionary/interior-structural-firefighting.

National Fire Protection Association. *National Fire Protection Handbook.* Quincy, MA: NFPA, 2008.

National Fire Protection Association. Standard 251, Standard Methods of Fire Tests of Building Construction and Materials.

National Fire Protection Association. Standard 1710, Standard for the Organization and Deployment of Fire Suppression Operations, Emergency Medical Operations and Special Operations to the Public by Career Departments. 2020. https://www.nfpa.org/codes-and-standards/all-codes-and-standards/list-of-codes-and-standards/detail?code=1710.

National Fire Protection Association. Standard 1720, Standard on Volunteer Fire Service Deployment. 2020. https://www.nfpa.org/codes-and-standards/all-codes-and-standards/list-of-codes-and-standards/detail?code=1720.

Naum, Christopher J. "Buildings on Fire: Collapse Zones: Building Characteristics & Occupancy Risks." *Firehouse Magazine.* June 1, 2015. Accessed January 30, 2023. https://www.firehouse.com/home/article/12069221/collapse-zones-building-characteristics-occupancy-risks

New England Institute of Technology. "What Are the Different Types of Construction?" April 6, 2021. Accessed March 6, 2023. https://www.neit.edu/blog/what-are-the-different-types-of-construction.

Quintiere, James G. *Principles of Fire Behavior*, 2nd ed. Boca Raton, FL: CRC Press, 2016.

Roarty, Dan. "An Overview of the 5 Types of Construction." Samuels Group. January 21, 2021. Accessed March 6, 2023. https://www.samuelsgroup.net/blog/5-types-of-building-construction.

Roberts, Bill (Fire Chief, City of Austin). "The Austin Fire Department Staffing Study." March 1993.

Smoke, Clinton. *Company Officer*, 3rd ed. Clifton Park, NY: Delmar Cengage Learning, 2009.

Snyder, Michael D. "Demystifying Building Code Occupancy Classification." American Institute of Chemical Engineers. February 2021. Accessed January 30, 2023. https://www.aiche.org/resources/publications/cep/2021/february/demystifying-building-code-occupancy-classification.

Solano County Fire Chiefs Association. *Solano County Fire Chiefs Association Guidelines.* April 30, 2018. Accessed January 30, 2023. http://www.solanofirechiefs.org/chiefs/documents/file/SCFC%20Policies/SCFCA%20Guideline%2010%20Rapid%20Intervention%20Crew%20%28RIC%29.pdf.

Thiele, Timothy. "The Function of the Drip Loop in an Electrical Service." The Spruce. Updated June 26, 2022. Accessed January 30, 2023. https://www.thespruce.com/electrical-service-drip-loops-1152352.

U.S. Department of Agriculture. *ICS Glossary.* n.d. Accessed January 30, 2023. https://www.usda.gov/sites/default/files/documents/ICSGlossary.pdf.

U.S. Department of Labor, Occupational Safety and Health Administration. "Safety and Health Topics." n.d. Accessed January 30, 2023. https://www.osha.gov/topics.

U.S. Fire Administration. *High-Rise Office Building Fire, One Meridian Plaza, Philadelphia, Pennsylvania.* Technical Report No. 49. 1991. Accessed May 3, 2023. https://www.interfire.org/res_file/pdf/Tr-049.pdf.

Ziavras, Valerie. "Occupancy Classifications in Codes." National Fire Protection Association. May 7, 2021. Accessed January 30, 2023. https://www.nfpa.org/News-and-Research/Publications-and-media/Blogs-Landing-Page/NFPA-Today/Blog-Posts/2021/05/07/Occupancy-Classifications-and-Model-Codes.

CHAPTER

6

Special Concerns in Firefighting

LEARNING OBJECTIVES

After studying this chapter, you will be able to:

- Explain prefire and postfire planning processes and describe how they ensure safe, efficient, and effective firefighting.
- Describe fire behavior in very small spaces with and without ventilation.
- Explain the various methods of ventilation and how each method impacts fire behavior.
- Describe the Governors Island study and explain how it changed firefighting methods.
- Explain salvage and overhaul, their roles in putting out fires, and methods used to reduce property loss.

CASE STUDY

Special Nursing Facility Fire

It is 2:00 PM on a Saturday. You receive an alarm. There is a report of a fire at a local nursing facility that houses 100 patients. The fire is reported to be in the laundry room.

On arrival of the first alarm units, they find the staff is helping to remove the patients from the building. Many of these patients cannot walk by themselves.

The laundry room is in the southwest part of the building. It is adjacent to the recreation and physical therapy rooms. There is a lot of smoke in the halls, and the alarm panel shows that the sprinkler system is offline.

1. Would this emergency be considered a great risk to life and property? For what reasons?
2. The first alarm consists of three engines, a truck, and one battalion chief. Are these resources enough for a proper action plan?
3. What other outside resources might be needed for the safety of the residents?
4. What tactical mode should be used for the fire attack?

Introduction

This chapter explores areas of special concern to firefighters. These areas, such as prefire and postfire planning, ventilation, salvage, and overhaul, show the importance of understanding fire behavior. Understanding fire behavior improves the efficiency and safety of firefighting. During prefire planning, it allows firefighters to predict how a fire may behave. At a postfire conference, firefighters can compare fire behavior principles to actual fire behavior. This comparison teaches staff and improves future firefighting.

Ventilation improves rescue and firefighting operations. It is a way to remove smoke and heat from a building and allow for the entry of cooler air. Ventilation receives special review in this chapter because incorrect ventilation can allow fire to spread to other areas of a building. This increases the threat to people inside the building and firefighters. It can lead to deaths.

We will review the Governors Island, New York, study on how outside ventilation can stop the spread of a fire inside a building. Ventilation can also help those trapped inside a burning building survive.

Salvage operations are the protection of property while fighting a fire. These are actions the firefighters take to prevent damage to an owner's property and belongings by fire, smoke, and water. Salvaging is important in firefighting because community members and peers view it as a measure of skill.

Fire service skill begins with an understanding of the combustion process. It continues by reducing the fire and smoke losses as much as possible. This is done by knowing how smoke, fire, and water move within buildings. This basic information can increase fireground safety.

Overhaul is the process of searching the fire scene for hidden fires or sparks that may rekindle. This process also helps find the origin and cause of the fire. Sometimes, fires thought to be out can rekindle and seriously injure firefighters. For this reason, we review the overhaul process and some of the methods to ensure that fires have been put out safely.

Advance Preparation for Firefighting

Firefighters should preplan for the community areas in their jurisdiction. In most cases, this area will be the first-in or first-due jurisdictional area of a fire department. The **first-in jurisdiction** is the area where a fire department has a duty to respond. To preplan, firefighters find areas and structures that pose the greatest threats to life and property. These places are then named **target hazards**.

Some examples of target hazards include:

- Buildings where there are groups of people (e.g., churches, schools)
- Large public areas (e.g., theaters, stadiums, and libraries)
- Properties with a hazardous manufacturing process or storage area (**FIGURE 6-1**)
- Properties that hold valuable items

FIGURE 6-1 A target hazard may include a facility or property with a hazardous storage area.

© jutawat Rawichot/Shutterstock

The Prefire Plan Inspection and Review

Prefire planning is the process of studying and preparing for firefighting activities at the scene of a hazard or structure (**FIGURE 6-2**). Often this involves first-alarm companies. Firefighters visit the structures or areas posing the greatest threats to life or property. All staff becomes familiar with these locations.

During prefire planning, firefighters think of tactics, strategies, or other actions that might be needed at a certain place in an emergency. Firefighters cannot wait for these emergencies to arise before deciding on a course of action. A preplanning session allows quiet thought on a fire problem before an emergency. Firefighters then create detailed plans to address what could happen there.

FIGURE 6-2 Firefighter conducting a building prefire plan.

© Jones & Bartlett Learning

Availability of Water

Water supply is a crucial part of prefire planning. There are several factors of the water supply to consider. First is the location of the fire hydrants. Fire crews also need to know how much water they can deliver and the size of the water main. In some departments, each engine company jurisdiction is divided into zones. A **fire management zone (FMZ)** is a zone within an engine company's area where similar hazards are grouped. The groups have about equal needed fire flow and number of hazards. **Fire flow** is the water supply needed to put the fire out. The type of hazard the building presents impacts the level of fire flow. Once the hazards are grouped, they receive a hazard severity ranking. The hazard severity ranking given to each zone can determine the personnel and equipment response.

Backup water sources are important, too. Firefighters are responsible for being familiar with and trained to use any backup water source. A prefire discussion with the water supplier should establish if other systems could be available in emergencies. Firefighters can preplan relays and water shuttles using water department equipment or other resources available.

In some areas, static water sources can be used to supply needed water. Some examples of static sources are lakes, ponds, rivers, small streams, or even swimming pools. Firefighters should note the location of any dry hydrants to help get water from these water sources. Fire crews can discuss the resources, supplies, methods, and any issues or needs before the incident (**FIGURE 6-3**).

Built-in Fire Protection

During prefire planning, firefighters examine systems inside buildings. They note all built-in fire protection equipment at the target hazard. Crews should learn and review all the processes needed to start and shut off the systems as well as the locations of key parts. It is valuable to know where to find **fire department connections (FDCs)**. FDCs are typically located

FIGURE 6-3 Static dry hydrant. During preplanning, firefighters should note the availability of any dry hydrant at static water sources.

Courtesy of H2o In Water Services LLC.

FIGURE 6-4 Firefighters should examine inside control valves during prefire planning.

© GolF2532/Shutterstock

outside of the building and can be either a stand-alone fixture or part of the outside wall of the building. These connections allow firefighters to supply additional water to the fire suppression sprinkler system or the standpipe system. The location of Siamese connections should also be noted. A **Siamese connection** is a hose fitting with a double hose line junction that allows two hoses to connect to one line. The one line is often larger than the supply lines. It is also good to know where to find control valves. **Control valves** control the flow of water in sprinkler and standpipe systems. The fire department will be required to connect to and operate these systems in an emergency (**FIGURE 6-4**).

Automatic and Mutual Aid Resources

The use of neighboring community resources must be preapproved and planned. It is helpful for agencies to share their preplanning information for locations where resources might be provided. Firefighters need to know what resources are available, including staffing. Crews should know the assisting staffs' abilities and fire agencies' procedures. This is so that any clash between systems can be worked out before an incident.

These resources can be provided as automatic aid or mutual aid. **Automatic aid** is a contract between agencies to provide resources to any incident on dispatch. These are usually based on the closest available resources. **Mutual aid** is a contract between agencies to provide resources on request. Most often, you will see mutual aid at large-scale incidents, such conflagrations, wildland fires, floods, or other natural disasters.

Problems can arise when this type of aid is provided. For example, in California, some fire departments use hose threads, which do not work with national standard threads. This has led to problems when those departments tried to use fire equipment and hose threads from nearby communities. This is one example of an issue that needs attention before an incident.

Prefire Planning and Fire Behavior

It is helpful for firefighters to think of how fire behaves as they review structures and their features. The layout of a building and the locations of stairwells, roof hatches, skylights, and windows can impact how firefighters ventilate a burning structure. Some structures may be challenging for firefighters if they have contents made from materials that will burn hotter or create toxic fumes.

Local Code Applications

Firefighters should know how local codes apply to fire suppression systems. Often cities will adopt standards for installing these systems. These systems may include smoke removal, air refill for self-contained breathing apparatus (SCBA) during a fire attack, and communication. The communication systems may be permanently wired and allow firefighters to contact each floor. The density of the concrete and steel in these structures often makes radio transmission impossible.

The building may also have a fire suppression systems control room. This area is kept up by the building engineer. It is useful for firefighters to locate the fire systems control panel and smoke removal controls. Firefighters should look for any system that would help in an effective fire attack or victim rescue.

Postfire Activities

Postfire planning meetings must be run in a positive way. Negativity in meetings will cut off communication. Negative feelings may even flow over into future emergency responses. Critical comments about an operation should be approached as a training process and not a way to place blame. Open, positive meetings will foster a learning atmosphere where all members can build on the success of their past actions and improve other areas.

Postfire Conference

Fire leaders hold postfire conferences to improve future fire operations. While the name of this meeting may vary regionally, the **postfire conference** is a meeting or meetings held after an incident to talk about what fire crews saw, what actions they took, and what they learned to improve their emergency response on future incidents.

The conference includes all responding companies. Several meetings may be needed to cover various work shifts. They may also be held at different locations. All members should have a chance to participate in the discussions. It takes tact and diplomacy from leaders to get participation by most of those in attendance.

Group participation is boosted by having each member describe their actions during the incident. This allows officers to understand what occurred. The officer chosen to record what happens at the meetings attends all of them. The officer can then present a summary of the findings to all members. This summary describes the firefighters' actions, both good and bad, along with any recommendations.

Postfire Form

One way to direct the focus of the conference is to use a form. The form includes standard questions to be answered during the conference. The form can be edited to meet the needs of every department. Some departments require all members to fill in the information. The form is then used as a training aid.

Cause

- Was the possibility of arson thoroughly investigated?
- Does the cause of the fire imply the need for increased fire prevention?
- If so, should the fire prevention inspections be done by station members or the fire prevention division or bureau?

Detection

- Would automatic fire detection devices benefit the property?
- Are there laws to require an automatic detection system?
- Are the automatic detection devices cost-effective?

Alarm

- Was the alarm reported, received, and sent correctly?
- If not, what was the problem? Can it be fixed so it will not happen again?

Equipment and Staffing Response

- Did the dispatcher provide a clear description of the fire/emergency when forwarding the information?
- Were the proper equipment and human resources dispatched?
- Were sufficient resources available?
- Were correct routes to the scene chosen?

Extent of the Fire

- What was the extent of the fire on arrival of the fire department?
- Did the fire continue to spread after firefighters started to put out the fire?
- Did special circumstances delay a fast application of water?

Size-Up

- During the size-up, did the commander use the acronyms RECEO-VS, SLICE-RS, or another department-approved method to consider all factors?
- What was the first-arriving officer's report? Did it accurately reflect the situation? If not, how could it have been improved?

Preplanning

- Did past preplanning activities and documents help the responders develop the best and safest strategy?
- Does the plan for the structure need to be updated?

ICS

- Was an incident command system put into operation?
- Was it correctly applied and used?
- Was the incident used as a training session for future fire officers?

RIT/RIC

- Was a rapid intervention team or rapid intervention crew (RIT/RIC) put into place?
- Were they equipped and ready to respond?
- Were higher levels of teams needed? If so, were they available?

Forcible Entry

- Were there signs of a break-in before fire department arrival?
- Was the building opened in a manner to minimize damage?
- Were special tools needed that were not available?

Rescue

- What were the conditions on arrival that impacted the rescue efforts?
- Was there enough staffing on the first alarm for rescues?
- Was an accountability system put in place?
- Were rescue efforts successful?
- If so, what were the factors contributing to the success?
- If not, describe factors that contributed to the failure of the rescue efforts.
- Can these be resolved for future rescue efforts to be successful?

Ventilation

- Was the building ventilated in a timely and correct manner?
- If not, what recommendations can be made?
- Was the size of the ventilation hole(s) or location a problem or were there other issues?

Hose Streams

- Were hose streams properly directed and applied?
- Were hose streams the proper size and pressure?
- Were there any suggestions for improving application or ventilation issues?
- Were there problems advancing the hose line?

Salvage

- Were salvage operations effective?
- Is there an estimate of property saved or protected?
- Are there suggestions for steps that can be taken to improve salvage operations?

Overhaul

- Was the fire put out completely?
- Was the building cleared of debris and excess water? Was sprinkler protection restored? Was the building properly closed?
- If arson was suspected, did the fire department stand by until the arson unit arrived?

Scene Security and Traffic Control

- Did the police department help with traffic control?
- Was the street opened for traffic as soon as possible?
- Were bystanders controlled?
- Were fire suppression personnel protected?

Utilities

- Were gas, water, and electrical services properly handled by the fire department?
- Was the utility service promptly secured?

Incident Communications

- Were the communications on the incident clear and concise?
- Were the radiofrequency assignments specific and logical? Were they assigned per the incident communications form?

By reviewing fire operations in a meeting, the members of the fire department who were not present at the fire can benefit from the experiences of those who were present. Meetings are also educational for those members who were at the fire but whose activities were limited to one area. It permits them to learn the firefighters' overall strategy and operations and the behavior of the fire.

Finally, a postfire conference is a chance to discuss how the mission could have been handled better and how future missions can be improved. The following materials are suggested for a postfire conference:

- A plan for a discussion and explanation to decide the order of events and operations on the fire ground
- A precise drawing of the fire, with photos if possible

FIGURE 6-5 shows an on-scene, informal postfire conference being conducted. The events are still fresh in everyone's minds, allowing the analysis to be as accurate as possible.

Other Important Postfire Meetings

In recent years, the fire service has held other postfire meetings. These meetings address the impact of incidents on firefighters. Memories of incidents can affect a firefighter's mental health. Critical incident stress management (CISM) is a protocol to help fire crews and others at traumatic events cope with these experiences.

There are a few types of CISM interventions. Two interventions, called defusing and debriefing, use group formats. **Defusing** is the process of talking about the experience. It occurs immediately after the incident. This type of meeting is recommended when there is a notable impact on one or more members after the incident. **Debriefing** is a voluntary group discussion that is used when there is no urgent need to discuss the incident. While similar, these meetings have different goals. Defusing is meant to ease the immediate trauma to the firefighter and decide the need for additional meetings. Debriefing is meant to promote resiliency, resistance to stress, recovery, and return of group members to a healthy functioning.

FIGURE 6-5 A quick on-scene postfire conference.

Courtesy of Los Angeles County Fire Department.

Ventilation

Ventilation is vital to successful and safe firefighting. It must be a high priority during a size-up. After fire crews decide how and where to ventilate, they should begin immediately (**FIGURE 6-6**). Ventilation usually requires more time than the placement and deployment of hose lines. When truck companies are primarily used for ventilation, other engine companies usually arrive

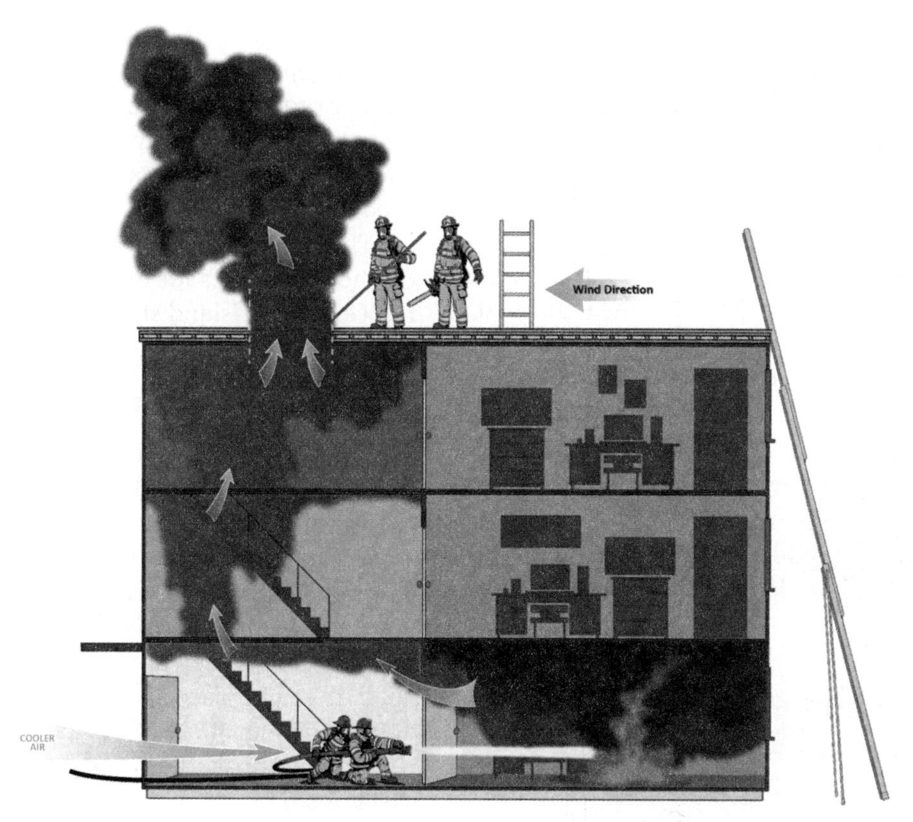

FIGURE 6-6 Tactical vertical ventilation to reduce fire spread.

Courtesy of FIRE - From Knowledge to Practice.

and conduct other parts of the size-up, request a water supply, and deploy fire attack lines.

In jurisdictions without truck companies, these important assignments must be carefully worked out during the prefire planning sessions. During this process, an engine company is often chosen for building ventilation. Companies given the ventilation function should be very familiar with fire behavior. They should also know the effect ventilation can have on the combustion process.

TIP

The type and amount of ventilation required are not always the same. Each fire is different, and the fire conditions vary. Where and when possible, vertical ventilation (Figure 6-6) is preferred to horizontal ventilation (**FIGURE 6-7**) because smoke rises. The height and construction of the building, along with the volume of smoke and the location of the fire, help determine the type of ventilation required or if any is needed at all. The choice between vertical or horizontal ventilation can be made based on a variety of issues. Are there sufficient time and resources for vertical ventilation? Often horizontal ventilation may be completed faster because it may only involve breaking ground-floor windows. In a multistory building, the floor the fire is on may demand horizontal ventilation. However, vertical ventilation can be more effective when combined with an interior attack.

Governors Island Study

For many years, the standard way to attack a fire was for firefighters to enter the building from the unburned side and push the fire back onto the burned side. This tactic proved very effective on the fire loads and building construction of the past. With today's modern buildings, the heat generated by fire is contained. In addition, the fuel from the many hydrocarbon-based products in these buildings creates a high fire load. We need to review the old tactics for fire suppression. New studies are needed to measure conditions caused by modern products. The basic chemistry of fire does not change (fuel, heat, and an oxidizing agent), but the fuel element has greatly changed. This element can be tested in live burn experiments. The results of these tests may be measured, and new tactics developed to prevent firefighter and civilian deaths.

FIGURE 6-7 Tactical horizontal ventilation.

Courtesy of FIRE - From Knowledge to Practice.

In 2012, the National Institute of Standards and Technology (NIST) and Underwriters Laboratories (UL), along with the New York City Fire Department (FDNY), ran live burn experiments to challenge conventional fire tactics.

On Governors Island in New York, test fires were set in old Coast Guard barracks. The tests proved that attacking a fire from outside the building increased the likelihood of survival for people inside the building. This also made it safer for firefighters to enter and continue suppression efforts. New standards for ventilation were issued.

The Vent–Enter–Isolate–Search acronym was developed from these tests. **Vent–Enter–Isolate–Search (VEIS)** is a firefighting technique that calls for venting outside the building for a safer entrance and a reduced fire flow path. This new tactic was proven effective through repeated tests on the structures at this site.

The process of venting the fire floor allows firefighters to keep the fire away from the rest of the structure. They can then enter a safer environment and perform a search.

The time-honored beliefs and methods are no longer the best tactics for modern buildings and fire load. The Governors Island study led to a more effective way to launch an offensive fire attack.

Roof Ventilation

Roof ventilation can be challenging because it varies in difficulty. Ventilating a building can be as easy as removing an attic cover or as complex as opening a large area by hand (**FIGURE 6-8**). One of the first things to do when preparing to ventilate a roof is to look for natural openings to use. Unfortunately, these openings are only of value if they are in the right place. If they are not in the right place and are opened, they will pull fire to the opening. Some of the natural openings used for ventilation are scuttle holes, skylights, staircases, and elevator shafts.

On small, single-story malls, fire crews may a need to cut a ventilation hole the width of the building. This

FIGURE 6-8 Vertical ventilation: Example of cutting a vertical ventilation open above the area of origin.

Courtesy of District Chief Chris E. Mickal/New Orleans Fire Department, Photo Unit.

operation involves multiple cutting crews. The purpose is to stop the progress of the fire through a common attic. This draws the heat and smoke out of the building. Hose lines would be placed inside the building below the ventilation cut to direct smoke and heat up using hydraulics.

Scuttle Holes

A **scuttle hole** is an opening that allows entry into the attic area. The hole is fitted with a cover that is sometimes locked. Scuttle holes can be used to ventilate only the attic because they usually do not go through the ceiling area. A wooden cover can be removed by prying it open with an axe, a pike pole, or a rubbish hook. Some holes have spring–loaded metal covers that may be locked from the inside. These covers are harder to open. In many cases, it is easier to find another place to ventilate.

Skylights

Skylights provide sunlight to areas below the roof. Their value in the process of ventilation is the same as other building openings. They allow the heated fire gases, heat, and smoke out of the building. Their effectiveness depends on the area they are ventilating. In hotels and apartments, they can be used to ventilate the hallway of the top floor. In commercial buildings, they will ventilate the storage or warehouse areas. Often this is where the seat or origin of the fire exists. Skylights can usually be removed intact because the frames are only lightly secured at the corners. However, if it is difficult to remove them, and time is important, a firefighter should not hesitate to break the glass.

In older structures, skylights are made of Plexiglass and wired glass. The Plexiglass skylights can provide automatic ventilation because Plexiglass melts at about 400°F (204°C) and will create an opening for ventilation to occur. Due to the intense heat at the ceiling of these buildings and the low melting temperature of Plexiglass skylights, they often self-ventilate before the arrival of the first-due fire companies.

Today, it is common for skylights to be made of tempered or laminated glass. These newer skylights may be opened by firefighters with the proper ventilation tools.

Solar Panels

Solar panels are a popular addition to commercial, industrial, and residential construction today. These panels can be mounted to the roof of a building or atop a freestanding structure to form an array for providing solar-generated electricity. These panels consist of photovoltaic cells that are generally rectangular and connected to one another.

Solar panels can be a hazard to firefighters during ventilation operations because they are slippery and often unsafe to walk on or to cut into. They can also prevent proper ventilation over the fire.

Staircases and Elevator Shafts

A stairway that leads to the roof can be quite valuable for ventilation. However, it can also prove risky if it is not used correctly. Using a stairwell for ventilation is easy because one has to only open a door. Prior knowledge of the building can be priceless in this situation. It is important to know what parts of the building below the roof will be threatened by opening a door. If the building does not have a complete compartmentation fire protection system, the doors on the floors below the fire may not close automatically. This would allow smoke rising in the stairway from the floor of origin to be pulled into the lower floors as it rises to the roof door to ventilate.

Some elevators have an equipment room located directly over the elevator shaft. In some cases, these rooms have a door that can be quickly opened or removed to release smoke through the shaft to the outside. These equipment rooms should be found and discussed during prefire planning.

FIGURE 6-9 Vertical ventilation: Example of tactical strip ventilation or trench cut.
Courtesy of Harry Garvin.

TIP

Firefighters must always keep in mind that ventilation operations can threaten the safety of tenants or other working firefighters.

Sounding the Roof

The ventilation officer must evaluate the roof to ensure that it is safe before firefighters walk on it. As firefighters enter the roof area, they should sound the roof ahead of them to find any weak spots. **Sounding the roof** means using a **pike pole**, a device with a sharp, pointed head and a curved hook, vent hook, or ax to tap the roof ahead to see if it is solid enough to support the weight of firefighters. A knowledge of roof construction is important to make this evaluation. Fire safety concerns about truss roof construction should be discussed ahead of time.

Some roofs by their nature are hazardous and require extreme caution when working on them. An example is a roof covered with 0.5-inch plywood in a commercial building that gives the appearance of being sturdy. Most roofs installed before 1960 are solidly made and well supported. They are usually safe, although they might appear somewhat springy. Firefighters should take extra care when working on buildings made after 1960 that are lightweight construction. **Lightweight construction** uses lumber that is smaller in dimension. Often 2 × 4 wooden studs, rafters, and ceiling joists will be found in structures made with lightweight construction. Today, this type of construction can include metal framing and engineered components.

Performing Roof Ventilation

There are several special considerations when cutting a hole in the roof. The ventilation hole needs to be cut directly over the fire. Sometimes, this location is readily found by observation. One can find it by looking for discoloration in the roof or by sounding the roof with a pike pole, vent hook, or ax to find soft or hot spots.

Cutting the hole over the hottest spot means a firefighter is on the most dangerous place on the roof (**FIGURE 6-9**). If there is any doubt about the safety of staff, then this spot should be avoided. A safer location should be found to make the cut.

A roof ventilation hole should be cut at the peak of the roof. It should be just below the ridge line and as close to above the fire as possible. Traditionally, the minimum size of the ventilation hole was 16 square feet, a 4 ft × 4 ft opening. The newer standard for ventilation holes is for them to be at least 36 square feet (6 ft × 6 ft) to allow for the increased number of British thermal units (BTUs) produced from plastics in modern products.

On most large building fires, the hole can be cut within 20 feet of the hottest spot. This will not seriously redirect the hot gases and fire through uninvolved parts of the building. Pulling the fire through uninvolved parts of the building should be avoided. However, if there are lives being threatened, firefighters will not hesitate to pull the fire away from the trapped people.

Many firefighters have fallen through the roof and been killed during ventilation operations. When conditions are right, an elevated platform might be a safer option. A line can be lowered from the platform and tied to the person opening the roof. The firefighter can be pulled to safety in case of roof collapse.

Louvering

Louvering is a method of roof ventilation that reduces the firefighters' exposure to smoke and heat. Initially it prevents cutting the supporting roof joists. It is an effective method for ventilating a plywood roof (**FIGURE 6-10**). It can be used effectively on sheathing but is most effective on plywood paneled decking.

The method consists of making two cuts lengthwise on either side of a single rafter. The cuts should be parallel to each other. The lengthwise cuts are then intersected with cross cuts. Once the cuts are made, one side of the cut panel can be pushed down, which results in a louvering effect. This method of opening the roof will ventilate the area below and direct most of the heat and smoke away from the working area. Cuts should be made so that firefighters making the cuts have the wind at their backs. The louvered panel should be used to direct the smoke away from those making the cut.

FIGURE 6-10 Louvered roof cuts: A tactical vertical ventilation using the louvered roof cut.

Horizontal or Cross-Ventilation

Cross-ventilation is a way of using windows, doors, and other horizontal openings for ventilation (**FIGURE 6-11**). This method can be very effective whenever there are many large openings. It limits the smoke to the floor being ventilated. This is particularly beneficial when there is a victim or valuable goods in other areas of the building that could be harmed by the smoke.

The wind direction should be assessed before starting horizontal or cross-ventilation operations. This is especially true whenever an entire floor or a large area has to be ventilated. The operation will be most successful when the smoke and gases are channeled out the leeward side of the building. Once the direction of the wind is decided, double-hung windows on the leeward side should be opened from the top. Sliding aluminum or wooden windows should be fully opened. Firefighters should remove all drapes, shades, screens, and other objects that would stop the airflow. Double-hung windows on the windward side should be opened from both the top and bottom or from the top only after all the leeward windows have been opened. Again, sliding aluminum or wooden windows should be fully opened and all blocks to airflow should be cleared.

It is a good practice when cross-ventilating to figure out where the heated gases and fire will go once they leave openings. They could extend upward and through open windows on the floors above. They could also go through openings into adjoining buildings. If there is a chance of this happening, then all exposed unprotected openings should be closed. Firefighters can also

FIGURE 6-11 Horizontal ventilation.

Courtesy of FIRE - From Knowledge to Practice.

spray streams to cover these openings to prevent any extension of fire.

Sometimes cross-ventilation works when the windows cannot be opened. Under these conditions, one may need to break the windows. Fire crews need to be careful when breaking windows because both civilian onlookers and firefighters can be injured by flying glass.

Smoke fans can be useful in cross-ventilation. They can be set up to create either a negative or positive airflow. There are two ways that firefighters use fans to create a positive airflow through a building. One uses smoke ejectors and the other uses positive-pressure ventilation.

Smoke ejectors are mechanical smoke fans that draw heated air and smoke outside of a structure. Positive-pressure ventilation is a process that blows air into a structure.

Negative-Pressure Ventilation

Negative-pressure ventilation is a method of forced ventilation that pulls air and smoke out of a structure. It uses mechanical fans to drain or pull the heated smoke and gases from inside of the burning building and move them outside. Negative-pressure ventilation has been used for several years. However, it has not proven to be an effective way to remove smoke. The fans never seem to fit the openings correctly. As a result, the openings around the sides of the unit need to be blocked. If the fit is not good enough, the fan will cause the air to churn. **Churning the air** is pulling air into the building from the outside and blowing it out again (**FIGURE 6-12**).

Other problems with this ventilation method include the fan sucking in debris, curtains, and other materials. Materials become trapped against the fan screen and need to be pulled off.

These fans were designed to not explode. However, overuse has been known to damage the electrical connections and the housing unit. This makes them dangerous in an explosive environment. For these reasons, the fire service tends to use positive-pressure ventilation.

Positive-Pressure Ventilation

As mentioned earlier, **positive-pressure ventilation** is a process that uses mechanical fans to blow air into a structure to remove smoke and gases through openings, like doors and windows. This allows firefighters to gain entry. This method combines the use of openings in the burning structure and mechanical blowers. These blowers may be powered either by electricity from the onboard generator on the fire truck or by gasoline-powered motors attached to the fan.

To conduct successful positive-pressure ventilation, the outlet opening must be controlled. If too many doors and windows are opened, it will not work. There will not be enough pressure in the smoke area to move the smoke quickly in the correct direction.

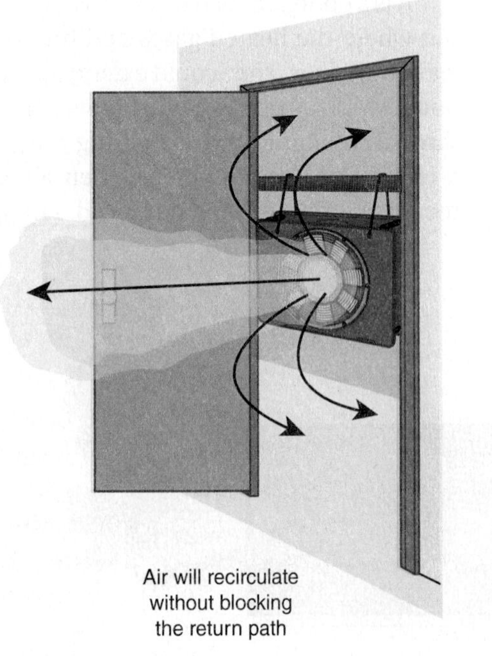

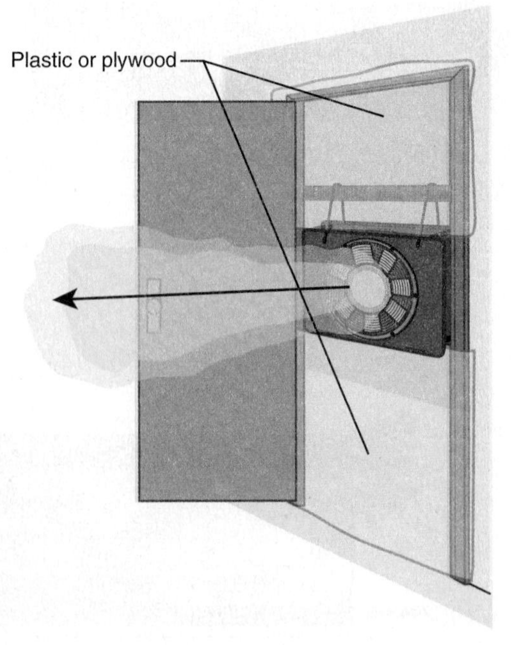

FIGURE 6-12 While using smoke ejector fans for ventilation, firefighters must cover the entire opening to avoid churning the air, which reduces the value of smoke removal.

FIGURE 6-13 Positive-pressure ventilation.

For best results, good timing, the right amount of opening space, and enough pressure are needed. Positive-pressure ventilation is seen in **FIGURE 6-13**.

TIP

Firefighters must minimize the movement of fire and smoke to unburned areas of the building.

- Do not use positive-pressure ventilation to enter the building if the fire is in the later stages of the combustion process. A backdraft may result.
- Do not push the smoke and fire into escape routes.
- Keep the openings for smoke exhaust as close to the fire area as possible.

FIGURE 6-14 Example of proper salvage cover application.

Salvage and Overhaul

Historically, in the United States, firefighting was based on bringing very large hose lines in the front of the building. Firefighters would push the fire from the front of the building back out the rear of the structure. Very little consideration was given to the amount of water and smoke damage or whether the fire department added to the fire damage. This resulted in the larger fire insurance companies funding their own salvage companies. They could respond to fires and work to keep the losses to a minimum. By providing this service, they could offset their losses and still pay for the costs of the salvage companies. However, as their operating costs increased, they phased out most of the salvage companies. They removed points to the insurance grading schedule for those fire departments that did not provide salvage work. **FIGURE 6-14** shows one of the few remaining fire underwriters salvage patrols.

For years, fire departments did not recognize salvage and overhaul work as their duty. These situations resulted in delays in salvage and overhaul training and the purchase of salvage and overhaul equipment. The fire department now accepts the duty of reducing water and smoke damage as much as possible.

Overhaul Operations

Overhaul operations require special attention be paid to the atmosphere and air quality in the area where firefighters are working. Firefighters should wear full personal protective equipment (PPE) with SCBA when the oxygen readings are below 19%.

The first step in the overhaul process is to watch the behavior of the smoke, heat, and water from firefighting operations and see how it affected the building and its contents. Some materials readily soak up water. This greatly increases the live load in the building and threatens the structure. **Live loads** are items inside a structure that are not attached to the structure or permanent. Often these live-load items can be product inventory, or they can be seasonal and change throughout the year. Smoke and fire can damage or destroy clothing, draperies, and furniture, which are part of the live load. Water can destroy drywall materials and some flooring. These are dead loads because they are part of the building. A **dead load** is static and associated with the full weight of the building.

The sturdiness of a building's walls and chimney is important for the safety of the firefighters who work inside and near the building. If needed, the walls or chimneys must be shored up before any overhauling work is done. In some cases, firefighters must get rid of extra water and water-soaked materials as quickly as possible. They can remove water by using the building's plumbing or diverting the water by cutting or removing walls or floors. They can also build water chutes or containment areas.

To develop the best strategy, firefighters must survey the damaged area before beginning overhaul operations. Part of the work survey should include a review of any tools or special equipment needed. The crew should request them early to avoid a delay in getting the resources.

Once the survey is complete, the incident commander should contact the owner of the building. This is especially important if the fire is in a commercial building. It is always best if the owner is on the property before overhaul starts. The owner can answer any questions about the inventory and business records. This is also the best way to protect the property because the owner may have either insurance coverage for securing the building or a preferred company.

The next step is to give work to teams. The teams are chosen to keep the work of each group from interfering with the work of the other groups. This is also to make sure all the work is done on time. Maintaining locations of crews working on the fire ground is essential to ensure the safety and direction of fire activities. This is called fire-ground management.

Teams must try to save all burned or partially damaged records. Such records may contain enough information to figure out the inventory, the amount of insurance carried, and the financial state of the business. The information in the records may be vital to the owner to prove the loss and may be important information for arson investigators.

Debris Handling

When handling debris from the fire, all regulations and polices regarding PPE should be followed. Firefighters should wear gloves to prevent cuts and burns.

Teams should also be careful when removing burned items that were in a deep-seated fire from the upper floors of a building. If the teams use an elevator, special policies should be put in place. These policies would direct actions to be taken when using the elevator. Firefighters should also use a special access key to control the elevator from the inside, selecting floors to access and those to avoid. Teams should bring a fire extinguisher into the elevator. Firefighters have been surprised by a rekindle of a fire as more oxygen was supplied to the debris when the elevator went down. A handheld, water-based extinguisher can put out the fire enough to allow firefighters to remove the burning material from the building.

Inside large industrial or commercial buildings, crews need to choose a safe place to put out the remains of a fire. Materials moved into this area may be smoldering. Any added oxygen can accelerate the fire.

Firefighters need to wet down and sift through all burned material thoroughly to check for hot spots. In residential buildings, it is better to move the debris outside of the building. Fire crews need to choose the area where the debris will be piled before moving any of the materials. It is a waste of time and effort to handle the burned material several times. Sometimes materials need to be piled on the sidewalk or street. If so, crews must provide enough barriers and notify the appropriate public agency.

Water Removal

One goal of the overhaul operation is to remove water from a building to prevent more damage. A building can become so waterlogged that the floor and walls could collapse.

A building may collapse at large fires if heavy streams have been aimed at the building. This could also happen at fires where the sprinklers have been on for some time. Sometimes, broken sprinkler risers or broken water pipes in the building cause flooding. If this happens, there is additional water damage to the building and its contents, plus the threat of collapse.

Stairways can be used to remove water from multiple-story buildings. The removal is easy if the stairs are concrete without a covering. If they are wooden or carpeted, they can be protected with salvage covers. Salvage covers should be opened to half their width to ensure a tight water seal.

Cast iron sewer pipes run both vertically and horizontally through buildings. This piping can be used to remove water. In many cases, the sewer piping system has a cleanout that can be opened to allow entry into the sewer system. If there is no access to the cleanout, the piping can be broken open for entry. It is good to have a screening device so that the system does not become clogged with debris.

One way of removing water is to cut a hole in the floor. The hole should be cut close to an outside window and be big enough to handle a large amount of water. Fire crews should remove all stock and other material below the spot where the hole will be cut before cutting the hole. A drain should then be set up to divert the water to the outside of the building. A ladder and salvage cover can be used to do this (**FIGURE 6-15**).

A wall breach may be needed when a large amount of water has collected on a floor and there are no other ways of removing it. This is done by choosing an outside window with a windowsill close to the floor area. The flow of water to the outside should not do further damage or reenter the building through the floors below. Firefighters should remove the window frame and glass and then remove the wall below the window opening to allow the water to drain outside the building.

Another method to rapidly remove large amounts of water from building areas with a bathroom is to remove the toilet from the floor. The removal of a toilet can be done by removing the nuts that fasten the bowl to the floor and lifting the toilet up. The large opening can then allow water to leave the floor via the sewer system drainage for the building.

Checking for Lingering Fire

The infrared **thermal imaging device**, also known as a thermal imaging camera (TIC), discerns between objects or areas with different temperatures and gives a visual display of those areas. The device is so sensitive that it may be able detect a heat source even if it is behind a wall. The device does have certain limitations. It may not be able to detect heat behind certain fabrics and types of glass. It is, however, often used during searches for fire victims and is also an effective tool during overhaul operations.

A firefighter should be sent into the attic to look for hot spots. This check should include looking at the areas close to the eaves. Birds, rats, and squirrels may build

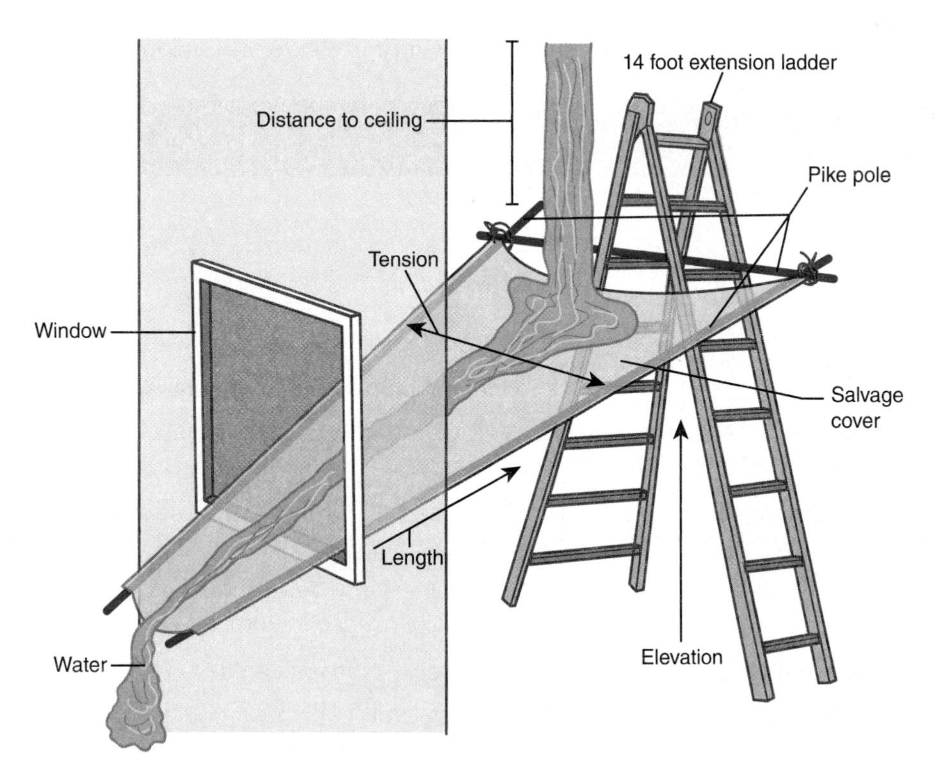

FIGURE 6-15 A ladder with a plastic salvage cover or plastic sheet can serve as a chute to divert water out of a building.

nests and store materials in these areas. These nests may have started burning without direct flame contact.

If fire is suspected between floors, it is best to go below the suspicious spot and pull the ceiling down. A pike pole or rubbish hook is a good tool for pulling ceiling away from the ceiling joists. Ceilings are easier to remove and less costly to repair than floors. If fire is suspected under the floor and it is impossible to get below the floor, the firefighters should not hesitate to cut a hole in the floor to gain access to the fire.

Fire tends to travel under window casings, around doors, and under trim at floor level. If fire is suspected in these areas, firefighters should remove the baseboards and window casings. When upper floors have been involved in fire, the bottoms of all windows, doors, and elevators should be checked for burning debris.

FIGURE 6-16 Building boarded up after a fire.

© rSnapshotPhotos/Shutterstock

Securing a Building

If possible, the building should be turned over to the owner when the fire department is ready to leave. If the owner is not there or cannot be reached, the fire department must lock up the building. The department should also board up all windows and doors broken during firefighting operations. Wood can usually be found on site to do this. If not, inside doors can be removed and nailed over openings.

Before returning to the fire station, the building should be protected from the weather. If holes were cut in the roof or the fire burned through the roof, the holes should be covered with plastic. **FIGURE 6-16** shows a building that has been boarded up after a fire.

At some fires, a firefighter needs to stay on as a fire watch. This task requires a firefighter to remain on the scene with a portable radio or cell phone to notify dispatch if there is a problem. The time spent for this task would be determined by the incident. Some fires may require an all-night watch to ensure the fire does not rekindle and to protect against vandalism. Others may only require a periodic return to the scene. If this is not practical, the jurisdictional engine company should be assigned to check on conditions periodically. The assigned company must always stop and reenter the building to check for lingering fire.

CASE STUDY
CONCLUSION

1. **Would this emergency be considered a great risk to life and property? For what reasons?**

   This facility is a great risk to life and property because most occupants cannot walk on their own. A threat to life is one of the first considerations in a target hazard classification.

2. **The first alarm consists of three engines, a truck, and one battalion chief. Are these resources enough for a proper action plan?**

   There are enough resources for an initial assignment to this incident. Crews should be able to begin a fire attack, limit the fire, and remove the patients. Additional fire support will be needed to complete this task.

3. **What other outside resources might be needed for the safety of the residents?**

   A key resource to ask for would be more emergency medical staff. This staff could address the various degrees of health of the patients.

4. **What tactical mode should be used for the fire attack?**

   Initially, an offensive fire attack would be put in motion. Hose lines would be taken to the area of origin. Crews would begin to put out the fire and prevent exposures from becoming involved.

WRAP-UP

SUMMARY

- Prefire and postfire planning, ventilation, salvage, and overhaul are all areas of special concern to firefighters.
- To preplan, firefighters find areas and structures that pose the greatest threats to life and property.
  - These places are then named as target hazards.
- Prefire planning is the process of studying and preparing for firefighting activities at the location of a hazard or structure.
- The postfire conference is a meeting or meetings held after an incident to talk about what fire crews saw, what actions they took, and what they learned to improve their response on future incidents.
- Ventilation should be a high priority during size-up. After fire crews decide how and where to ventilate, they should begin immediately.
- The Vent–Enter–Isolate–Search (VEIS) acronym was developed from the Governors Island Study. VEIS calls for outside venting for a safer entrance and a reduced fire flow path.
- Some of the natural openings available for ventilation are scuttle holes, skylights, staircases, and elevator shafts.
- Cross-ventilation, or horizontal ventilation, is a way of using windows, doors, and other horizontal openings for ventilation.
- The fire department now accepts the duty of reducing water and smoke damage as much as possible.

KEY TERMS

automatic aid A contract between agencies to provide resources to any incident on dispatch.

churning the air Pulling air into the building from the outside and blowing it out again.

control valve A valve used to control the flow of water in sprinkler and standpipe systems.

dead load The full weight of the building and its static parts.

debriefing A voluntary group discussion that is used when there is no urgent need to discuss the incident; part of critical incident stress management.

defusing The process of talking about the experience that occurs immediately after the incident; part of critical incident stress management.

fire department connections (FDCs) Fixtures, either stand-alone or part of the building's outside wall, that allow firefighters to supply additional water to the fire suppression sprinkler system or the standpipe system.

fire flow The water supply needed to put the fire out.

fire management zone (FMZ) A zone within an engine company's area where similar hazards are grouped because they have roughly equal needed fire flow and number of hazards.

first-in jurisdiction The area where a fire department has a duty to respond.

lightweight construction Construction that uses lumber that is a smaller dimension; often 2 × 4 wooden studs, rafters, and ceiling joists are found in these structures.

live load Items inside a structure that are not attached to the structure or permanent.

louvering A method of roof ventilation that reduces the firefighters' exposure to smoke and heat and initially prevents cutting the supporting roof joists.

mutual aid A contract between agencies to provide resources on request.

negative-pressure ventilation A method of forced ventilation that pulls air and smoke out of a structure.

overhaul The process of searching the fire scene for hidden fires or sparks that may rekindle; also helps find the origin and cause of the fire.

pike pole A device with a sharp pointed head and a curved hook; used to pry open scuttle holes and sound the roof.

positive-pressure ventilation A process that uses mechanical fans to blow air into a structure to remove smoke and gases.

KEY TERMS CONTINUED

postfire conference A meeting or meetings held after an incident to talk about what fire crews saw, what actions they took, and what they learned to improve their emergency response on future incidents.

prefire planning The process of studying and preparing for firefighting activities at the scene of a hazard or structure.

salvage operation The protection of property while fighting a fire.

scuttle hole An opening that allows entry into the attic area.

Siamese connection A hose fitting with a double hose line junction that allows two hoses to connect to one line.

smoke ejector Mechanical smoke fan that draws heated air and smoke outside of a structure.

sounding the roof Using a pike pole, vent hook, or ax to tap the roof ahead to see if it is solid enough to support the weight of firefighters.

target hazard Structure or area that poses the greatest threats to life and property.

thermal imaging device A device that discerns between objects or areas with different temperatures and gives a visual display of those areas.

Vent–Enter–Isolate–Search (VEIS) A firefighting technique that calls for venting outside the building for a safer entrance and a reduced fire flow path.

ventilation A way to remove smoke and heat from a building, allow for the entry of cooler air, and improve rescue and firefighting operations.

REVIEW QUESTIONS

1. How does the need to learn and understand the fire combustion processes apply to prefire and postfire planning?
2. What is the importance of preplanning? Please include safety concerns in your answer.
3. Why is it important during a postfire conference for firefighters to participate?
4. When cutting ventilation holes, it is sometimes necessary to make louver cuts. What is a louver cut, and when and why it is used?
5. What are some fire combustion concepts that are important when conducting salvage and overhaul operations?

DISCUSSION QUESTIONS

1. What are the specific differences between automatic aid and mutual aid?
2. When deciding on ventilation methods for a two-story building, what important fact should be considered, and how does this affect the type of ventilation used?
3. When a water supply is not immediately available, what other resources may be requested for water supply?

APPLYING THE CONCEPTS

In reference to the Chapter 6 Case Study and lessons learned from Chapter 6:

- When conducting a postfire evaluation of this incident, who would you include?
- Should salvage operations be considered?
- Should overhaul operations be considered?
- When should tactics/standard operating procedures be changed?

REFERENCES

A & C Plastics. "Skylights and Plastic Sheeting." n.d. Accessed February 6, 2023. https://www.acplasticsinc.com/informationcenter/r/plastic-sheeting-for-skylights.

Angle, James S., Michael F. Gala, Jr., T. David Harlow, William B. Lombardo, and Craig M. Maciuba. *Firefighting Strategies and Tactics*, 4th ed. Burlington, MA: Jones & Bartlett Learning, 2020.

Brassell, Lori D., and David D. Evans. *Trends in Firefighter Fatalities Due to Structural Collapse, 1979–2002.* NIST Interagency/Internal Report (NISTIR). Gaithersburg, MD: National Institute of Standards and Technology, 2003. https://tsapps.nist.gov/publication/get_pdf.cfm?pub_id=861268.

CISM International. "What Is CISM: Understanding CISM." n.d. Accessed February 6, 2023. https://www.criticalincidentstress.com/critical_incidents.

DelBello, Chris. "Vertical Ventilation: A Firefighter's Ladder-to-Roof Guide." FireRescue1. November 26, 2021. Accessed February 6, 2023. https://www.firerescue1.com/fire-products/ventilation/articles/vertical-ventilation-a-firefighters-ladder-to-roof-guide-shezGumWxQoJFSKU/.

Delmar Cengage Learning. *Firefighting Handbook: Firefighting and Emergency Response*, 3d ed. Clifton Park, NY: Delmar Cengage Learning, 2008.

Diamantes, David. *Fire Prevention: Inspection and Code Enforcement*, 4th ed. Scotts Valley, CA: CreateSpace Independent Publishing Platform, 2015.

Federal Emergency Management Agency (FEMA). *Rapid Intervention Teams and How to Avoid Needing Them*, FEMA TR-123. 2003. http://tkolb.net/tra_sch/RIT/RIT-123.pdf.

Gagnon, Robert M. *Design of Special Hazard & Fire Alarm Systems*, 2nd ed. Clifton Park, NY: Delmar Cengage Learning, 2007.

Klinoff, Robert. *Introduction to Fire Protection and Emergency Services*, 6th ed. Burlington, MA: Jones & Bartlett Learning, 2019.

Mittendorf, John H. "Strip Ventilation Tactics." *Fire Engineering* 149 no. 3 (1996).

National Institute of Standards and Technology. "Live Fire Tests with FDNY Will Guide Improvements in Fire Department Tactics." July 11, 2012. Accessed February 6, 2023. https://www.nist.gov/news-events/news/2012/07/live-fire-tests-fdny-will-guide-improvements-fire-department-tactics.

Offthecouchwellness.com. "Critical Incident Defusing & Debriefing." n.d. Accessed February 6, 2023. https://www.offthecouchwellness.com/critical-incident-defusing-debriefing.

Smoke, Clinton H. *Company Officer*, 2nd ed. Clifton Park, NY: Delmar Cengage Learning, 2005.

University of Wisconsin Green Bay. *CISM-Defusing*. n.d. Accessed February 6, 2023. https://www.uwgb.edu/UWGBCMS/media/bhtp/files/CISM-Defusing_1-(1).pdf.

Zimmerman, Don. *Fire Fighter Safety and Survival*, 3rd ed. Burlington, MA: Jones & Bartlett Learning, 2021: 262–76.

CHAPTER

7

High-Rise Building Fires

LEARNING OBJECTIVES

After studying this chapter, you will be able to:

- Explain why high-rise buildings present special concerns and problems.
- Identify the differences in construction methods of high-rise buildings.
- Explain how different construction materials and designs impact fire behavior in these buildings.
- Discuss the special problems at high-rise fires, such as communications issues, the stack effect, ventilation concerns, evacuation issues, and elevator control.
- Explain the purpose of the special fire protection equipment that may be found in high-rise buildings.
- Identify when a stairwell support procedure may be needed.
- Discuss why the greatest need for property protection in a high-rise is downward.

CASE STUDY

The Winecoff Hotel Fire

The Winecoff Hotel fire of December 7, 1946, occurred in the early morning hours. It was the deadliest hotel fire in U.S. history. It killed 119 people, including the hotel's owners. When the hotel was built, it was hailed as a fireproof building. While its steel structure was fireproof, the interior decor was not. This contributed to the fire. The hotel was also designed with only one stairwell. It went from the ground floor to the top floor, which was a total of 15 floors. The stairwell was open and there were fire doors at each floor landing.

The rising smoke and heat trapped those on the floors above the fire. To compound the spread of fire, the rooms included wooden transoms. Transoms are small rectangular windows above entry doors that add ventilation to rooms. Those that were open at the start of the fire helped to spread the smoke and heat into the rooms on the fire floor and those above the fire floor.

Because there was only one stairwell, trapped people were forced to jump. Some landed on the firefighting nets, while others fell to their deaths. Firefighters were only able to reach the 10th floor with water. Their ladder trucks could only reach the fifth floor.

1. Would the Winecoff Hotel meet the definition of a high-rise?
2. What fire suppression construction/building regulations would have reduced the number of deaths?
3. What would have been the best method to control and remove the smoke?

Introduction

The development of vacant land in cities has increased property values. As the amount of land becomes scarce, the cost of the land increases. Building developers then build upward, rather than buying land to build horizontally. This upward growth leads to what firefighters call high-rise buildings. The National Fire Protection Association (NFPA) Standard 101 *Life Safety Code* defines **high-rise buildings** as "Buildings where the floor of an occupiable story is greater than 75 ft (22.9 m) above the lowest level of fire department vehicle access."

There are variations in the code definitions of high-rise buildings throughout the country. For firefighters, the best policy is to check locally for the appropriate code. This will help them to better understand the code specifics. Almost all definitions of a high-rise imply that these buildings have floors higher than the reach of fire department ground ladders. This means fighting the fire is done from inside the building. Upper-floor fires present firefighters with different problems than those in buildings that can be easily and quickly reached from the ground level.

High-rise fires can be compared to single-story fires. They are just many floors above the ground. While the firefighting tactics used on high-rise fires are the same as those used on other structure fires, these buildings present special concerns and problems.

- The building may have multiple occupancies, such as apartments, stores, restaurants, and parking areas.
- The height of the building allows for more smoke movement within the building.
- All firefighting equipment, supplies, and human resources must be sent to the upper floors. This is a lot of work and takes added staff.
- Access to and exit from high-rises can be difficult in an emergency. People fleeing the building and firefighters must sometimes use the same stairwells.
- The HVAC systems can help to spread fire and smoke into other parts of the building.

Because of these differences, the fire problems and fire behavior in high-rise buildings are of special concern to firefighters.

High-Rise Buildings

According to the National Fire Incident Reporting System (NFIRS), for the years 2014 to 2018, there were 13,400 high-rise fires in the United States. These fires led to $204 million in annual property losses and 464 injuries, including 39 deaths (NFPA, 2021). **FIGURE 7-1** shows the dangers that are present in a high-rise fire.

People who live in high-rise buildings face several fire behavior and safety issues. These issues include the

FIGURE 7-1 Fire in high-rise buildings can present firefighters with unique challenges.

© Gorloff-KV/Shutterstock

FIGURE 7-2 A highrise building constructed before 1920.

© ChicagoPhotographer/Shutterstock

construction manner, design, materials, as well as the building's contents. Firefighters group high-rise construction into three time periods.

Early Fire-Resistive Buildings (1870 to 1920)

Structures built from 1870 to 1920 were constructed with little or no concern for fire safety. At the time, there were no standards for the protection against heat for the building's steel parts. When heated to a little above 1,000°F (537.8°C), steel quickly loses its strength. Steel also expands at a rate of 0.06% to 0.07% for each 100°F (37.8°C). This means that at 1,000°F (537.8°C), it will expand about 9% for each 100 ft in length. In a masonry building where the steel beam is blocked by a wall, the steel can cause a partial or total wall collapse.

Some of these early buildings were built with the floors supported by concrete piers. The piers created open void spaces where fire and heated gases could move under the structure. To prevent fires from spreading from these open spaces, some building contractors used terra cotta tile for fire resistance. The tiles were effective as long as they did not have holes or poorly cemented joints. Shoddy workmanship and cheap materials led to fires that would spread quickly through buildings.

Some of these buildings were equipped with standpipes, but most water supply lines were too small to deliver the water needed in a fire. Prefire planning is advised for areas that have older buildings. A risk assessment should be done as well. **Risk assessment** is the evaluation or comparison of risks. It is used to develop successful approaches to an incident. **FIGURE 7-2** shows a pre-1920 high-rise.

High-Rise Construction (1920 to 1960)

Buildings built from 1920 to 1960 have fire safety features that were lacking in earlier buildings. They also have some features that are not found in newer high-rise buildings.

These buildings were steel framed and tiled with concrete or masonry. This was used for protection from fire. Concrete steel-reinforced columns that were spaced about 30 feet apart supported the floors. Each floor had several widely spaced concrete-encased steel stairwells. The windows on each floor could be opened, allowing firefighters to ventilate. Many of the windows leaked, so air moved within the immediate floor area rather than throughout the building.

A good example of this type of building is the Empire State Building, which was built in 1930 (**FIGURE 7-3**). It has thick masonry walls, which is why the building weighs about 23 pounds per cubic foot. Newer high-rise buildings are built with drywall, which weighs about 7 to 8 pounds per cubic foot. The added mass of the older buildings absorbs more heat from fire than the lighter buildings. The heat not absorbed is reflected back into or contained within the room area. This results in a hotter inside environment for firefighters.

High-Rise Buildings Constructed after 1960

Buildings built after 1960 are the modern steel and glass buildings found in most cities. These buildings differ from the earlier high-rise buildings. They have newer construction materials, methods, and systems. They use gypsum wallboard and lightweight concrete to protect against fire. They also have central core

FIGURE 7-3 Empire State Building.

FIGURE 7-4 Application of spray-on insulation.

construction and use central heating units, ventilation, and air-conditioning systems for heating and cooling.

Gypsum Wallboard/Plaster Board

Gypsum board is primarily used in inside finishes for walls and ceilings. There are newer gypsum board coverings that are now being used for exterior finishes on buildings. **Gypsum board** is usually referred to as drywall or plasterboard. It is made with a core of gypsum, starch, water, and other additives between two heavy paper sheets (4 ft × 8 ft or 4 ft × 12 ft) to give it shape. The board can be nailed to wood studs or fastened with screws onto metal studs. It is light in weight compared to other concrete products. Its strength quickly fades when the material is wet or struck by an object.

On September 11, 2001, in the World Trade Center, the impact from an airplane crash knocked sheets of wallboard off the stairwell walls. The displaced Sheetrock blocked the stairwells. People were trapped. The National Institute of Standards and Technology (NIST) study of this tragic event urged that buildings be designed with more entrances and exits. Also, the construction materials used to protect building systems should offer a haven from fire inside the building.

Lightweight or Blown-on Concrete

Builders use a coating of concrete called lightweight, or blown-on, concrete to protect steel structural parts from fire. **Lightweight concrete** is a light-in-weight concrete mixture that is blown onto metal. This provides a layer of material that resists fire. The fire rating varies with the material and its thickness. The thickness and the ratio of concrete to trapped air determine the resistance to fire. A variety of these materials are sold under various trademark names. After the 9/11 terrorist attacks, as with gypsum board, questions were raised about the application processes. There were also questions about the poor adherence of the concrete used to protect the steel floor truss parts from fire and heat. **FIGURE 7-4** shows the application of spray-on insulation.

Central Core Construction

In pre-1960 construction, exits were on every floor, and in most buildings, there was one exit in every corner. These exits opened to a stairway to the first floor or the floor that exited onto the street. Utility chases were put in rooms next to bathrooms, electrical control rooms, and HVAC control rooms.

The post-1960s high-rises used the central core method of construction. In **central core construction**, elevators, stairs, restrooms, and support systems are located in the center of the building. Thus, all the plumbing, telephone, and electrical systems could be channeled within utility chases (**FIGURE 7-5**). This reduced the cost of materials and installation because all the plumbing and wiring were in one place. Office and meeting areas were placed around the perimeter of the building.

Likewise, exit stairways could be built at less cost if some of the walls could be common walls for elevators and other exits. The idea was to add these common needs to the center of the building, leaving the outer portions of the building available for lease (**FIGURE 7-6**). One problem with this type of construction is that the vertical channels allow fire and toxic gases to rise.

FIGURE 7-5 Plumbing, telephone, and electrical systems can be channeled within utility chases.

Courtesy of Joseph Marranca.

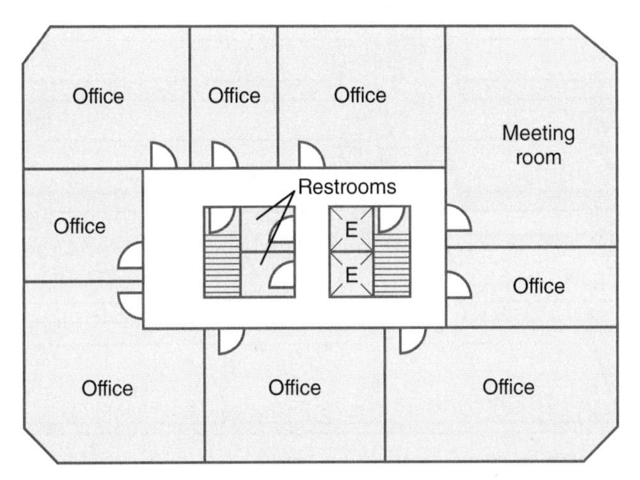

FIGURE 7-6 Central core floor plan for high-rise building.

HVAC Systems

Since the late 1960s, architects and building owners have realized that energy is not being used efficiently. It is costly to draw air in from outside, filter out dust and pollutants, heat or cool it, and move it to chosen areas. In effect, this air is a valuable item. If it can be kept within the building and recycled, the cost of operating the building can be greatly lowered.

With this concept in mind, architects design high-rise buildings to be sealed. Windows no longer open. As an added feature, the air conditioning for several floors can be done from one central floor (e.g., mechanical equipment room).

This presents severe problems in the event of a fire. First, windows must be broken to provide horizontal ventilation if it is needed. The HVAC system must be shut down during a fire or designed to prevent the system from moving the fire from floor to floor.

In these systems, the circulating air is either heated or cooled and then forced by mechanical fans through the building or, in higher buildings, within a zone of the building. They use return ducts or a plenum space to return the air to the driving fan. A **plenum space** is a space above a suspended ceiling or hallway that is kept under negative pressure for return air. On the air return above the suspended ceiling, the air is moved by negative pressure through the ducts or plenum to the heating or cooling element. There it is cooled or heated and returned via the ducts again by the positive pressure created by the fans.

Following the MGM Grand Hotel and Casino fire, it was determined that a number of incident deaths were a result of smoke and fire gases traveling up through the HVAC ducting into the hotel guest rooms. Due to this finding, fire codes were changed to require fire dampers to be installed in the return ducting. Fire dampers prevent fire and heated gases from being spread to other parts of a building. A **fire damper** is a device that, when turned on, closes the air flow in the HVAC system ducting. Firefighters should know the locations of fire dampers. They should follow up to make sure fire and smoke are not being spread to other parts of the building when they are activated. After the fire is out, firefighters must ensure that fire dampers are reset.

Only 15% of the total air volume is drawn into the system each air circulation cycle, which reduces the total amount of heating or cooling needed. This is a marked cost savings over the older heating and cooling systems in which none of the air was reused.

In some buildings, the systems can be set up so that the total system or zone can be used to exhaust smoke from some areas and supply fresh air to other areas in the event of a fire. This requires a working knowledge of the specific building system and a thorough knowledge of how these systems work. Conversely, some systems must be shut down immediately. Information about the HVAC system should be discussed with the building maintenance service. Written operating instructions should be obtained during prefire planning.

Stack Effect

The **stack effect**, or chimney effect, is the natural movement of air within a tall building. It is caused by the temperature difference between the outdoor and indoor air. This temperature difference causes the air in the building to rise or fall and varies with the degree of difference. The air movement becomes very noticeable in buildings over 60 feet high. It becomes stronger as the building gets taller and the temperature difference becomes greater.

For example, if the temperature outside a 60-story office building is 100°F (37.8°C) and the temperature inside is 70°F (21.1°C), then there is a 30° difference. This difference will drive air currents in the building downward. On the other hand, if it is 40°F (4.4°C) outside and 70°F (21.1°C) inside, the temperature difference is 30° again, but now the direction of the air in the building is upward.

The pressure difference forces heated air in a structure at lower levels to the upper levels. This lighter heated air eventually reaches a point where the air temperatures inside and outside the building are equal. The weight of the air is balanced. This is called the neutral plane or **stratification location**. For firefighters, this means that at this location, heated smoke and fire gases will move horizontally, resulting in a sideways fire spread. In most cases, it occurs at or close to the middle of the building (**FIGURE 7-7**).

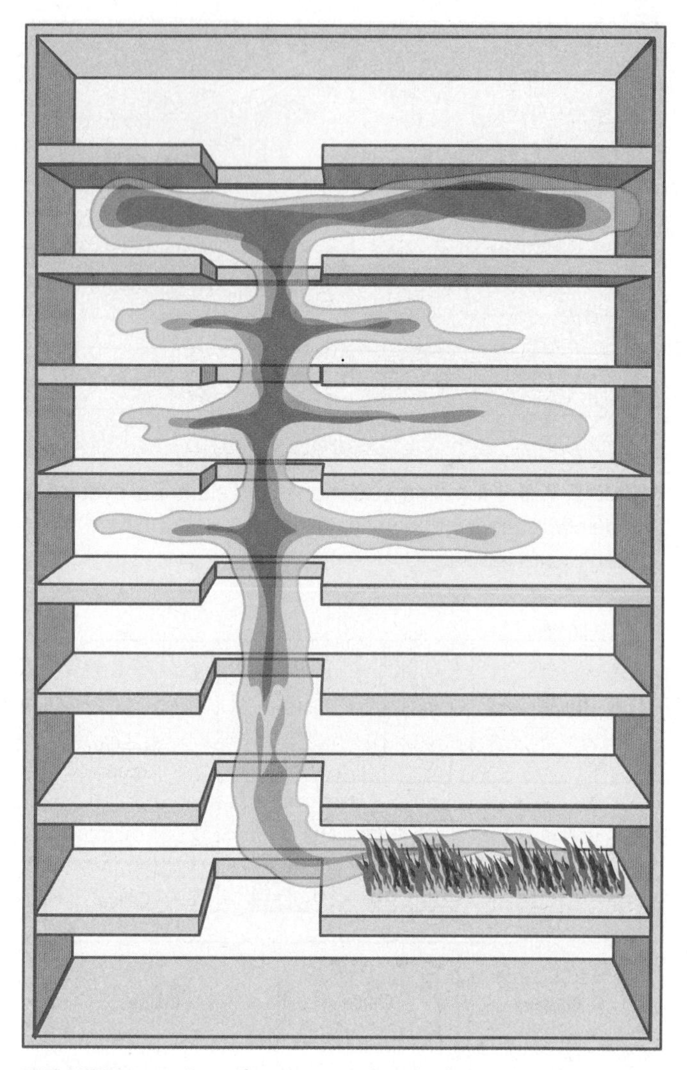

FIGURE 7-7 Stratification inside a high-rise building.

Smoke and heated gases can also be impacted by the stack effect. In a fire in the lower levels (below the neutral plane), the internal pressure created by the stack effect draws the smoke and heated gases toward any shafts or stairway openings. In fires at higher levels (above the neutral plane), air draws smoke and heated gases away from shafts or stairs toward the outside of the building. In fires within the neutral plane, the stack action has little effect on smoke movement. **FIGURE 7-8** shows how temperature differences impact the stack effect.

Ventilation

Ventilation is the planned removal of smoke, heat, and gases from a structure. Firefighters ventilate to get rid

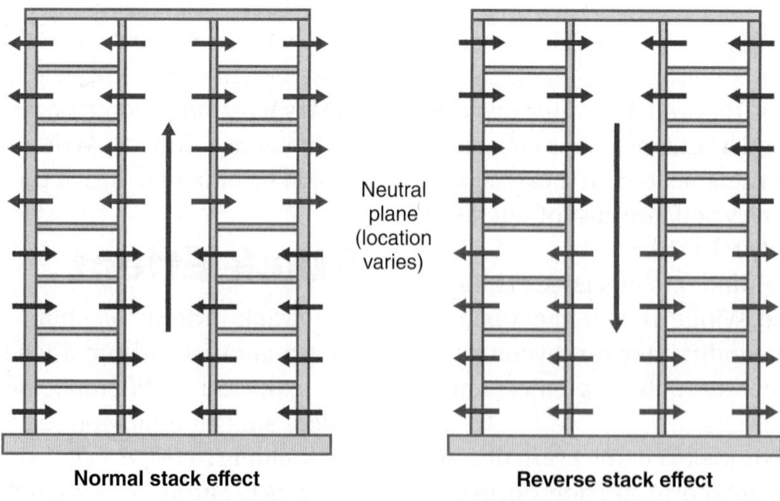

FIGURE 7-8 Temperature differences impact the stack effect in high-rise buildings.

of toxic gases and hazardous health conditions for firefighters and those in the building. It also allows firefighters to find the fire more quickly and reduce fire and smoke damage.

The traditional methods to ventilate at fires are opening or breaking windows, bulkheads, and scuttles. Firefighters can also cut a hole in the roof. These methods do not always work at high-rise building fires. In older high-rise buildings with windows that work, firefighters can ventilate by opening windows on the fire floors and using positive or negative ventilation to draw out the smoke and heated gases. In high-rises built after 1960, ventilation is more complex because most buildings are sealed. In addition, the buildings built in the late 1960s and 1970s will probably not have HVAC systems set up to exhaust fire gases and smoke. If horizontal ventilation is not an option (e.g., it is too dangerous to break windows), then an enclosed stairwell can be used. The stairwell should have an opening to safely exhaust smoke. It also should not be used for people leaving the building or firefighter access.

If a stairwell is available, then the door on the fire floor into the stairwell is propped open. An exhaust fan directs the smoke up through the enclosed shaft and out of the building. All doors and other openings to intervening floors not affected by fire must be closed so that fire and smoke do not spread. This operation must be planned and checked continually.

Some newer buildings are equipped with fans that create positive air pressure in the stairwell. Firefighters should use these systems with caution because the pressurized air can move smoke and fire into uninvolved parts of the building. This is an important feature to learn more about from building engineers and maintenance staff during prefire planning visits.

Elevators

A key part of good high-rise firefighting is the safe use of elevators to haul firefighters to and from the fire in the building. However, we know from the past that elevators sometimes do not work during emergencies. This results in the deaths of tenants, building staff, and firefighters. Firefighters will attack a high-rise fire using the stairwell with fires on the first six floors. They typically do not use elevators on the lower floors unless the elevators can be inspected.

Most fire departments allow elevators to be used for incidents higher than the sixth floor after proof of the elevator's safety. In 2013, a landmark study by the NIST looked at 14 critical tasks performed by firefighters when fighting fires in high-rises. This research was conducted in an empty 13-story office building. As part of the study, fire protection engineers reviewed the use of fire service access elevators to move firefighters and equipment up to the staging floor. The study revealed that most tasks began 2 to 4 minutes faster when elevators were used compared to when stairs were used. This finding confirms the importance of developing methods and procedures for firefighters to safely use elevators.

Some departments allow elevators away from the fire to be used if there is separation between the elevators and the fire area. Firefighters using these elevators move the staff and equipment two floors below the fire. They then go up the stairs and travel horizontally to the fire. This approach is much safer and easier than going up multiple stairs. It also keeps firefighters who are loaded down with heavy equipment out of the stairwells where people may be exiting the building.

An understanding of local fire department policies and procedures is important when deciding if and when elevator use in high-rise firefighting is safe and will be allowed.

Use of Elevator Systems

In 2008, due to firefighter deaths and injuries, the American Society of Mechanical Engineers (ASME) adopted new codes for elevator safety. These codes called for the installation and retrofitting of fire service override systems on elevators in high-rise buildings. These systems allow manual control over all elevator controls in the cars. This is safer for firefighters and people who live in the building. The codes provide standards throughout the country. They apply to new systems during installation and when older elevators are upgraded.

The ASME has continued to update the safety codes for elevators and escalators with the latest edition (ASME A17.1/CAS B44-2019) released in 2019. Among the most recent updates were requirements to ensure communication with trapped passengers, including people who are hearing impaired. Another update increased the door protection on passenger elevators.

To control the elevator system, the firefighter must first find the elevator key. It should be in a key box or another predesignated location in the building. The manual control system has three phases.

Phase 1 is often called the firefighter's recall. Firefighters can control the elevator by inserting the key into the fire control panel. This brings the cars nonstop to the lobby or predesignated area and opens the doors. In some systems, phase 1 can be triggered by a smoke detector. The aim is to make

FIGURE 7-9 Elevator panel showing the fire department key control functions.

Reproduced from Bigda, Kristin. 2017. "NFPA 1: Standardized Fire Service Elevator Keys, #FireCodefridays." NFPA Today. https://www.nfpa.org/news-and-research/publications-and-media/blogs-landing-page/nfpa-today/blog-posts/2017/03/31/nfpa-1-standardized-fire-service-elevator-keys-firecodefridays

sure that people in the building use the stairways. It is also to prevent the car from stopping on the fire floor. Finally, it does not allow the doors to open on floors other than the predesignated floor.

Phase 2 allows firefighters to control the elevator from inside the car. The key in phase 1 cannot override the key operation in the car. Many departments have a standing policy forbidding the use of any elevator car that does not have phase 2 fire service controls.

Phase 3 is linked to the building fire/smoke alarm system. It applies when a fire or smoke alarm is triggered. If the smoke detector goes off in the lobby, the elevator cars will be recalled to a predesignated floor other than the lobby. Firefighters have to go to the designated floor to access the car. The location of the car can be found by viewing the panel in the lobby. During the prefire planning walk-through, the panel should be checked. The location should be logged in the prefire planning notes. **FIGURE 7-9** shows an elevator panel with fire department control functions.

The following are safety rules for firefighters for the use of elevators.

- Use the elevator key to place and keep all elevator cars in multiple hoist ways under fire department control.
- Never use an elevator that does not feel or look safe. Do not overcrowd the cars.
- Never take the elevator directly to the fire floor. Always stop two or more floors below the fire and then walk to the fire floor. Department policies often dictate on which floor to exit the elevator.
- Stop the rising elevator every fifth floor to check for smoke. This is done by opening the elevator escape hatch and checking with a flashlight for smoke.
- Always wear protective clothing, including SCBA, forcible entry tools, and a safety line.

On some fires, firefighters have been trapped in elevators that stopped in between floors as the power was interrupted. For this reason, firefighters must always bring tools, ropes, and extra air bottles on elevators.

Building Fire Protection Systems

One tactical consideration with fires in high-rises is the reliance of fire departments on built-in systems to help with fire control and extinguishment. Building systems available to firefighters are:

- standpipes;
- sprinklers and water supply systems;

- fire pumps;
- firefighter air replenishment systems;
- fire communications and command systems; and
- pressurized stairwells.

These systems are varied and can be complex. Fire departments must learn how to use them for successful fire operations. A closer look during prefire planning at some of these systems is needed so they can be used correctly during fire operations.

Standpipes

Most building codes require standpipe systems in high-rise buildings. Some buildings are not considered high-rise buildings, but they should be defined as such for firefighting purposes. As discussed earlier, this definition includes buildings with stories beyond the reach of ladders.

One way to understand standpipes is to view them as fire water mains with outlets that are part of the building. These standpipes are found in the stairwell of every floor. On the outside of the building is a **fire department standpipe connection**, which is where the fire pumper connects to pressurize the building standpipe system. It is a common practice to connect to the standpipe at the fire floor. Fire crews then advance the hose line upstairs. Once the hose line is charged, it is easier to pull the hose line down from the floor above than to pull it up from the lower floor. Prefire training exercises are an effective way to see if a building has a working standpipe system. **FIGURE 7-10** shows a fire department connection and postindicator valve.

Pressure relief valves regulate nozzle pressure for firefighters. They ensure a constant water pressure for fire attack by preventing backpressure from the upper floors.

Sprinklers and Water Supply Systems

Most new high-rise buildings have fire sprinklers. Because of the height of these buildings, many of the systems are installed in zones. This is so fire pumps can provide enough water pressure.

When properly installed and kept up, fire sprinkler systems can quickly control and put out most fires with water from one or two heads. As an example, in 1991, the One Meridian Plaza office high-rise in Philadelphia, Pennsylvania, was undergoing sprinkler retrofitting. It was not yet completed. A fire started on the 22nd floor and burned to the 29th floor (the floors without sprinklers). On the 30th floor, 10 sprinkler heads put out the fire.

There is a water supply tank inside each zone. The tank is filled from outside water supplies using a pump or a series of pumps. During a fire, the tank and a fire pump are used to supply water for the fire sprinkler system and standpipes within that zone. Firefighters need to be familiar with how these water tanks and pumps work and the location and coverage of the various zones in high-rise buildings.

Fire Pumps

All high-rise buildings are required to have backup fire pumps. A **fire pump** is a specially designed pump that

FIGURE 7-10 Fire department connection and postindicator valve.

increases the water pressure serving a fire protection system. These pumps can be found alone or in pairs. A sprinkler head or standpipe valve will open, decreasing the pressure in the pipe. The pressure decrease causes a switch on the pump to note the drop in pressure and start the pump.

Because most fire sprinklers and valves are rated at 175 pounds per square inch (psi), a pump relief valve might be needed to prevent overpressurization, which could damage the sprinkler system. There are also pressure-regulating valves on the discharge side of the fire pump. These valves protect sprinkler systems in high-rises where the pressure needed at the base is more than the rating for the system parts on the lower levels.

Although some systems may be more complex, firefighters should know how to use fire pumps. They should check the inlet and discharge gauges. They should also make sure the bypass valve is closed after pump testing. Fifty percent of the pump capacity can be lost with an open bypass valve. A **bypass valve** is a tap that allows a water supply to travel around the pump and provide fire protection. It is used when taking the fire pump out of service for maintenance and repairs.

Usually, fire pumps are turned on using instructions posted on the control panel. Once the firefighter starts the pump, that person should remain nearby to make sure that the pressure is maintained. The person must also ensure that no one stops the pump without permission from the incident commander.

Firefighter Air Replenishment Systems

Firefighter air replenishment systems (FARS) are networks of pipes and air storage that deliver air via refilling stations in a high-rise (**FIGURE 7-11**). These systems are standpipes for air that remove the need for firefighters to leave a burning building to refill their SCBAs. Fire codes recommend FARS placement on the fifth floor and every three floors up. There has not yet been a code requirement. Each jurisdiction can choose whether to include this section of the fire codes.

FIGURE 7-11 Firefighter air replenishment systems (FARS).

Used with permission of Johnson Controls.

Fire Communications and Command Systems

In postfire reviews for high-rise building fires, the issue of poor fire-ground communications is often raised. The reason is that high-rises are built with steel columns and beams enclosed with concrete. In many cases, the steel and concrete beams block emergency communications.

To solve this problem, newer high-rises have communications systems built into them. These communications systems use hard wiring that enables communication inside the building. A series of speakers and telephones are located throughout the building. The system allows the fire department to direct people exiting the high-rise. The incident commander can speak directly to the operations and division chiefs.

Stairwells

In most high-rise buildings, the people can use the stairwell to exit. The stairwells can have a 2-hour fire protection rating, as well as assigned areas of refuge in the building for people to seek temporary protection from the fire. However, firefighters moving upward while people exit interferes with not only the firefighters' movement but also the tenants' escape. One way to avoid this problem is to choose one or more stairwells for those exiting and one or more for firefighting operations.

If elevators cannot be confirmed safe, a stairwell support procedure must be put in place. A **stairwell support procedure** is a system used to move resources to the fire attack area when the elevators are unsafe. Firefighter teams position themselves two floors apart in the stairwell to carry resources up two floors. The team descends and rests while waiting to carry and pass on more equipment. This relay system provides a steady flow of resources to the fire area. While this method demands a large number of firefighters, it is an option to supply needed resources to the fire attack area.

Pressurized Stairwells

Newer high-rise buildings have pressurized stairwells designed to provide a smoke-free environment. The principle of a pressurized stairwell is to enclose the stairwell. One or more fans are then added to increase the air pressure. When a door or doors open, the air pressure inside the stairwell will push the smoke back into the floor area. In some systems, if too many doors open at the same time, the pressure falls to a point that the stairwell fills with smoke. Fire department fans can supplement the pressure in the stairwell if needed by adding positive pressure.

Search and Rescue

The key to successful search and rescue in high-rise buildings is the same as in other buildings. Firefighters need to use an organized search method to make sure all areas are checked for people left in the building.

When searching a high-rise for victims, firefighters need to be aware of convergence cluster behavior, which was first observed during the 1980 Las Vegas MGM Grand Hotel fire. **Convergence cluster behavior** is when people who feel threatened gather as a group to feel safe. This leads to a group in one area or room rather than people waiting for rescue in separate rooms. During some fires, firefighters have wasted time and energy searching empty rooms only to find all victims in one room. Firefighters need an accountability system to confirm the location of all occupants.

Relocation of Occupants

Total evacuation of a high-rise building during firefighting is neither practical nor doable. The main effort depends on the number of people to be moved, how they can be moved, and if there is safe shelter in the building. Hospitals provide safe areas or zones so that patients can be moved horizontally. With building code additions, safety zones have become required. The safety zones provide safe areas for people to gather during a fire. Firefighters need to identify these areas during the prefire planning phase.

TIP

During 9/11, gypsum board and areas of refuge did not protect the people in the building. Concrete has proven to be a suitable material to provide a safe area. While it adds weight and cost to the construction of the building, its durability is without question. Efforts are under way to find ways to reinforce building construction in safe refuge areas.

Salvage

High-rise office buildings, hotels, and apartments usually have valuable contents, such as computers, office equipment, business records, and personal items. During an emergency, the owners and tenants may ask firefighters to help them carry these items out of the building. For good community relations, it is important to help these people.

In a fire, people and property above the fire are at the most risk. Fire, smoke, and heat will first affect materials on the fire floor and the floors above. However, the greatest need for property protection is downward. Water flows through stairs, walls, utility chases, electrical fixtures, and other openings, damaging and destroying more property under the fire. Firefighters should protect valuables with covers. They should also direct water down the stairs or through drains. Like life-safety and fire-ground operations, preserving property must be well planned and carried out to reduce loss.

Overhaul

Overhauling high-rise buildings is a lot of work. A good prefire plan helps to determine where to apply overhaul efforts. The plan can direct firefighting forces to hidden shafts to check as well as false ceilings that should be either pulled down or opened. Crews should be assigned to every floor above the fire to check for smoke and fire spread. Crews should also search areas below the fire to confirm that the fire has been put out.

CASE STUDY

CONCLUSION

1. **Would the Winecoff Hotel meet the definition of a high-rise?**
   To meet the definition of a high-rise, the building would have to be at least 75 feet in height.
2. **What fire suppression construction/building regulations would have reduced the number of deaths?**
   During the design of the building, a sprinkler system, fire doors at the stairwell, and additional stairwells would have helped firefighting efforts. This would have also allowed the people in the hotel to escape.
3. **What would have been the best method to control and remove the smoke?**
   In multistory buildings, the use of positive-pressure ventilation is required.

WRAP-UP

SUMMARY

- Almost all definitions of a high-rise imply that these buildings have floors higher than the reach of fire department ground ladders.
- Fighting the fire is done from inside the building.
- High-rise construction is grouped into three time periods:
  - Early fire-resistive buildings (1870 to 1920)
  - High-rise construction (1920 to 1960)
  - High-rise buildings constructed after 1960
- The stack, or chimney, effect is the natural movement of air within a tall building. It is caused by the temperature difference between the outdoor and indoor air.
- Ventilation is the planned removal of smoke, heat, and gases from a structure.
- Most fire departments allow elevators to be used for incidents higher than the sixth floor after proof of the elevator's safety.
- The manual control system of an elevator has three phases.
  - Phase 1: Firefighter's recall
  - Phase 2: Firefighters control the elevator from inside the car
  - Phase 3: Applies when a fire or smoke alarm is triggered
- One tactical consideration with fires in high-rises is the reliance of fire departments on built-in systems to assist in fire control and extinguishment. Building systems available to firefighters are:
  - Standpipes
  - Sprinklers and water supply systems
  - Fire pumps
  - Firefighter air replenishment systems
  - Fire communication and command systems
  - Pressurized stairwells
- Newer high-rise buildings have pressurized stairwells designed to provide a smoke-free environment.
- Convergence cluster behavior is when people who feel threatened gather as a group to feel safe.
- In salvage, the greatest need for property protection is downward.
- A good prefire plan helps to determine where to apply overhaul efforts.

KEY TERMS

bypass valve A tap that allows a water supply to travel around the pump and provide fire protection.

central core construction Method of construction where the elevators, stairs, restrooms, and support systems are located in the center of the building; also called center core construction.

convergence cluster behavior People who feel threatened gather as a group to feel safe.

fire damper A device that, when turned on, closes the air flow in the HVAC system ducting.

fire department standpipe connection Where the fire pumper connects to pressurize the building standpipe system.

fire pump A specially designed pump that increases the water pressure serving a fire protection system.

firefighter air replenishment systems (FARS) Standpipes for air that remove the need for firefighters to leave a burning building to refill their SCBAs.

gypsum board Board made with a core of gypsum that is primarily used in inside finishes for walls and ceilings; also called drywall or plasterboard.

high-rise building A building where the floor of an occupiable story is greater 75 ft (22.9 m) above the lowest level of fire department vehicle access (NFPA 101).

lightweight concrete A light-in-weight concrete mixture that is blown onto metal and provides a layer of material that resists fire; also called blown-on concrete.

plenum space A space above a suspended ceiling or hallway that is kept under negative pressure for return air.

risk assessment The evaluation or comparison of risks to develop successful approaches to an incident.

stack effect The natural movement of air within a tall building caused by the temperature difference between the outdoor air and indoor air; also called chimney effect.

stairwell support procedure A system used to move resources to the fire attack area when the elevators are unsafe.

stratification location The point in a high-rise building where heated air has risen and is the same temperature and weight as the surrounding air; at this location, air moves horizontally.

REVIEW QUESTIONS

1. Why is there an increase in high-rise buildings in the United States?
2. What is the definition of a high-rise building in your jurisdiction? If you are not in the fire service, provide the NFPA definition of a high-rise building.
3. What are some of the fire behavior problems you might expect to encounter during a high-rise fire?
4. How does more mass or more concrete in a high-rise building impact firefighting and fire behavior?
5. What is the stack effect? What is its impact on the movement of fire and heated fire gases in a high-rise building?

DISCUSSION QUESTIONS

1. What major issue needs to be addressed in prefire planning for mid-rise and high-rise buildings?
2. What two areas of concern affect the arriving firefighters and the occupants of a high-rise building? How might these be addressed in the prefire plan?

APPLYING THE CONCEPTS

Marco Polo Condominium Fire

On July 14, 2017, a fire broke out in the Marco Polo Condominium in Honolulu, Hawaii. It was on the 26th floor of a 36-story building. The fire quickly spread to three more floors, making it a 7-alarm fire. This was the deadliest high-rise fire in Hawaii's history. More than 200 of the 568 units were damaged.

APPLYING THE CONCEPTS CONTINUED

The Marco Polo building complex was built in 1971 and had no sprinkler system. Four years later, sprinkler systems became mandatory for new construction in Honolulu.

In response to the Marco Polo fire, in May 2018, the City and County of Honolulu passed Ordinance 18-14. It requires all buildings 10 stories or higher to conduct safety evaluations and retrofit needed improvements or automatic sprinklers. Of 102 buildings that had conducted safety evaluations by April 2021, only 6 passed. Buildings are required to complete safety evaluations or sprinkler retrofits by the spring of 2024.

1. What building and fire suppression design additions would have helped firefighters to combat the Marco Polo condominium fire?
2. How would better exits from the hotel floors have helped to evacuate the residents?
3. What would be the best fire suppression sprinkler and standpipe system for this type of building?

REFERENCES

Ahrens, Marty. *2014–2018 High-Rise Fire Estimates and Selected Previously Published Incidents.* March 2021. National Fire Protection Association (NFPA). Accessed February 17, 2023. https://www.nfpa.org/-/media/Files/News-and-Research/Fire-statistics-and-reports/Estimates/osHighRiseFireEstimates.ashx.

American National Standards Institute. "ASME A17.1-2019: Safety Code for Elevators and Escalators." February 4, 2020. Accessed April 10, 2023. https://blog.ansi.org/2020/02/asme-a17-1-2019-safety-code-elevator-csa-b44/#gref.

Angle, James S., Michael F. Gala Jr, David Harlow, and William B. Lombardo. *Fire Fighting Strategies and Tactics,* 4th ed. Burlington, MA: Jones & Bartlett Learning, 2021.

East Coast Power Systems. "Strict Building Codes for Elevator Panelboards and the Effects on Safety and Inspection." n.d. Accessed February 17, 2023. https://www.ecpowersystems.com/resources/panelboards/strict-building-codes-for-elevator-panelboards-and-the-effects-on-safety-and-inspection/.

FEMA. "Highrise Fires," *Topical Fire Research Series* 2(18). 2002. Accessed February 17, 2023. https://nfa.usfa.fema.gov/downloads/pdf/statistics/v2i18-508.pdf.

Fire Engineering. "Throw Back to Basics: Fire Service Operations with Elevators." February 25, 2016. Accessed February 17, 2023. https://www.fireengineering.com/firefighter-training/fire-service-operations-with-elevators/#gref.

Hall, John R. *High-Rise Building Fires.* Quincy, MA: NFPA, 1997.

International Fire Code. "Appendix L. Requirements for Fire Fighter Air Replenishment Systems." 2022. International Code Council. https://codes.iccsafe.org/s/IFC2018/appendix-l-requirements-for-fire-fighter-air-replenishment-systems/IFC2018-Pt07-AppxL

National Construction Safety Team. *Final Report of the National Construction Safety Team on the Collapses of the World Trade Center Towers.* Gaithersburg, MD: National Institute of Standards and Technology, 2005.

National Fire Protection Association (NFPA). Standard 101 *Life Safety Code.* 2021. https://www.nfpa.org/codes-and-standards/all-codes-and-standards/list-of-codes-and-standards/detail?code=101.

National Fire Protection Association (NFPA). "Determining Sprinkler Requirements for High-Rise Buildings." August 21, 2020. Accessed February 17, 2023. https://www.nfpa.org/high_rise_sprinklers.

National Institute of Standards and Technology. *NCSTAR 1: Final Report on the Collapse of the World Trade Center Towers.* 2004. https://nvlpubs.nist.gov/nistpubs/Legacy/NCSTAR/ncstar1.pdf.

National Institute of Standards and Technology. "Landmark NIST Study Evaluates Effectiveness of Crew Sizes in High-Rise Fires." April 17, 2013. Accessed February 17, 2023. https://www.nist.gov/news-events/news/2013/04/landmark-nist-study-evaluates-effectiveness-crew-sizes-high-rise-fires.

Norman, John. *Fire Officer's Handbook of Tactics, Fire Engineering,* 5th ed. Saddle Brook, NJ: Penwell Publishing Company, 2021.

Rescue Air. "Firefighter Air Replenishment Systems (FARS)." n.d. Accessed February 17, 2023. https://rescueair.com/firefighter-air-replenishment-systems/.

RKS Plumbing & Mechanical Inc. "Firefighter Air Replenishment Systems (FARS)." n.d. Accessed February 17, 2023. http://www.rksplumbing.com/Fire%20Fighter%20Air.html.

Smoke, Clinton. *Company Officer,* 3rd ed. Clifton Park, NY: Delmar Cengage Learning, 2009.

U.S. Fire Administration. *High-Rise Office Building Fire, One Meridian Plaza, Philadelphia, Pennsylvania,* Technical Report 49, 1991. https://www.usfa.fema.gov/.

U.S. Fire Administration. *New York City Bank Building Fire, Compartmentation vs. Sprinklers, New York,* Report 71, 1993. https://www.usfa.fema.gov/.

U.S. Fire Administration. *Multiple Fatality High-Rise Condominium Fire, Clearwater, Florida,* Report 148, 2002. https://www.usfa.fema.gov/.

U.S. Fire Administration. *Risk Management—Planning for Hazardous Materials: What It Means for Fire Service Planning,* Report 124, 2003. https://www.usfa.fema.gov/.

CHAPTER

8

Wildland Fires

LEARNING OBJECTIVES

After studying this chapter, you will be able to:

- Explain the basic fire combustion principles as they apply to wildland fires.
- Identify the differences between wildland fire behavior and structural fire behavior.
- Examine how weather conditions impact wildland fuels and the behavior of wildland fires.
- Identify the various parts of a wildland fire.
- Explain the unique behavior of wildland fires.
- Identify the method used to classify resources for wildland fires and how fire behavior impacts the type and number of resources needed to suppress these fires.
- Identify the various resources and tools used to extinguish wildland fires.
- Explain the attack methods and actions firefighters use to suppress wildland fires.

CASE STUDY

The Camp Fire

The Camp Fire was one of the deadliest and most destructive wildland fires in California. Named after Camp Creek Road, its place of origin, the fire started in November 2018 in Northern California's Butte County. The number of residential units in the area had recently increased because of its popularity among retirees. The cause of the fire was faulty PG&E electrical lines.

The fire raged through the towns of Butte Creek Canyon, Concow, Paradise, and Magalia. It burned wildland, destroyed homes and businesses in the towns, and caused a massive evacuation. The heavily wooded areas had small and narrow roads, and only 2 to 3 roadways, which made evacuation perilous and limited the firefighters' access to the fire. The fire resulted in 85 civilian deaths, with injuries to 12 civilians and 5 firefighters. In total, 153,336 acres were burned, and 18,804 structures were destroyed.

California had been plagued by several years of low rainfall, which created drought conditions and limited water supply. Many environmental groups had lobbied to stop the removal of trees from the forest, logging in certain areas, and yearly prescribed burns that reduced the amount of dead vegetation on the forest floor. By not reducing the vegetation, the medium and heavy fuels are more likely to burn, and the heat generated from wildland fires increases. With proper forest management, these fires would not have been so intense and could have been extinguished much sooner.

1. What effects did the lack of roadways and the increased residential construction in the wildland/urban interface have on the wildland fire?
2. How does the lack of forest management affect the rate of fire spread?
3. How would an adequate fire suppression force be deployed to such a fire during wildland fire season?

Introduction

A wildland fire is a nonstructure fire that takes place in areas of vegetation or in natural fuels. By the end of 2022, the number of wildland fires reported in the United States was almost 69,000, with approximately 7.6 million acres burned. Humans caused 89% of the wildland fires that burned 3.4 million acres. While most wildland fires are caused by humans, they are also frequently caused by lightning strikes. In 2022, lightning sparked 7,500 wildland fires and burned 4.2 million acres (National Interagency Fire Center, 2023).

There is a vast difference between wildland firefighting and structural firefighting. Although there are differences in these types of fires that impact fire behavior, all fires follow the natural laws of physics. Once these laws are understood, the fire service can apply the laws to both types of fires. This chapter reviews the differences and similarities between fighting fires in wildland and structural fires.

We begin with an overview of the combustion process. The fire tetrahedron described in Chapter 3 discussed the parts needed for a fire: heat, fuel, oxygen, and the chemical chain reaction. Removal of any side of the fire tetrahedron will put out the fire. Most wildland agencies use the fire tetrahedron model to explain the processes of wildland fires.

The Fire Tetrahedron for Wildland Fires

In wildland firefighting, firefighters stop the process of combustion by cooling the heat, removing the fuel, or excluding the oxygen needed to light a fire and keep it lit. Firefighters use hose lines, hand tools, or machines or use dirt or foam to smother, cool, and/or exclude the oxygen. Chemical applications can also help put out a fire. **FIGURE 8-1** illustrates the fire tetrahedron.

Heat Removal

One side of the fire tetrahedron is heat. This side can be removed or reduced so the ignition temperature of the fuel is lowered. Once the fuel is cooled below the temperature at which it gives off vapors, combustion will no longer take place. The fire is out. Water is the most effective fire-suppressing agent that is readily available. In the wildland scenario, water is not as available as in a structure fire setting. There are no hydrants placed along the roads or trails. The key to using water is to use only the amount needed to do the job. Firefighters use

FIGURE 8-1 The fire tetrahedron shows the wildland fire combustion process.

a fine spray or steam to break up the mass of the water so the maximum amount of heat will be absorbed by as little water as possible. For deep-seated fires in forests, a straight stream is needed to penetrate and cool the material below its ignition temperature.

Because water is not available in abundance at wildland fires, firefighters use other means of extinguishment. For example, small amounts of water from a hose line or backup pump cool the fire so firefighters can get close to the fire's edge. They then use tools to remove the burning fuel from the path of the fire. They will clear an area that is twice as wide as the fuel is tall. This creates a fuel break, commonly referred to as a **fire line**. A fire line removes the fuel side of the tetrahedron.

In the 1988 Yellowstone National Park fire, firefighters successfully used a compressed air foam system to coat and protect dwellings with Class A foam. The fire service has used compressed air foam systems since the late 1970s.

The **compressed air foam system (CAFS)** is a fire suppression system that contains a water supply, foam supply, pump, and hose line. It adds compressed air to the foam and water mixture with an air compressor to propel the extinguishing agent farther than the standard nozzle application. The foam is made by mixing 0.1% to 0.3% foam solution with 10% water and compressed air. This method significantly reduces the amount of water needed, thus enhancing its application to wildland fires. **FIGURE 8-2** shows a firefighter applying Class A foam to a wildland fire. Because a large amount of water has been replaced by air in the hose lines, firefighters can move the lines more easily.

The foam can be made wetter if needed, for use on a hot, advancing fire, or drier to adhere to vegetation. The foam is given a rating of 1 for dry to 5 for wet and provides a protective blanket from the fuel. The added detergent in compressed air foam allows water

FIGURE 8-2 Class A foam applied to a wildland fire.

to break the surface tension and penetrate fuels more easily and quickly. **TABLE 8-1** identifies the relationship between foam expansion rate and drain time. This principle of wetter (deeper penetration) to drier (lightweight protective covering) is shown in **FIGURE 8-3**.

Firefighters can slow a large fire using foam and fire retardants. A **fire retardant** is any substance that is used to slow or stop the spread of fire or to decrease a fire's intensity. Fire retardants are often applied by aircraft in wildland fires. These aircraft apply the fire retardant directly on the head or leading edge of the fire to stop its advancement. Retardant can also be dropped ahead of the fire's path to prevent fire spread. The retardant combined with water provides added moisture to the fuels, which helps suppress the fire. Fire retardants can also be applied by special foam application–equipped vehicles, such as the CAFS.

Aircraft used on wildland fires can drop either water or flame-retardant solutions to suppress the combustion process by cooling and smothering the fuel. To be approved for use, the flame-retardant material must degenerate or be biodegradable generally within 30 days of application.

Fuel Removal

In the United States, there are 13 classifications of vegetation fuels. The fire service assigns wildland fuels into three categories: light, medium, and heavy.

- **Light fuels** are fuels less than 2 ft tall.
- **Medium fuels** are fuels 2 to 6 ft tall.
- **Heavy fuels** are fuels over 6 ft tall.

In wildland fire suppression, removing fuel from the path of the fire is the most common method to put

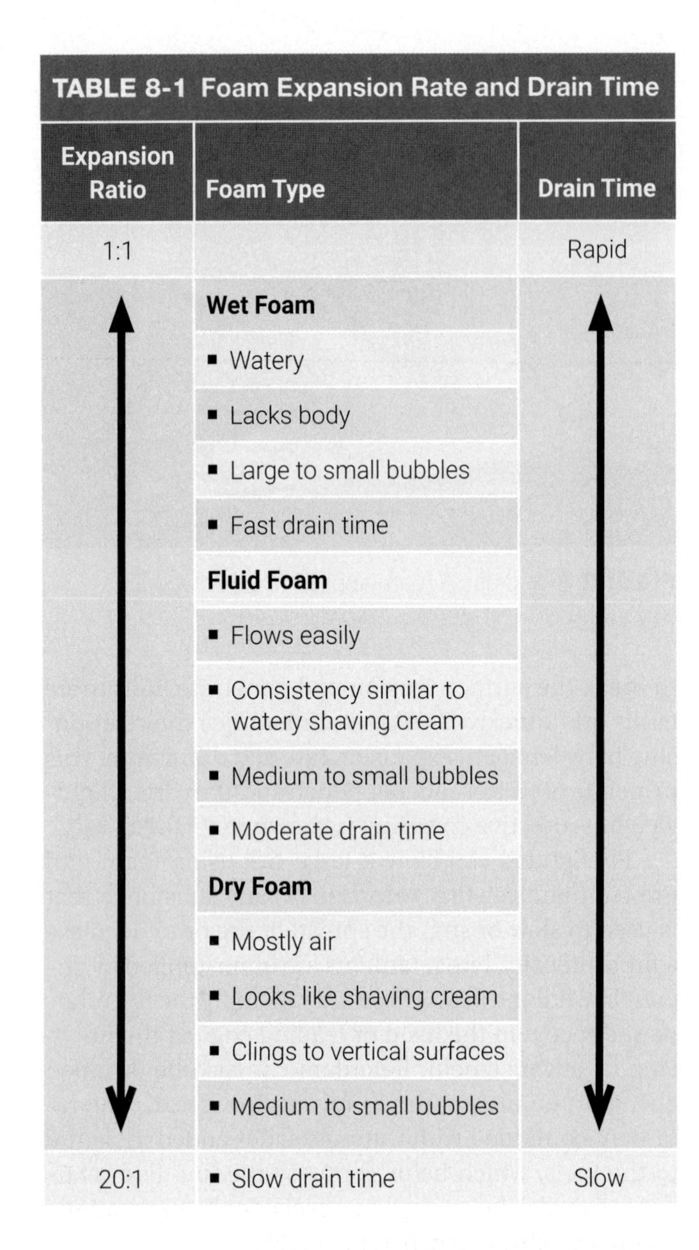

TABLE 8-1 Foam Expansion Rate and Drain Time

Expansion Ratio	Foam Type	Drain Time
1:1		Rapid
↑	**Wet Foam**	↑
	▪ Watery	
	▪ Lacks body	
	▪ Large to small bubbles	
	▪ Fast drain time	
	Fluid Foam	
	▪ Flows easily	
	▪ Consistency similar to watery shaving cream	
	▪ Medium to small bubbles	
	▪ Moderate drain time	
	Dry Foam	
	▪ Mostly air	
	▪ Looks like shaving cream	
	▪ Clings to vertical surfaces	
↓	▪ Medium to small bubbles	↓
20:1	▪ Slow drain time	Slow

out the fire. These operations include cutting or scraping, using bulldozers and/or hand crews, and clearing the fuel in front of the fire. This action removes the burnable fuel from the path of the fire.

Hand crews can operate directly along the fire's edge when the flame height is less than 2 ft tall in light fuels. Fire crews use hand tools, Pulaskis, McLeods, and shovels to remove fuels to the bare dirt level. The fire line breaks are usually 1.5 times the height of the surrounding fuels. For example, in light ankle-high grass, a fire break of 12 to 24 in. would be constructed.

Oxygen Removal

Oxygen can be removed or excluded from the fuel on a wildland fire by throwing dirt directly at the base of the flames or at the area on fire. This action is followed by other extinguishing actions to ensure that the fire is out and cannot spread to any unburned fuel. If there is limited water available, this method reduces the total amount of water needed; however, the penetrating ability of water is needed to put out any deep-seated fire.

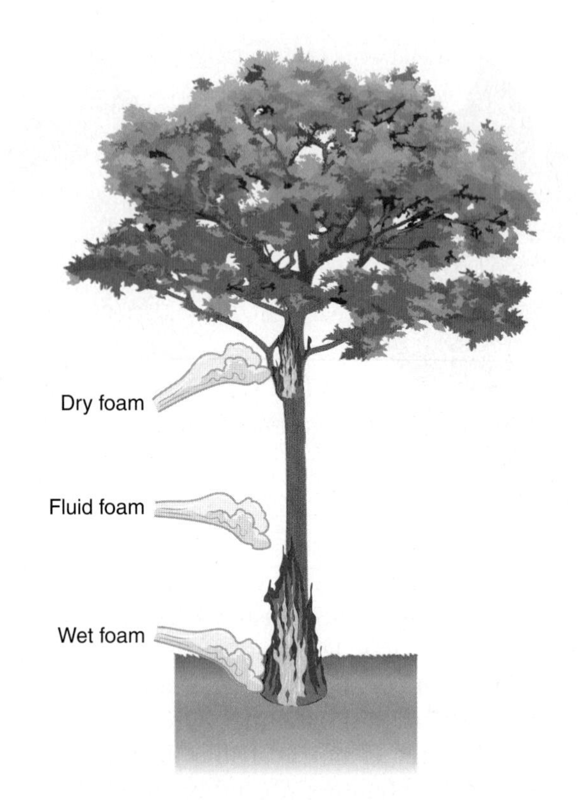

FIGURE 8-3 Dry, fluid, and wet foam applications.

Wildland Heat Movement

Wildland fire heat output is transferred the same way as in structural fires. The heat moves in four ways: conduction, convection, radiation, and direct flame impingement.

Conduction

Conduction is the transfer of heat within the material itself. Firefighters know that fire can travel via metal fences and other metal objects found in wildland areas. Although not conduction by definition, a fire in deep-seated vegetation litter underneath standing trees can transmit heat through the underground root system. In some cases, the smoldering roots can carry fire across a prepared fire line. Cold trailing is needed to detect and control this unseen fire. **Cold trailing** is a method to determine whether a fire is still burning by inspecting the ground to detect any heat source.

Firefighters remove all fuel from a ground surface area in a strip at least 18 in. wide, and then check the ground with their gloved hands to ensure that underground fuels are not burning. Crews also dig out every live ember and trench any live edge.

Convection

Convection is the transfer of heat by a liquid or a gas. In a wildland fire, the fire extends itself through heated air, which arises from the burning fuel. This movement of heat through the air transfers the heat into the cooler unburned fuel. If sufficient heat is present, it will raise (preheat) the fuel to its ignition temperature.

Radiation

Fire radiates heat waves, or energy, ahead of the fire. In doing so, it preheats the fuel and increases the chance that the area will catch fire. **Area ignition** is the accumulation of heated gases from individual fires. Once a fuel is preheated to its ignition temperature, all fuel in the area catches fire either simultaneously or in quick succession.

Direct Flame Impingement

Direct flame impingement occurs on wildland fires when the flame front is moving upslope and the flames lay down or move at an angle to the slope so that the flame front burns directly into the exposed fuel. This increases the fire's burning rate because it not only preheats the fuel, but the flames also provide a direct source of ignition. **FIGURE 8-4** illustrates how a wildland fire can move uphill.

Wildland Fire Size-Up

To size up a wildland fire, the fire service uses the same five-step decision-making process described in Chapter 5. The acronyms RECEO-VS and SLICE-RS are not needed unless structures are involved or exposed. The five-step process is applied to each of the following factors.

1. **Factors impacting life safety.** Life safety concerns are those factors that may lead to the death of civilians and/or firefighters.
2. **Factors impacting property safety.** Keeping property safe is a concern because fire can rapidly spread from a hot wildland fire into structures, which may lead to a conflagration.

FIGURE 8-4 A wildland fire can move uphill quickly when the flames lean at an angle and heat the fuel by convection currents and direct flame. A rule of thumb is that the fire's rate of spread will double when it transfers from flat land to a slope.

3. **Factors that may harm the environment.** One factor that may harm the environment is the damage and loss of the soil after burning away the vegetation because the roots work to hold the soil in place. Without the roots to hold the soil in place, the vegetation will be washed away during the rainy season.

   A second environmental issue is the toxic particulates and ash given off during the burning of wildland fuels. Some smoke and ash carry acids and other toxic gases, which can affect people who are sensitive to these types of materials.

TIP

It is important to keep fire retardants away from reservoirs used for drinking water. In the past, fire retardants have not always broken down in the environment. The chemicals have been toxic to fish and wildlife. During the Rim wildland fire in California in 2013, the San Francisco water district denied the use of any fire retardants close to the edge of their largest water supply, the Hetch Hetchy reservoir.

4. **Factors that may harm wildlife.** The loss of vegetation results in the loss of food supply for many forest creatures.
5. **The availability of needed firefighting resources.** The nonavailability of aircraft and other resources can seriously hamper the efforts of firefighters as they try to control a fire in its early stages. In

2019, a study by the California Governor's Office of Emergency Services recommended the advanced positioning of fire resources and staffing to be prepared to respond to wildland fires. The expectation is that these fires can be controlled if proper and sufficient resources are available during the first 20 to 30 minutes of the ignition. Many uncontrolled fires are pushed by the wind or eventually spread into structures that were exposed by brush. **Brush** is dense vegetation that consists of shrubs or small trees.

Factors That Affect Wildland Fires

Fire behavior is also affected by weather, topography, and fuels. These three main factors are often referred to as the "fire behavior triangle." One method that firefighters and wildland management personnel use to measure a fire's intensity and determine staffing levels is the burning index. The **burning index** is an estimate of the potential difficulty of containing a fire as it relates to the flame length at the most rapidly spreading part of a fire's edge. The index takes into account the fuel moisture content, wind speed, relative humidity, temperature, and recent precipitation.

Weather

Understanding weather characteristics is essential for the safe and effective extinguishment of wildland fires. A change in weather conditions can either add to or subtract from the fire's progress.

Temperature

In fire chemistry, fuel temperature is essential to light a fire and keep it lit. Temperature is one component of the weather that affects the behavior of wildland fuels. The range of ignition temperature in woody fuels varies between 572°F and 752°F (300°C and 400°C). Typically, the woody fuels will burst into flame at approximately 540°F (282°C) (Rocky Mountain Research Station, n.d.). However, the time required to bring the wood to its ignition temperature will vary with the amount of moisture in the fuel, because the heat must drive the water from the fuel to bring it up to its ignition temperature.

The highest temperature the sun can provide on a slope with no wind is 150°F (66°C), which is far below the required temperature for fuel to catch fire. The sun's heat plays an essential role because it preheats the fuel, which raises its temperature. At the same time, the temperature elevation reduces the amount of outside heat needed to bring the fuel to its ignition temperature. At night, the air is cooler than the fuel. This causes heat to transfer from the fuel to the cooler air.

Stable and Unstable Air Masses

Atmospheric stability determines whether air will rise and form clouds and storms (instability) or sink and create clear skies (stability). If the temperature of an air mass is greater than the temperature of the surrounding air, the air mass will rise. This air movement produces an unstable condition, which is dangerous, as is any shift of vertical and horizontal currents. At the same time, this creates an unpredictable wind situation.

Temperature Inversion

A **temperature inversion** occurs when air temperature increases with height. Cooler air is found close to the ground, and warmer air is above the cooler air. This often occurs along the coasts when cool, moist ocean air flows inland and settles in low areas (marine inversion). If a fire in this low area heats the cooler low-lying air to a temperature that is greater than the warm air overlying the area, the heated air breaks through the inversion layer, resulting in a quick rush of cooler air. An **inversion layer** is comparatively warm air overlaying cool air. In this stable atmospheric condition, the atmosphere will resist the vertical motion of air.

Firefighters should be concerned when an inversion layer is present because conditions can lead to a rapid change in the behavior of the fire. The inrush (entrainment) of a fresh supply of oxygen near the ground surface will increase a fire's intensity.

Relative Humidity

Water vapor is water in a gaseous state and is a significant ingredient in fire behavior. Air picks up water vapor as it moves over the surface of water. This process is called evaporation. **Evaporation** is the process by which molecules in a liquid state spontaneously become gaseous. The oceans are the primary source for adding water vapor to the atmosphere. Lakes, rivers, snow, and vegetation provide lesser amounts.

Relative humidity is the amount of water vapor in the air to the maximum amount that the air can hold. If the relative humidity is 100%, the air is saturated. It can hold no more water. If the relative humidity measures 50%, then the air is only half saturated and is able to hold 50% more water vapor. Low relative humidity draws moisture from fuels, allowing them to heat more rapidly. A wildland fire is possible when the relative humidity is below 30%. The National Weather Service issues a **red flag warning** when relative humidity is below 15%, which causes fuel to become very dry and makes conditions favorable for a wildland fire. During red flag warnings, fire departments prepare for wildland fires by organizing resources for initial fire responses.

Wind Speed and Direction

Wind is the result of the uneven heating of the Earth's surface. Heating the air causes it to become lighter; then it rises and draws in cooler air from below. Over the surface of the Earth, several major forces are at work to stimulate air movement. There are vast areas of heating near the equator that are much closer to the sun. There the heat produces rising air currents that return to Earth in the cooler regions. This movement sets up a circulation pattern of rising and descending air currents. The pattern is affected by the Earth's rotation, which deflects the rising air currents and causes them to move in a horizontal direction.

The changing seasons alter the wind pattern as well, because the hot and cold regions of the Earth shift with the Earth's tilt. This tilting results in changes to the upward vertical, downward, and horizontal movements of the rising and descending currents. Firefighters must prepare for the wildland fire season by attending training sessions, checking the local weather report, and holding preplanning sessions.

Foehn Winds

Winds called foehn (pronounced "fane") are **gravity winds**, which occur when air is pushed over high elevations and flows downhill. The **foehn wind** is a dry wind with a strong downward component. Locally, this wind is called the Santa Ana, North, Mono, Chinook, or East Wind. Gravity winds occur when a high-pressure system is located near mountain ranges where the airflow around the high-pressure system causes some of the air to spill over the higher elevations. This results in a strong wind moving downhill at high speed. As the foehn wind flows downslope, it is compressed and becomes warmer and drier. This movement of warm air dries the fuels and reduces the fuel's moisture. During these conditions, winds can reach 70 mph or higher.

Although the foehn wind is responsible for a drying influence on fuels and high rate of fire spread, the real hazard for firefighters comes from the day of transition. The **day of transition** is the first day when offshore winds subside and cool, moist onshore flows begin gradually returning. On the transition day, fire spread changes from downslope to upslope. This change may come at any time, and as a result, firefighters must always be on guard to not get trapped.

Some of the costliest fires in the western part of the United States occur when gravity or foehn winds push wildfires into populated urban interface lands. The **wildland/urban interface** lands are areas where structures are built close to or in the wildland area (defined as structures within 400 ft of the wildland area).

Dust Devils, Mirages, and Fire Whirls

Dust devils, mirages, and fire whirls indicate unstable surface conditions that will cause erratic fire behavior. These conditions result from uneven heating that causes ground temperatures to vary. Heated air rises and is pushed horizontally by the prevailing wind, and in some cases, it begins to whirl, or spin. The problem for firefighters is that these whirling winds throw burning materials and create spot fires. **Spot fires** jump over the control line and burn the vegetation ahead of the main fire front. This causes fuel preheating and results in rapid fire spread.

Fire Tornadoes

Fire tornadoes are exceptional atmospheric conditions. Previously referred to as fire whirls, these have recently been identified during large fire storms. The vortex created by the **fire tornado** is comparable to what we associate with a standard land-based tornado. The U.S. Forest Service now recognizes fire tornadoes.

Topography

Topography includes elements such as slope, aspect, elevation, and configuration, or the lay of the land. Topography can be considered static because the forces that change it, like rain and wind, generally work very slowly. However, topography causes changes in fire behavior as the fire moves across the terrain. Erratic fire behavior is prompted by changes in the wind pattern. These changes can be caused by the slope, aspect, elevation, canyons, saddles, ridges, and chimneys found on the land surface.

Slope

Slope affects the spread of fire in two ways. First, the fuel is preheated by direct flame impingement and convective air currents; second, a draft effect is created. A fire will run faster uphill than it will downhill as long as the wind is not strong enough to change the direction of the spread. On the uphill side of the fire, the flames are closer to the fuel, which causes preheating and faster

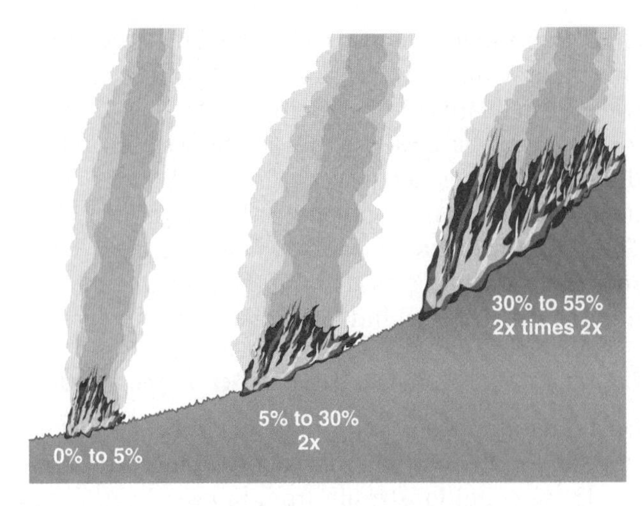

Slope/percentage	
0%–5%	x = spread of fire in feet per second
5%–30%	2x = spread of fire in feet per second
30%–55%	2x times 2x = spread of fire in feet per second

FIGURE 8-5 Speed of increase as fire travels upslope.

ignition of the fuel. The heat rises along the slope, causing a surge of air and increasing the rate of spread. A fire burning on level ground or up to a 5% slope will spread twice as fast when it reaches a 30% slope. The rate of fire spread will double again when the slope reaches a 55% slope. **FIGURE 8-5** depicts the speed at which you can expect fire to move upslope.

TIP

Firefighters cannot outrun a fire moving upslope. They must move in a horizontal direction or downslope.

Fires that start at the bottom, or toe, of a slope become larger because they have greater amounts of fuel available. The **bottom** or **toe** of the fire is the rear-most area of a fire. This is used as the anchor point, or safe location, to conduct fire suppression operations. Preheating is a result of direct flame impingement, radiative heat, and convective air currents. These make fires running upslope the common denominator in most fatal wildland fires.

Steep slopes also affect the ability of firefighters to work efficiently. The incline makes constructing fire lines and moving hose lines difficult. In addition, mid-slope fires present firefighters with great danger because the fire can move below them, suddenly erupt, and move rapidly uphill and trap them. Always check for fire below where you are positioned because gravity can cause burning materials to roll downhill and cross control lines. When one is working on a mid-slope or mid-slope road, it is wise to post lookouts to watch for any spot fires that begin above the mid-slope position. Spot fires can be started by burning embers traveling upslope ahead of the fire front. They can cross the mid-slope road, igniting the vegetation above the position of the firefighters.

Wildland fires have claimed the lives of many firefighters in the United States. This is due to lack of proper lookouts, firefighters working with fire below them on an upslope position, and changes in fire conditions. Ignoring either one of the 10 Standard Fire Orders or one of the 18 Watch Out Situations can also lead to fire deaths. Refer to Appendix B for the lists of standard fire orders and watch out situations.

According to the National Fire Protection Association (NFPA), the fires listed in **TABLE 8-2** are the top 10 deadliest wildland firefighter fatality incidents.

Aspect

The direction a slope faces to the sun is called the **aspect** or **exposure**. This direction impacts the spread of fire based on rising and falling temperatures. North-facing slopes will reach their warmest temperature at noon. South-facing slopes are exposed to more radiant heat by the sun, resulting in temperatures that are normally higher with lower humidity. East-facing slopes heat up first with the rising sun. Eastern slopes begin to cool down after noon. Western slopes continue to increase in temperature until sunset. Drier fuels are lighter and ignite more quickly than wetter fuels. Fuels on western slopes are usually heated more than eastern ones and as a result are dry because of their aspect position. The warming changes hourly as the sun moves across the sky, resulting in different fire behavior that is affected by the angle of the sun and its heating of the fuels.

Aspect also influences the type of natural vegetation. Fuels on a southwestern exposure typically present the greatest fire hazard because they are exposed to direct sunlight for longer. The fuels are smaller and drier. Fuel on the northern slopes receives more moisture and is heavier. Fires on the northern aspect move more slowly because of the moisture content of heavier fuel; however, once burning, these fuels generate more BTUs. As a result, they present a greater resistance to fire control.

Elevation

Elevation affects fire behavior. It influences how air moves from the valleys that are warming to the cooler

TABLE 8-2 Top 10 Deadliest Wildland Firefighter Fatality Incidents

Event	Date	Number of Firefighter Deaths
The Devil's Broom Wildland Fire St. Joe Valley, Idaho	August 20, 1910	78
The Griffith Park Fire Forest Fire Los Angeles, California	October 3, 1933	29
Yarnell Hill Fire Yarnell, Arizona	June 30, 2013	19
Rattlesnake Fire Mendocino National Forest Fire Willows, California	July 6, 1953	15
Blackwater Forest Fire Shoshone National Forest, Wyoming	August 21, 1937	15
Wildland Fire South Canyon Glenwood Springs, Colorado	July 6, 1994	14
Mann Gulch Fire Helena National Forest Helena, Montana	August 5, 1949	13
The Loop Fire Disaster Forest Fire Los Angeles, California	November 1, 1966	12
Wildland Fire Hauser Canyon Fire Cleveland National Forest, California	October 1, 1943	11
Wildland Fire Inaja Fire Cleveland National Forest, California	November 24, 1956	11

Reproduced from NFPA's Fire Incident Data Organization database and NFPA archive files. Updated June 2017.

ridges (warm air rises). It can also affect the positioning of warm/cool air masses in thermal belts (the layering of air masses). Elevation affects the length of the fire season (the lower the elevation, the longer the season) because the light flash fuels are drier and quick burning. **Flash fuels**, such as grass, leaves, ferns, tree moss, and some types of slash, ignite readily and are consumed rapidly when dry. Elevation will affect types of fuel (types of vegetation and fuel loading change with higher elevation) and placement. Notice the grass to brush to timber change as the elevation increases in **FIGURE 8-6**. Fuels at sea level are more compact.

Elevation leads to the fuel ladder expansion of fire. The **fuel ladder** is the progression of the fire from the lighter to heavier fuels. As the elevation increases, fuels are spaced farther apart. Light grasses and low shrubs are not present at higher elevations. Large trees such as conifers are found in these rocky areas.

Canyons, Saddles, and Ridges

The topography of the land affects wind patterns and fire spread. Canyons, saddles, and ridges directly influence how a fire burns. Wide canyons will have little effect on the wind pattern, while narrow canyons may

FIGURE 8-6 Elevation changes fuel types.

significantly impact the wind as it is driven down the canyon. In narrow canyons with steep walls, a fire can jump across by radiant or convective spotting. This is due to the short distances involved. These canyons are extremely dangerous for firefighters because the fire can easily cross over the canyon and get behind working fire crews.

TIP

Take special precautions when firefighting in narrow canyons with intersecting canyons. The wind eddies that occur in these intersections can cause erratic fire behavior.

Saddles are the low topography between two high points and are the point of least resistance for wind and wind eddies. They provide the potential for rapid fire spread because fires are pushed through saddles faster during upslope fire runs. Firefighters should not use saddles as a safety zone.

Ridges are raised land masses that divide the terrain. They may not only divide the terrain, but also have different wind conditions on each side. This is especially true along coastal regions when the weather patterns change due to the temperature difference from the ocean to the warmer land areas.

Chimneys

A **chimney** is a topographic feature that has three walls and a steep, narrow chute. As discussed earlier, radiant or convective heating can occur across narrow canyons. Chimneys draw fire up through them (a drafting effect) in the same fashion as a flue draws heat up the chimney of a fireplace.

Fuels

Understanding and identifying the type of fuel is critical to firefighter operations and safety. The composition of the fuel, including its chemical makeup, density, and moisture, determines its flammability. The level of moisture is the most important consideration.

The characteristics of the fuel will also affect fire behavior, and the decisions and actions that firefighters take on the fire line.

Fuel Loading

Fuel loading is the amount of fuel available to burn in a given area. When deciding on the strategy and tactics to use when fighting a structure fire, the firefighter calculates how much fuel is in the structure by the construction type and contents, or **fire load**. Based on this information, fire crews can determine the number of resources needed.

The fuel loading of an area in a wildfire is determined by tons of fuel per acre, the fuel type, height, density, arrangement and shape of fuel, its moisture content, and temperature. **TABLE 8-3** shows how the fuel type relates to tons per acre. When firefighters compare the fuel to the needed suppression activities, the fuel is classified as light, medium, or heavy.

Light Fuels. Grass and other small plants grow on the floor of all forests. These light flash fuels influence

TABLE 8-3 Tons of Fuel per Acre

Fuel Type	Average Estimated Tons per Acre
Grass	0.25 to 1
Medium brush	7 to 15
Heavy brush	20 to 50
Timber	100 to 600

Data from U.S. Forest Service. 2006. *Fire Management Notes*. Boise, ID: U.S. Government Printing Office; Xanthopoulos, Gavriil, and Miltiadis Anthanasiou. 2020. "Crown Fires." In *Encyclopedia of Wildfires and Wildland-Urban Interface (WUI) Fires*, edited by Samuel L. Manzello, 1–15. Cham, Switzerland: Springer. https://doi.org/10.1007/978-3-319-51727-8_13-1.

the rate of fire spread. If there is a continuous cover of dry grass on the forest floor, the rate of fire spread will be much faster. These fine, light fuels ignite easily and provide an avenue for carrying the fire from one area to another. They provide the heat necessary to move the fire to the heavier, denser fuels of the forest.

Medium Fuels. Medium fuels consist of brush that is 6 ft in height or lower, growing in fairly thick stands. In many cases, the medium fuels are those areas between the light and heavy fuels that move the fire from light to heavy fuels.

Heavy Fuels. Fires in the heavier (more mass) fuels produce more BTUs and as a result are more resistant to fire suppression activities. They require more water to cool, extinguish, and control them. They consist of brush taller than 6 ft, timber slash, and standing conifer and hardwood trees.

Fuel Shape and Arrangement

A fuel's shape may determine how the fuel can affect the ignition and spread of a fire. The smaller the fuel (the lower its mass), the faster it will preheat, ignite, and spread. The larger the fuel (the more its mass), the more BTUs it will produce.

Fuels that grow and spread fire from the ground surface into higher or taller vegetation are called **ladder fuels**. During suppression activities, firefighters should be aware of their surroundings to include what is happening above them.

Aerial Fuels

The **aerial fuels** include all green and dead materials in the upper forest canopy. The main aerial fuels are tree branches, crowns, snags, moss, and high brush. These can provide a path for fire travel to fuels higher in the air. These fires are inaccessible, so they are difficult to extinguish.

Crown fires are fires that have traveled from the ground to the forest canopy and spread through it. They burn horizontally across the tops (crowns) of the trees or tall brush. These fires occur in fuels that are at least 6 ft (1.828 m) tall. When the fire is burning, the flame length is between 30 and 200 ft (9.14 m and 60.96 m). For a crown fire to occur, the vegetation must be close together (dense), continuous (one plant close to another), and have ground or ladder fuels to carry the fire into the tops of the trees.

These types of fires usually move quickly and independently of the surface fire. They can spread fire in advance of the fire's head by dropping burning material to the forest floor below.

Fuel Moisture

The amount of moisture in the fuel will affects how easily it will ignite and how intensely it will burn. The moisture content of fuel is classified in two ways. One is dead fuel moisture, which is the content of moisture in fuel that is not living. The second is live fuel moisture. The live fuel moisture is the amount of fuel moisture in living, growing fuels, which gain moisture primarily from the plant root system. Live fuel moisture is changed by the growing cycle of the vegetation and varies greatly between species and seasons. Dead fuel moisture is changed only by the moisture content of the air. Both are measured as a percentage of the weight of the fuel.

The **time lag** is the time it takes for the moisture content of fuel and the surrounding air to equalize. Time lag is usually expressed in hours. With afternoon rain, light grass would absorb moisture; however, if the afternoon were warm, these fuels could dry out in an hour. This would allow them to ignite. Grass is considered a 1-hour time lag fuel. It takes about 1 hour of air exposure to change the fuel moisture of grass up or down. This means that grass can go from 100% fuel moisture to a combustible fuel in 1 hour.

Small- to medium-size brush is classified as 10-hour fuel, and logs or downed trees are classified as 100-hour fuel. The large timber and large downed trees are 1,000-hour fuels. This time relates to how long firefighters have until the fuel is dry enough to ignite.

Fuel Temperature

The warmer the fuel, the less heat that is required to ignite it. This is particularly important when dealing with light fuels. Their small size and mass allow them to heat up more quickly and change their moisture content faster.

There can be a 50°F to 80°F difference between the surface temperatures of fuel in the sun and those in the shade, with a corresponding difference in fuel moisture. Fire is a chemical reaction. For every 18°F rise in temperature, the chemical reaction of fire doubles in speed. This influences both the fire's intensity and the rate of fire spread.

Fire Behavior

This section first identifies the parts of a wildland fire. Firefighters must know the parts of a fire where it is safest to work. It then explores spotting fires, which start from embers drifting away from the main fire, and area ignition, which is much like a flashover in a building, where a large amount of fuel ignites. The section briefly touches on surface fires and ground fires.

Identifying Parts of a Wildland Fire

Wildland fires are identified by a left flank (left side of the fire) and a right flank (right side of the fire). The left side is always the left side of the fire when looking at a map. In other words, the left hand becomes the left flank of the fire. The **head** of the fire is the outermost portion of the fire that is moving from the rear.

- The **anchor point** is a safe location, usually a barrier to fire spread, where fire crews can start building a fire line.
- **Islands** are areas inside the fire that have not been burned out.
- **Pockets** are the areas on the edge of a fire that have not been burned. A pocket is surrounded on one side by a finger.
- **Fingers** are narrow strips of fire that have separated from the main body. A finger creates a pocket between it and the main body of the fire.
- The **rear** is the portion at the edge of a fire opposite the head. It is the point of origin in most fires.

FIGURE 8-7 identifies the critical parts of a wildland fire. Note the spot fires and hot spots over and near the fire line. Firefighters always find a safe working area either using the anchor point or a safe location from which they start working on the fire line.

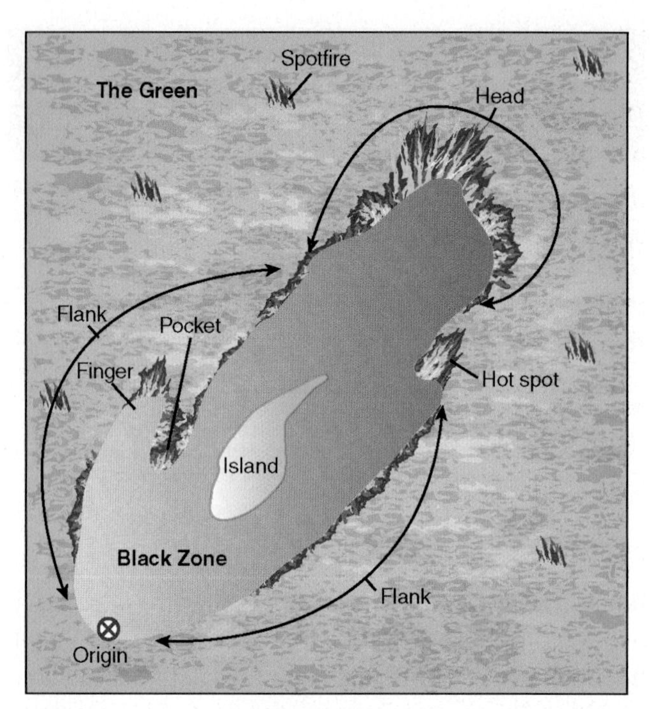

FIGURE 8-7 Parts of a wildland fire.

Spotting

The appearance of spot fires depends on the terrain, weather, and the type of fuels in the main fire. Spotting occurs when wind and **convection columns**—thermally produced columns of gases, smoke, and debris produced by a fire—spread hot fire brands into the unburned fuel ahead of the main fire. This is the **green area**, the unburned area next to vegetation fires. Firefighters always watch for and protect against uncontrolled fires moving into the green.

Crown fires, with their intense heat and strong convection currents, frequently start spot fires. They can throw burning embers (preheating the brush) far ahead of the main fire front.

Large Fires

Wildland fires can become so large that the burning process creates its own indraft of oxygen. This indraft serves to continue the burning process. Firefighters use this principle to burn out the fuel near the fire's edge to separate the fire from the fuel. This fire suppression method works well when a fire is lit to be entrained into the main fire. This action removes the fuel from the path of the main fire.

Area Ignition

As mentioned earlier, area ignition is the accumulation of heated gases from individual fires that ignite. This happens either simultaneously or in quick succession. The fires are close enough to each other that they soon support each other to produce a hot and fast-moving fire. Area ignition can occur when spot fires flare up in an area of unburned vegetation or when the spot fires are burning in a canyon or an area where the heat from the spot fires is trapped. In some cases, an inversion layer will hold the heated gases within the canyon. The spot fires preheat the unburned vegetation to a point where it reaches its ignition temperature. All at once, the preheated vegetation explodes into flames. If firefighters are working in the canyon, their chances of survival are slim. Similar conditions exist in a confined room when a flashover occurs, and the entire room explodes into fire.

If a fire has moved through a forested area without burning the crowns, a reburn in the form of a crown fire is still a possibility. In this situation, the crown fire may be very intense because the first fire through the fuel dried and scorched the aerial fuels.

Surface Fires

Surface fires occur when leaf matter, fallen branches, and dead or downed trees burn above the ground. These fires are common in wooded areas that are not maintained, and the fallen vegetation has built up over the years.

Ground Fires

Ground fires occur when the fire progresses underground through the root systems or the ground surface material. These types of fires are most often seen in peat fields.

Fire Resources

Firefighters use engine companies, water tenders or trucks that carry water, hand crews, aircraft (both fixed wing and rotary wing), and bulldozers for wildland fires. The National Incident Management System (NIMS) classifies all resources based on their design, capability, and inventory. Using this standardized system, all agencies can quickly identify the type of equipment to request or send to another agency.

The classification of all resources used on wildland fires is available in a field operations guide. The **field operations guide (FOG)**, available in print, online, or as an app, is a system that classifies units into strike teams and task forces. A strike team is a set number of resources of the same type with minimum staffing levels. Engine strike teams include five engines and a strike team leader. The team leader rides in a separate vehicle for mobility and communications purposes.

A task force is any grouping of resources with a task force leader, assembled temporarily for a specific mission. A task force can be any size and include a variety of resources.

The NIMS system classifies the engines within a strike team or task force into types, which identify their capabilities. The engine types are not supposed to be mixed on strike teams; however, if they are in a strike team configuration, the team takes on the classification of the least capable unit.

Recently, the addition of staged strike team resources has proven beneficial in reducing the zero impact time these resources have on the fire. **Zero impact time** is the time from dispatch to the incident to when firefighters are on scene to take action.

Engine Types

The fire engine is the most versatile of the various apparatus in the air and on the ground (**FIGURE 8-8**). It can fight fire day and night and is not restricted by visibility, unlike an aircraft. It can deliver water (attack the heat side of the tetrahedron) to a fire at various capacities and transport a crew to a fire with their equipment. Its crew can construct hand lines (attack the fuel side of the tetrahedron) and the unit can be used to fight structural, vehicle, or other types of fire. **TABLE 8-4** classifies the standard types of fire engines.

Hand Crew Types

On any sizable wildland fire, the biggest portion of suppression personnel consists of hand crews (**FIGURE 8-9**). A hand crew is simply a group of firefighters formed into a suppression team. They use hand and power tools to reduce the fuel side of the tetrahedron (**FIGURE 8-10**).

Hand crews usually comprise 12 to 20 firefighters who are assigned to actively suppress low-flame-production fires. They also construct hand lines along

FIGURE 8-8 Wildland type 3 engine. This type 3 engine is designed for off-road fire suppression.

TABLE 8-4 Type Classification of Engines

Engine Companies	Pump Gallons per Minute (gpm)	Water Tank Gallons (gal)	2 1/2 Hose	1 1/2 Hose	1 Hose	Ladders	Staff
Type I engine	1,000	400	1,200	400	200	20 ft	4
Type II engine	500	400	1,000	500	300	20 ft	3
Type III engine	120	300		1,000	800		3
Type IV engine	70	750		300	300		3
Type V engine	50	500		300	300		3

Modified from U.S. Fire Administration/National Fire Academy. 2016. *Field Operations Guide: ICS 420-1*. https://www.usfa.fema.gov/downloads/pdf/publications/field_operations_guide.pdf.

FIGURE 8-9 Wildland hand crew cutting a fire suppression line.

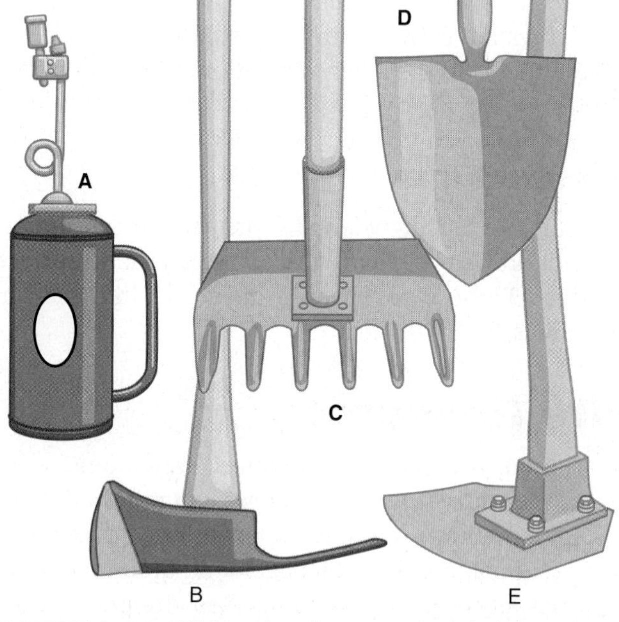

FIGURE 8-10 Wildland hand crew tools: **(A)** drip torch, **(B)** Pulaski, **(C)** McLeod, **(D)** wildland shovel, **(E)** Green Grubber.

the fire's edge or ahead of the fire for future holding action or for creating a backfire. These crews often get the toughest assignments because they can get to locations where vehicles, including bulldozers, are unable to travel.

A strike team of hand crews consists of a strike team leader with two or more supervisors and up to 40 crew members, all of the same type. The kind of fire line work that can be performed by the crew is based on the type of the hand crew. There are two classifications of hand crews:

Type I: These people are highly trained to work on all fires directly on the fire line.

Type II: These people have some training with fire line work restrictions.

Hand crews can work with engine companies to carry a hose line or assist a bulldozer by following and picking up any fire missed by the dozer blade. Hand crews also need to perform cold trailing to make sure the fire is completely out.

Bulldozer Types

Bulldozers in conjunction with other ground components provide a balanced fire suppression force. Bulldozers attack the fuel side of the tetrahedron. A strike team of bulldozers consists of two dozers, two

FIGURE 8-11 Firefighting bulldozer.

operators, two swampers, and a strike team leader. A **swamper** is a worker on a bulldozer crew who pulls the winch line and helps maintain equipment. They generally assist with suppression work on a fire. Bulldozers are divided into three broad classifications:

Type I: Heavy D-7, D-8, D-9—Heavy, thick brush, timber, or road construction

Type II: Medium D-6 and HD-11—Heavy to medium brush

Type III: Light D-4 and HD-6—Light fuels, grass, fast-moving fire

Bulldozers are efficient and effective in removing fuels along a fire line. They do their best work during a direct fire attack by cutting a fire line on the fire flanks. On an indirect attack, they can cut brush lines, escape routes, or safety zones in front of the fire line. They can also open roads and other access points in the wildland areas. Bulldozers often cut water bars to ensure that soil is trapped rather than washed off the side of the hill during the rainy season. The water bars look like ridges of piled-up dirt that run parallel to the hillside. They are generally at least 18 in. tall and will direct water moving down slope and prevent erosion. **FIGURE 8-11** shows a firefighting bulldozer.

Fixed-Wing Aircraft Types

Both fixed-wing and rotary-wing aircraft will attack the heat leg of the fire tetrahedron by cooling with water or retardants. Aircraft are typed on their ability to carry water or fire-retardant materials. Aircraft can drop the load from their tanks either once or in a series, thus spreading the load over a larger area. **TABLE 8-5** lists the classifications of fixed-wing aircraft and their carrying capacity.

TABLE 8-5 Type Classifications of Fixed-Wing Aircraft

Type	Carrying Capacity (gal)
I	3,000+
II	1,800 to 2,999
III	600 to 1,799
IV	100 to 599

Data from Firescope California. *2022 Field Operations Guide ICS 420-1*. https://firescope.caloes.ca.gov/fog-manual.

FIGURE 8-12 Fixed wing aircraft: Air tanker making a suppression drop.

Fixed-wing aircraft drop retardant or water on the head or front of a hot, fast-moving fire. This action will slow it down and allow engine companies, hand crews, and bulldozers to work the flanks. Eventually, the crews might catch the head once it has cooled and is moving slowly. In addition to slowing a fire, the aircraft can paint the fire line with fire retardant to confine the fire to a specific area. Because it carries much more retardant than a rotary aircraft, the fixed-wing aircraft can cover a larger area. This helps to reduce the total heat build-up and allows access for ground crews. **FIGURE 8-12** shows a fixed-wing aircraft assisting in fire operations.

Rotary-Wing Helicopters

The advantage of a helicopter is its ability to work from locations close to and on the fire line. Its ability to take off and land vertically makes it a valuable tool to support ground-based and other aerial operations. Helicopters have attachments that allow them to perform jobs that would be difficult, if not impossible, by any other means (**FIGURE 8-13**).

FIGURE 8-13 A rotary-wing aircraft can drop fire retardant.

TABLE 8-6 Type Classifications of Rotary-Wing Aircraft

Type	Personnel Capacity	Agent Capacity (gal of retardant)
I	16 seats; 5,000 lb	700
II	9 to 15 seats; 2,500 to 4,999 lb	300 to 699
III	5 to 8 seats; 1,200 to 2,499 lb	100 to 299
IV	3 to 4 seats; 600 to 1,199 lb	75 to 99

Data from Firescope California. *2022 Field Operations Guide ICS 420-1*. https://firescope.caloes.ca.gov/fog-manual

Rotary-wing aircraft cannot carry as much as fixed-wing aircraft, but they can work closer to houses. The fire service needs both fixed- and rotary-wing aircraft because each unit fulfills important but different fire protection functions.

Helicopters can apply liquids, such as water or retardant chemicals, over key fire line targets more precisely than fixed-wing aircraft. Because of their ability to hover, they can be directed precisely via radio communications with ground personnel. **TABLE 8-6** classifies rotary-wing aircraft and their capacities.

Helicopters can also transport injured firefighters to hospitals and burn victims to hospital burn wards. First responders often use helicopters to evacuate people stranded or threatened by fire.

FIGURE 8-14 Direct attack. Wildland hand crew cutting a fire suppression break in light fuel.

Wildland Fire Tactics

Just as in structure firefighting, for wildland fires there are multiple options for strategy and tactics. These can be initially chosen and then, after an evaluation of the suppression activity, they can be changed. The following are examples of tactics available to firefighters.

Direct Attack Method

The incident commander of the first-arriving units will decide which attack method to use based on the flame length of the fuel, the terrain, and weather conditions. These decisions are made during the initial size-up for extended suppression activity. For large-scale fires, a planning meeting is held before each operational period to define an **incident action plan (IAP)**. Operational periods usually last 12 or 24 hours. The planning meeting allows all the information on the fire that has been collected to be considered. Resources are then ordered, and an IAP is developed. The IAP is given to the firefighters battling the fire during the next operational period.

The **direct attack method** is when personnel and resources work at or very close to the edge of the burning fire. The direct method includes attacks with water, chemical agents and retardants, cooling, or separating the fire from the available fuel (**FIGURE 8-14**). For example, firefighters can spray water directly on the burning edge of the fire or a fire can be set to burn the fuel in front of the fire. This method of attack has some advantages and disadvantages for firefighters.

One advantage is that the firefighters can work and control the fire on the fire line. The line cut around the fire must be continuous enough to keep the fire to the burning side of the line. If the direct attack is successful, it prevents the fire from building momentum. It also eliminates the need for a backfire. A **backfire** is a fire that is intentionally set ahead of the progressing fire to remove fuel load. The direct attack method is typically used in the initial attack phase. Fire crews attack the flanks or head of the fire depending on the amount of heat the fire is generating. The objective of this attack method is to stop the fire spread as soon as possible.

Another advantage of the direct attack is that it holds the fire to a minimum acreage by taking advantage of line portions that have not burned. This can be the result of a lack of fuel or poor burning conditions. Other advantages include controlling a fire more quickly with less loss of wildland resources than a backfire would require.

There are also many disadvantages of the direct attack method. It may not be effective against an intensely hot or fast-moving fire. This type of attack is also risky and requires close coordination of all crews. Firefighters work directly on the fire line and are exposed to more heat and smoke. Fire crews can accidentally spread burning materials across the fire line. Because the direct attack follows the fire line, it does not take advantage of barriers, such as rock outcroppings, lakes, rivers, or other barriers. Finally, there is usually more cleanup and patrol work required and crews must follow behind to ensure that all fires are out.

Indirect Attack Method

The **indirect attack method** consists of constructing control lines or creating a backfire at a distance ahead of the fire. The fire service uses this method when the fire is burning too hot and rapidly to use direct attack methods. It is also used when sufficient suppression forces are not available. Firefighters use the indirect method on hot, intense fires, so it is often the first choice for large fires.

A fire line is constructed ahead of the fire and is kept in place by ground and air forces. If it can be held, it becomes the final control line. Fire lines can be reinforced with retardant drops to cool the fire as it approaches the line, or they can be protected with backfires on fuel in front of the main fire.

One advantage of the indirect attack method is that firefighters can construct control lines on favorable terrain. Such a line can often be completed with fewer suppression resources. In most situations, the total fire line cut is shorter than in a direct attack. Another benefit of this method is that firefighters can work away from heat and smoke.

The disadvantages are that while the control lines are being put into place, fire is not being extinguished. During this period, the fire consumes additional fuel. Also, any areas that are not burning are included in the control line perimeter. No advantage is taken of these lighter fuel areas and, in some situations, unburned fuel is left inside the control line. It also places firefighters in areas where burning out or backfiring is in progress. In some cases, secondary lines may need to be cut or burned out, which requires more work and exposes firefighters to additional danger.

Combination Attack

The **combination attack method** takes advantage of the direct and indirect attack methods by combining both methods. Firefighters work directly on the fire line in areas that are safe and can be reached quickly to contain as much of the fire as possible. At the same time, resources permitting, crews begin to work indirectly in those areas where they cannot directly attack the fire line to put in and burn out a clean line. Most incident commanders will use this method of attack as the opportunities on the fire line allow.

Application of Attack Methods

There is not a single method of fire attack to follow. Many times, a combination of two or more methods is used. In some situations, all methods are used simultaneously.

Tandem Action

A **tandem action** is a direct attack method with the attacking forces working in tandem (one unit following the other). The attack starts at an anchor point, with units following closely on the flank of the fire. The units move up as the first unit knocks down the fire. This attack method is used on a fast-moving fire because it is a direct attack on the fire line, and firefighters can always move from the fire line green area (the area not yet burned) into the **black zone** (the area already burned) to seek safety.

Flanking Action

A **flanking action** is attacking the fire on both flanks at the same time. The objective of a flanking action is to prevent the fire from spreading on a given flank and thus threatening an exposure. An exposure could be heavier fuel, a recreational area, or an area with structures. Any type of ground or air suppression resource can conduct a flanking action. A flanking action requires sufficient resources to cover the entire fire area (both flanks) and good, close resource coordination. Flanking allows firefighters to work away from the direct flame front, reducing their exposure to heat and smoke.

Pincer Action

A **pincer action** moves crews along both flanks of the fire to a point where flanking forces create a pinching action near the head of the fire. A pincer action may be conducted on any size fire; however, it is most often used on small fires. Resources include hand crews, bulldozers, engines, aircraft, or a combination.

Envelopment Action

An **envelopment action** involves suppressing a fire at many points in many directions simultaneously. It provides for a rapid attack and, if coordinated, can be highly effective on smaller fires. This attack method requires a large number of suppression forces to be committed to both flanks of the fire. Crews must establish close command and control. Units taking part in this action should be experienced and aggressive.

Parallel Action

The **parallel action** is another kind of attack. In this suppression method, the fire line is constructed parallel to and just far enough away from the fire edge to allow personnel and equipment to work effectively. The line may be shortened by cutting across unburned fingers. The intervening strip of unburned fuel is often burned out as the control line proceeds. It may be allowed to burn out unassisted when this happens without undue delay or threat to the line.

Backfire

As discussed earlier, a backfire is a fire that is intentionally set ahead of the progressing fire to remove fuel load. There will be times when the head of the fire is moving rapidly and setting up fuel breaks is not sufficient to stop the fire. In these situations, firefighters may need to burn a designated area or remove a large amount of fuel ahead of the fire.

Fire crews use drip torches to light the designated area to ensure it burns toward the oncoming fire (**FIGURE 8-15**). As the two fire heads burn, they will draw toward each other. Once they meet, the progression of each fire stops.

This technique will be in the operations plan to ensure everyone is aware of it. Backfires can be dangerous and require calm weather conditions for their success.

FIGURE 8-15 Application and use of drip torches to create a backfire.

CASE STUDY
CONCLUSION

1. **What effects did the lack of roadways and the increased residential construction in the wildland/urban interface have on the wildland fire?**

   The lack of roadways impacted the evacuation out of the town. It created traffic jams and overwhelmed the capacity of the roadway. It also limited the fire department's access to the fire.

2. **How does the lack of forest management affect the rate of fire spread?**

   The lack of forest management meant that dead and dying forest vegetation piled up on the forest floor. The amount of heat generated increases the rate of the burn intensity and thus the ability of the fire to ignite the 100- and 1,000-hour fuels.

3. **How would an adequate fire suppression force be deployed to such a fire during wildland fire season?**

   Resources are vital to any fire suppression plan. Having sufficient fire resources staged to reduce the zero impact time would be a positive measure.

WRAP-UP

SUMMARY

- In wildland firefighting, firefighters stop the process of combustion by cooling the heat, removing the fuel, or excluding the oxygen needed to light a fire and keep it lit.
- Wildland fire heat output is transferred the same way as in structural fires. The heat moves in four ways: conduction, convection, radiation, and direct flame impingement.
- To size up a wildland fire, the fire service applies a five-step decision-making process that includes each of the following factors:
  - Factors impacting life safety
  - Factors impacting property safety
  - Factors that may harm the environment
  - Factors that may harm wildlife
  - The availability of needed firefighting resources
- Fire behavior is also affected by weather, topography, and fuels, which are often referred to as the "fire behavior triangle."
- Wildland fires are identified by a left flank (left side of the fire) and a right flank (right side of the fire); the head of the fire is the outermost portion of the fire that is moving from the rear.
- Firefighters use engine companies, water tenders or trucks that carry water, hand crews, aircraft, and bulldozers for wildland fires.
- Firefighters use the direct attack method, the indirect attack method, and the combination attack method to combat wildland fires.
- To apply attack methods, the fire service can use tandem action, flanking action, pincer action, envelopment action, or parallel method, or create a backfire.

KEY TERMS

aerial fuel Fuel that includes all green and dead materials in the upper forest canopy.

anchor point A safe location, usually a barrier to fire spread, where fire crews can start building a fire line.

area ignition The accumulation of heated gases from individual fires that ignite either simultaneously or in quick succession.

aspect The direction a slope faces to the sun; also called exposure.

atmospheric stability The air's tendency to either rise and form clouds and storms (instability) or sink and create clear skies (stability).

backfire A fire that is intentionally set ahead of the progressing fire to remove fuel load.

black zone The fuel or vegetation that has been burned; it is considered a safe area because the fuel has been removed.

bottom The rear-most area of a fire, used as the anchor or starting point of the fire suppression activities; also called a toe.

brush Dense vegetation that consists of shrubs or small trees.

burning index An estimate of the potential difficulty of containing the fire as it relates to the flame length at the most rapidly spreading part of a fire's edge.

chimney A topographic feature that has three walls and a steep, narrow chute.

cold trailing A method to determine whether a fire is still burning by inspecting and feeling with gloved hands to detect any heat source.

combination attack method A method that takes advantage of the direct and indirect attack methods by combining both methods.

compressed air foam system (CAFS) A fire suppression system that injects air into the foam solution with an air compressor.

convection column The ascending column of gases, smoke, and debris produced by a fire.

crown fire Fire that has traveled from the ground to the forest canopy and spread through it.

day of transition The first day when offshore winds subside and cool, moist onshore flows begin gradually returning.

direct attack method A method of fire suppression in which personnel and resources work at or very close to the edge of the burning fire.

envelopment action A fire suppression technique that suppresses a fire at many points and in many directions simultaneously.

KEY TERMS CONTINUED

evaporation The process by which molecules in a liquid state (e.g., water) spontaneously become gaseous (e.g., water vapor).

field operations guide (FOG) A system that classifies units into strike teams and task forces; available in print, online, or as an app.

finger A narrow strip of fire that has separated from the main body of the fire.

fire line A fuel break that is created when firefighters clear an area twice as wide as the fuel is tall.

fire load Amount of fuel in the structure that includes the construction type and contents.

fire retardant Any substance that is used to slow or stop the spread of fire or to decrease a fire's intensity.

fire tornado A vortex comparative to a standard land-based tornado.

flanking action A technique in which the fire is attacked on both flanks at the same time.

flash fuel Fuel, such as grass, leaves, ferns, tree moss, and some types of slash, that ignites readily and is consumed rapidly when dry.

foehn wind A dry wind with a strong downward component; locally called Santa Ana, North, Mono, Chinook, or East Wind.

fuel ladder The progression of the fire from the lighter to heavier fuels.

fuel loading The amount of fuel available to burn in a given area.

gravity wind Wind that occurs when air is pushed over high elevations and flows downhill.

green area The unburned area next to vegetation fires.

head The outermost portion of the fire that is moving from the rear.

heavy fuel Fuel over 6 ft tall.

incident action plan (IAP) A plan for the firefighters battling the fire during the next operational period.

indirect attack method A method that consists of constructing control lines or creating a backfire at a distance ahead of the fire.

inversion layer A layer of comparatively warm air overlaying cool air.

island An area inside the fire that has not been burned out.

ladder fuel Fuel that grows and spreads from the ground surface into higher vegetation.

light fuel Fuel less than 2 ft tall.

medium fuel Fuel 2 to 6 ft tall.

parallel action A fire suppression technique in which the fire line is constructed parallel to and just far enough away from the fire edge to allow personnel and equipment to work effectively.

pincer action A fire suppression technique in which crews move along both flanks of the fire to a point where flanking forces create a pinching action near the head of the fire.

pocket An area on the edge of a fire that has not been burned.

rear The portion at the edge of a fire opposite the head.

red flag warning An alert issued by the National Weather Service when relative humidity is below 15%, which causes fuel to become very dry and makes conditions favorable for a wildland fire.

relative humidity The amount of water vapor in the air to the maximum amount that the air can hold; it is presented as a percentage.

ridge A raised land mass that divides the terrain.

saddle The low topography between two high points.

spot fire Fire that jumps over the control line and burns the vegetation ahead of the main fire front.

swamper A worker on a bulldozer crew who pulls the winch line, helps maintain equipment, and generally assists with suppression work on a fire.

tandem action A direct attack method in which the attacking forces work in tandem (one unit following the other).

temperature inversion Atmospheric condition in which air temperature increases with height; cooler air is found close to the ground and the warmer air is above the cooler air.

time lag The time it takes for the moisture content of fuel and the surrounding air to equalize; usually expressed in hours.

wildland/urban interface An area where structures are built close to or in the wildland area.

zero impact time The time from dispatch to the incident to when firefighters are on scene to take action.

REVIEW QUESTIONS

1. Do the basic fire and combustion processes found in wildland firefighting differ from structural firefighting? Explain your answer.
2. What are the similarities and differences between the strategy and tactics of wildland firefighting and structural firefighting?
3. While working to put out a wildland fire, how would you exclude the oxygen, reduce the heat, and remove the fuel?
4. What are the three main factors that impact the behavior of a wildland fire? How do each of these factors impact fire behavior?
5. Why are chimneys so dangerous to firefighters?

DISCUSSION QUESTIONS

1. What are the three areas of concern in wildland firefighting? Which would you consider the most important?
2. Which major parts of a wildland fire are of the greatest concern? How do they individually relate to fire attack?
3. How many methods of fire attack can be deployed on a wildland fire? Which is the best attack method in your opinion?

APPLYING THE CONCEPTS

During the Tunnel fire, in Oakland, California, in 1991, a small-acreage grass fire was fought and presumed safe to leave with unstaffed hose lines in place. The following day, high foehn winds and high temperatures ignited a rekindle of the initial fire and began to spread. The fire ultimately killed 25 people and injured 150 others. The 1,520 acres (620 ha) destroyed included 2,843 single-family dwellings and 437 apartment and condominium units. The economic loss from the fire was estimated at $1.5 billion.

The compactness of the vegetation in the area and the proximity of residences fed the fire and created a firestorm. It was reported the residential homes were exploding and on fire in an average of 11 to 13 seconds. Damage to electrical power lines in the area hampered the use of water tanks and their pumps for firefighting efforts.

High winds carried burning embers across an eight-lane freeway and pushed the fire further into residential areas. Embers and debris were carried as far as 8 miles away from the area of origin.

1. Explain which of the three main factors of the fire behavior triangle contributed the most to the fire spread.
2. Following a firestorm of this magnitude, what precautionary steps should be taken during the rebuilding of the area?
3. Based on the known origin of this fire and the subsequent expansion the following day, what standards of practice should be put in place?

REFERENCES

Barr, Robert C., and John M. Eversole, eds. *The Fire Chief's Handbook*, 7th ed. Salt Lake City, UT: Fire Engineering, 2015.

Clayton, B., D. Day, and J. McFadden. *Wildland Firefighting*. North Highland, CA: State of California, Office of Procurement, 1987.

COMET MetEd. "Unit 2. Topographic Influences on Wildland Fire Behavior." n.d. Accessed March 22, 2023. http://stream1.cmatc.cn/pub/comet/FireWeather/S290Unit2TopographicInfluencesonWildlandFireBehavior/comet/fire/s290/unit2/print.htm#page_4.7.0.

"Effectiveness Fire Pooled Fire Resources on Wildland Fires." Report to the Los Angeles County Board of Supervisors.1989.

FIRESCOPE. "2022 Field Operations Guide (FOG) Information." 2022. Accessed March 22, 2023. https://firescope.caloes.ca.gov/fog-manual.

Gavriil, and Miltiadis Anthanasiou. 2020. "Crown Fires." In *Encyclopedia of Wildfires and Wildland-Urban Interface (WUI) Fires*, edited by Samuel L. Manzello, 1–15. Cham, Switzerland: Springer. https://doi.org/10.1007/978-3-319-51727-8_13-1.

REFERENCES CONTINUED

Geographic Area Coordination Centers. "Common Terms: National Fire Danger Rating System (NFDRS)." n.d. Accessed March 22, 2023. https://gacc.nifc.gov/rmcc/predictive/fuels_fire-danger/drgloss.htm.

Government of the Northwest Territories. "Environment and Natural Resources: Wildfire Operations." n.d. Accessed March 22, 2023. https://www.enr.gov.nt.ca/en/services/wildfire-operations/suppressing-wildland-fires.

Klinoff, Robert. *Introduction to Fire Protection and Emergency Services*, 6th ed. Burlington, MA: Jones & Bartlett Learning, 2019.

Kutztown University. "Tutorial 18: Atmospheric Stability." n.d. Accessed March 22, 2023. https://faculty.kutztown.edu/courtney/blackboard/physical/17stability/stability.html.

Lowe, Joseph D. *Wildland Firefighting Practices*. Albany, NY: Delmar Cengage Learning, 2001.

Met Office. "What Is a Temperature Inversion?" n.d. Accessed March 22, 2023. https://www.metoffice.gov.uk/weather/learn-about/weather/types-of-weather/temperature/temperature-inversion.

National Institute of Standards and Technology. "New Timeline of Deadliest California Wildfire Could Guide Lifesaving Research and Action." February 8, 2021. Accessed March 22, 2023. https://www.nist.gov/news-events/news/2021/02/new-timeline-deadliest-california-wildfire-could-guide-lifesaving-research.

National Interagency Fire Center. "National Fire News." n.d. (continually updated). Accessed November 11, 2022. http://www.nifc.gov/fire_info/nfn.htm.

National Interagency Fire Center. "Statistics." n.d. (continually updated). Accessed April 11, 2023. https://www.nifc.gov/fire-information/statistics.

National Park Service. "Wildland Fire Behavior." n.d. Last updated February 16, 2017. Accessed March 22, 2023. https://www.nps.gov/articles/wildland-fire-behavior.htm.

National Park Service. "Wildland Fire Program: Wildfires, Prescribed Fires, and Fuels." n.d. Last updated January 12, 2022. Accessed April 11, 2023. https://www.nps.gov/orgs/1965/wildfires-prescribed-fires-fuels.htm.

National Weather Service. "Fire Weather Criteria." n.d. Accessed March 22, 2023. https://www.weather.gov/gjt/firewxcriteria.

National Wildfire Coordinating Group. *Instructor Guide: S190 Unit 2: Fuels*. n.d. Accessed March 22, 2023. https://www.nwcg.gov/sites/default/files/training/docs/s-190-ig02.pdf.

National Wildfire Coordinating Group. "Glossary: Wildland Fire; Fire Line." n.d. Accessed April 11, 2023. https://www.nwcg.gov/term/glossary/wildland-fire.

Oblack, Rachelle. "Atmospheric Stability: Encouraging or Deterring Storms." n.d. ThoughtCo. Accessed March 22, 2023. https://www.thoughtco.com/atmospheric-stability-and-storms-3444170.

Perry, Donald G. *Wildland Firefighting: Fire Behavior, Tactics & Command*, 4th ed. Stillwater, OK: Fire Protection Publications, 2016.

Queen, Phillip L. *Fighting Fire in the Wildland/Urban Interface*. Bellflower, CA: Fire Publications, 1993.

Safeopedia Inc. "Fire Retardant." June 29, 2019. Accessed March 22, 2023. https://www.safeopedia.com/definition/6602/fire-retardant.

Safer States. "Toxic Flame Retardants." n.d. Accessed April 10, 2023. https://www.saferstates.org/toxic-chemicals/toxic-flame-retardants/.

Teie, William C. *Firefighter's Handbook on Wildland Firefighting: Strategy, Tactics and Safety*, 4th ed. Stillwater, OK: International Fire Service Training Association, 2016.

The New Humanitarian. "Fire Retardant Use Explodes as Worries About Water, Wildlife Risk Grow." November 27, 2017. Accessed April 10, 2023. https://deeply.thenewhumanitarian.org/water/articles/2017/11/27/fire-retardant-use-explodes-as-worries-about-water-wildlife-risk-grow.

University of Missouri, MU Extension. "Green Horizons: Know the Fire Danger Levels." 2012. Accessed March 22, 2023. http://agebb.missouri.edu/agforest/archives/v16n4/gh10.htm.

U.S. Census Bureau. "Camp Fire—2018 California Wildfires." November 2018. Accessed March 22, 2023. https://www.census.gov/topics/preparedness/events/wildfires/camp.html.

U.S. Department of Agriculture Forest Service. "Rocky Mountain Research Station." n.d. Accessed April 14, 2023. https://www.fs.usda.gov/research/rmrs.

U.S. Department of Agriculture Forest Service. "Wildland Fire Terminology." n.d. Accessed March 22, 2023. https://www.fs.usda.gov/detail/r5/fire-aviation/management/?cid=stelprdb5396693.

U.S. Department of Agriculture Forest Service. *Federal Wildland Fire Policy*. 2022. https://www.fs.usda.gov/managing-land/fire.

U.S. Fire Administration and FEMA. *The East Bay Hills Fire Oakland–Berkeley, California (October 19–22)*. Technical Report No. 060. 1991. https://www.caloes.ca.gov/wp-content/uploads/Fire-Rescue/Documents/US-Fire-Admin-East-Bay-Hills-Fire-Report.pdf.

U.S. Forest Service. *Fire Management Notes*. Boise, ID: U.S. Government Printing Office, 2006.

CHAPTER 9

Transportation Fires and Related Safety Issues

LEARNING OBJECTIVES

After studying this chapter, you will be able to:

- Discuss fire behavior and safety-related problems in transportation vehicles.
- Explain actions that firefighters take to resolve safety issues.
- Define the different classifications of transportation vehicles.
- Explain fire behavior problems one might encounter with each of the classifications of transportation vehicles.
- Explain the importance of prefire planning for each of the categories of transportation vehicles.

CASE STUDY

MacArthur Maze Fire

On April 29, 2007, a speeding gasoline tanker slammed into a guardrail near the Oakland–San Francisco Bay Bridge. The accident created an intense fire that destroyed part of the MacArthur Maze freeway complex.

The tanker was carrying 8,600 gallons of gasoline, UN classification number 1203, which is a flammable liquid classification. The impact of the collision caused a gasoline leak from the tanks, and the heat and sparking metal ignited the liquid and started a fire. The heat from the fire was estimated at over 2,000°F. Structural steel will melt at 2,700°F. The fire's intense heat, coupled with the weight of the concrete and asphalt roadbed, led to the structural collapse of the overpass.

1. What classification of fire would this be?
2. What class extinguishing agent would be best to extinguish this fire?
3. If you had access to the tanker, how would you determine the name of the product being carried in the tanks?

Introduction

This chapter examines the types of fires and related emergencies that occur in vehicles. According to the National Fire Protection Association (NFPA), an estimated 212,500 vehicle fires caused 560 civilian deaths, 1,500 civilian injuries, and $1.9 billion in property damage in the United States in 2018.

There is a wide variety of transportation vehicles classified under the U.S. Department of Transportation's (DOT) Federal Highway Administration system. For this reason, this chapter covers fires and related emergencies only in the major vehicle classifications. This includes passenger vehicle fires, truck and recreational vehicle fires, rail transportation vehicle fires, marine vehicle fires, and aircraft fires.

We will identify specific fire problems, fire behavior, and safety issues first responders face. Fire principles and suppression tactics are similar to those used for structural and wildland fires; however, some unique and dangerous fire behavior situations arise from vehicle fires. **TABLE 9-1** shows vehicle deaths, injuries, and losses.

TIP

Responders should consider the presence of hazardous materials, such as composites, plastics, and lubricants, in every transportation emergency. They may not only be in the cargo but may also be in the materials used to build the vehicle.

Passenger Vehicle Fires

Before the 1960s, most cars were made almost entirely of steel. The materials used in the interiors contained fewer combustible materials than today's vehicles, which have a far greater fire load due to the use of plastics. Fires were put out with small booster lines, and the lines provided enough water for most fires. Occasionally, flammable liquid escaped from fuel tanks, creating fires that required more water than a booster line could supply. The booster lines have been replaced with a 1 3/4 hose line that provides an adequate water supply to extinguish today's fires.

Throughout the 1960s, the use of flammable plastics increased. Plastics boosted fuel efficiency and allowed manufacturers flexibility in vehicle design. Fuel tanks were also made of polyethylene or polypropylene plastic. These tanks were lighter and more durable than steel tanks. Unfortunately, they melted after a few minutes of flame contact. This use of plastics increased the fire danger without adding safeguards.

Vehicle fires burned hotter and produced more toxic smoke. These cars also had more systems operating under pressure. Some of these systems used polyethylene or polypropylene plastic fuel supply lines pressurized from 15 to 90 pounds per square inch (psi). Because the tanks were also pressurized, a rupture sprayed flammable liquid on firefighters working suppression lines. If sufficiently heated, these systems could explode.

In the late 1960s, NFPA estimated over 400,000 vehicle fires occurred in the United States per year, 25% of which started in the vehicle's interior. To address vehicle interior fires, the National Highway Traffic Safety

TABLE 9-1 Transportation Vehicle Deaths, Injuries, and Losses

Type	Fire	Percentage	Fire Deaths	Percentage	Fire Injuries	Percentage	Fire Loss (Billion $)	Percentage
Passenger	142,345	83.3%	345	56.2%	1300	59.8%	$692	55.8%
Freight road	37,100	8.9%	104	17.7%	345	14.7%	$183	14.7%
Rail transportation	650	0.02%	6	1.0%	12	0.5%	21.0	1.7%
Water transportation	1,540	0.4%	6	1.0%	73	3.1%	$20.9	1.7%
Air transportation	200	0.1%	38	6.5%	25	1.1%	$39.7	3.2%
Heavy equipment	6,260	1.6%	7	1.2%	65	2.8%	$77.2	6.2%
Special vehicle	2,200	0.6%	2	0.3%	29	1.2%	$9.3	0.7%
Unknown	56,810	14.2%	94	16.0%	395	16.8%	$197.2	15.9%

Data from the United States Fire Administration, 2018.

Administration developed a small-scale regulatory fire test in 1969. This test evaluates the flammability of vehicle interiors and is still used today. The **National Highway Traffic Safety Administration (NHTSA)** is an agency within the DOT that works to reduce injuries, deaths, and economic losses due to motor vehicle accidents.

TIP

An average vehicle fire requires a minimum 1 3/4-in. (44.45-mm) line to supply the water because of the increased use of plastics. Plastic- or hydrocarbon-related fires burn twice as hot as wood-based fires and produce a very dense, toxic smoke. The smoke further hampers firefighting activities. Firefighters must wear self-contained breathing apparatus (SCBA) and full protective gear to safely suppress vehicle fires.

Front Bumpers

The hydraulic cylinders on front bumpers can absorb a 5-mph crash test without damage. Firefighters need to be careful when working with extrication tools around bumpers because they can explode under the intense heat from a flammable liquid fire.

Air Bags

The accidental deployment of active air bags during a vehicle rescue is another problem facing first responders. Since the 1970s, vehicles have had supplemental restraint systems in addition to seat belts. The **supplemental restraint systems (SRS)** are additional passenger restraint systems, some of which are added to the vehicle roof, dash, or roof pillar.

The **side impact curtain (SIC)**, or side impact protection system (SIPS), was introduced in the 1990s as side impact and rollover protection. This system deploys more quickly and more forcefully than the frontal restraint systems because of the shorter distance to the passenger. The **inflator** is a pressurized container that releases cool/hot pressurized gases that fill the appropriate impact curtain or bag. The pressure vessel in hybrid inflators has a stored gas pressure ranging from 3,000 psi to over 4,000 psi. Firefighters could inadvertently rupture a hidden impact inflator while removing a roof or cutting through a vehicle's interior.

Safe Work Zones

At fires along roadways, firefighters are often in more danger of being hit by another vehicle than being injured by fire. To provide some protection, fire departments create

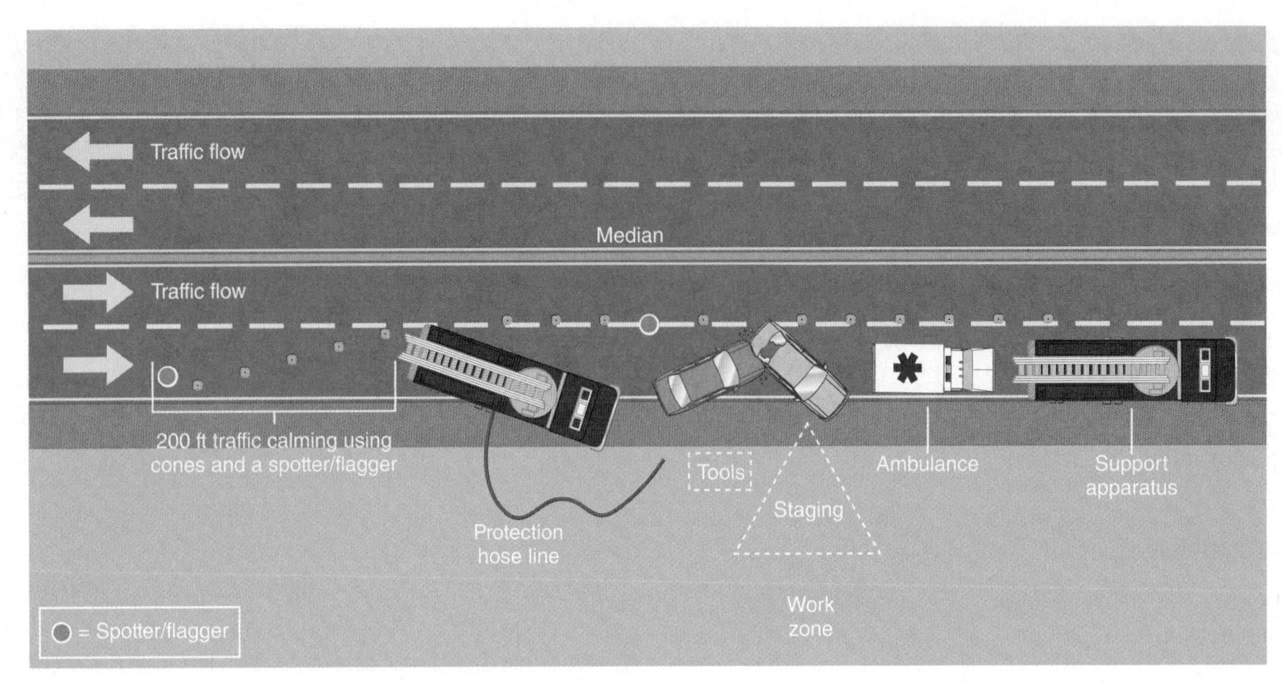

FIGURE 9-1 How a safe work zone is created on the incident scene.

FIGURE 9-2 Firefighters in full protective gear and SCBA attack a passenger vehicle fire with hand lines.

safe work areas while on roads or highways. These often involve positioning the apparatus, using warning devices to alter traffic flow, and using personnel to monitor and direct traffic flow. **FIGURE 9-1** shows how a safe work area is created at the incident scene. The fire apparatus provides a traffic barrier while cones and a spotter/flagger slow traffic. **FIGURE 9-2** shows firefighters in full protective gear and SCBA attacking a passenger vehicle fire with hand lines.

Alternative Fuel Vehicles

Today we see a large part of the public and private transportation industry using alternative fuel vehicles. An **alternative fuel vehicle** is powered by a fuel or energy source other than petroleum. These fuels include **compressed natural gas (CNG)**, liquefied petroleum gas (LPG), battery powered (electric), hydrogen, ethanol, and biodiesel; some vehicles use a combination of electricity and gasoline for power (hybrid).

CNG is a natural, nontoxic, lighter-than-air gas that rises and dissipates when released. In a vehicle, CNG is stored at high pressures in cylinders similar to SCBA cylinders. These cylinders are located in the trunk or cargo area. In a vehicle fire, firefighters must cool these as with any gas-filled cylinder.

LPG is stored in the same type of cylinder as those used for heating and cooking purposes. LPG is heavier than air, so the vapors pool and collect in low areas. **TABLE 9-2** shows the heat of combustion of commonly used fuels. CNG and LPG produce as much energy as gasoline without as much heat and toxic gases.

Electrical Vehicles

At the turn of the 20th century, the introduction of the automobile was at its inception, and manufacturers had many power sources. One of these was the steam engine. Pioneered by the Stanley twins, this practical means of power for the automobile was more refined than the newer form, the internal combustion engine. The internal combustion engine used gasoline for fuel—a product that had been widely used as a cleaning agent.

Also in production at this time were automobiles powered by lead acid batteries. These vehicles were

TABLE 9-2 Heat of Combustion of Commonly Used Fuels

Fuel	BTUs[a]/lb	kJ/kg
Hydrogen	61,600	141,800
Methane (natural gas)	24,100	55,500
Propane	21,500	50,000
Diesel fuel	20,700	48,000
Gasoline	20,000	46,800
Cooking oils	14,000	34,700
Methanol	9,900	22,680
Wood	8,500	19,700

Data from National Fire Protection Association. 2021. *Fire Protection Handbook*, 21st ed. Quincy, MA: National Fire Protection Association.

widely used in the New York City taxicab system. However, the original electric-powered vehicles did not have the desired speeds and were very costly. They did not have a long range on a single charge, and charging stations were not readily available or in production at the time. With the refinement of the internal combustion engine, powered by gasoline, the early electric vehicles (EVs) found their usefulness restricted to business where they were used for loading materials and for a short time in mass transportation.

Currently, international interest in slowing climate change through decreased carbon emissions and renewable energy has created a surge of improved battery designs. The introduction of the lithium-ion battery has propelled the manufacturing of today's electric vehicles. These batteries are linked together to form a battery bank that powers the electric motor and all of the accessories in the vehicle. The power rating of these battery packs is far greater than the 12 volts needed for an internal combustion engine. Often, lithium-ion car batteries can provide 600 to 800 volts. Doubling the voltage reduces the charging time. The specific voltage is determined by the manufacturer. The electric vehicles in **FIGURE 9-3** have high voltage and amperage, which causes these cars to be fire prone.

Lithium-Ion Batteries

A **lithium-ion battery** is made from lithium iron phosphate ($LiFePO_4$), lithium manganese oxide ($LiMn_2O_4$–), and lithium nickel manganese oxide ($Li_2Mn_3NiO_8$–). The batteries power both electric and hybrid electric vehicles. In these vehicles, the lithium-ion batteries are enclosed in a watertight, fire-resistant box made from steel, aluminum, or mixed materials. While this box keeps water out under normal conditions, natural disasters and floods do occur. If the vehicle is under water, water may leak into the battery box. Water in the box can cause the battery cells to decay. Decay can cause a cell to fail, leading to a thermal runaway event. A **thermal runaway event** is when a lithium-ion battery self-heats uncontrollably.

FIGURE 9-3 Electric vehicles at charging station.

Hydrogen Fuel Cell Electric Vehicles

Hydrogen fuel cell electric vehicles (FCEVs) are similar to EVs because they use an electric motor instead of an engine to power the wheels. However, while EVs run on batteries that must recharge, FCEVs generate their electricity onboard.

In FCEVs, hydrogen is stored in fuel tanks and transferred to the fuel cell stack. There it is combined with oxygen from the air and converted into electricity. The electricity is then stored in the vehicle battery pack. A converter changes the voltage from high-voltage direct current (DC) into low-voltage DC, which powers the electric motors.

Hydrogen is a very light chemical element, weighing only one-fifteenth as much as air. It has a wide flammable range (4% to 75%) and is one of the most flammable gases. It also has a high heat content per pound (see Table 9-2). It is colorless and odorless, and contains no carbon products, causing it to burn with a nonluminous flame. The energy released from hydrogen is three times that of gasoline without producing toxic by-products. This is one reason it was chosen for research as a fuel alternative to petroleum-based products. When released from containment, hydrogen presents both an

FIGURE 9-4 Hydrogen-powered cars and refueling station.

© Scharfsinn/Shutterstock

explosion and a fire hazard. There are over 50 hydrogen refueling stations in the United States, the majority of which are in California. **FIGURE 9-4** shows a hydrogen-powered car refueling station.

Hybrid Electric Vehicles

Hybrid electric vehicles (HEVs) are run by a variety of fuels and powering systems, but they often use an internal combustion engine and a lithium-ion battery system like EVs. Some hybrid vehicles use voltage systems that are between 100 and 600 volts.

Firefighters can usually put out a vehicle fire in a few minutes. In 2022, in Montgomery County, Maryland, however, it took firefighters more than 3 hours to put out a fire in a hybrid vehicle. It also took 4,000 gallons of water, about 40 times the water required for a car with a combustion engine. Firefighters must become familiar with these newer fueling systems because they present new safety issues and impact fire behavior. They also affect the strategies firefighters use to put these fires out. **FIGURE 9-5** shows a hybrid vehicle.

TIP

Although the standard for electrical systems in gasoline-powered cars is 12 volts, electric and hybrid electric cars have systems with voltages up to 600 volts. This boost in voltage increases the possibility that fuel released during an incident will ignite. If flammable fuels are involved, disconnect the batteries. The procedure for disconnecting the batteries should follow the manufacturer's recommendations. Keep other outside ignition sources from the area.

Strategy and Tactics for Alternative Fuel Vehicles

The first step in putting out a fire in a passenger vehicle is to examine the situation. Use the decision-making

FIGURE 9-5 Hybrid vehicle for commercial use.

© Grisha Bruev/Shutterstock

process described in earlier chapters. Gather the needed information to develop a safe strategy and tactics. The fire attack should be from the downwind side and upslope, if possible and safe.

Firefighters should prevent leaking fuels from spreading to other areas of the vehicle. Fuel can leak into low areas or outside the vehicle into storm drains. In some cases, foam can be effective at putting the fire out and containing the fuel. It can act as a vapor blanket.

When responding to a vehicle fire, firefighters should be able to recognize alternative fuel vehicles. Many vehicles and public buses use CNG or LPG. A fire attack on a vehicle powered by these fuels should be performed at a distance of at least 100 ft. This distance will vary based on the materials burning and the recommended fire procedures outlined in the Safety Data Sheet (SDS) for the materials (see Chapter 10). The tanks carrying these gases may explode.

Firefighters should use dry chemical extinguishment to prevent the spread and growth of the fire. Water is effective for reducing the heat from the fire. The use of water should be monitored to ensure it is not cooling the pressure relief valve, which would cause it to close. This would allow additional pressure to build up in the tank.

Electrical Vehicle Fires

Lithium-ion batteries are highly reactive, which makes them hazardous. When these batteries catch fire, they sometimes explode. Water used to fight these battery fires becomes contaminated. After use on such fires, the water is a hazardous material and must be disposed of by law.

In some emergencies, passengers become trapped and need the fire department to remove them from a vehicle. Firefighters must disconnect the battery before rescuing victims to ensure leaking flammable fluids do not catch fire.

Cables connect the batteries to the electric motors that power the wheels. Cutting these cables may result in injury or death because of the high voltage of the car battery. The cables extend from the battery bank in the rear to the front of the car via the undercarriage. They pass directly under the center of the driver's seat. Firefighters should take extra care when using rescue tools on these vehicles to avoid cutting the high-voltage cables.

TIP

Firefighters should avoid cutting into any electrical wire harness when possible because damage could allow an outside electrical charge to ignite the inflator.

Unfortunately, there is no effective conventional way to put out an EV fire. It typically takes many hours on scene and tens of thousands of gallons of water. Even if crews can stop a thermal runaway event, the vehicle might still catch fire again at the salvage yard.

While not ideal, the best strategy is to move the vehicle away from exposures and allow it to burn itself out, which should take about an hour. Firefighters should not pry, cut, or remove any part of the battery case to access the fire. Doing so would be very dangerous.

If the vehicle has been partially or fully submerged, it is best to assume that the vehicle's batteries have been compromised. When cleaning up after a flood or submersion in water, fire crews should transport hybrid and electric vehicles on a flatbed truck. If an EV is towed with the drive wheels on the ground, the motor can back-feed power into the batteries. Ask the tow truck driver to store the vehicle outside, at least 50 ft from all structures or other potential fuel sources, such as trees or brush, that could spread fire if the battery ignites.

Motor Homes, Buses, and Recreational Vehicles

A recreational vehicle contains living space for travel or camping. These vehicles range from small pop-up tent trailers to large buses with all the amenities of home. Cooking, bathing, and lighting pose the same hazards as in homes. The only difference is that recreational vehicles and motor homes are mobile. The fuel lines and electrical connections are subject to vehicle accidents and vibration as the vehicle moves.

Motor homes are divided into three categories: Class A, B, or C. The Class A units are built as motor homes. Class B units have been converted from vans. Class C units use the cab and motor from a vehicle and add an over-the-cab sleeping unit.

As stated earlier, these vehicles are homes on wheels and contain the same risks as stationary homes. Many motor homes have heaters, stoves, air conditioners, and appliances. To run these appliances, the vehicles use batteries, ranging from 12 to 24 volts DC, with switches to use a 115-volt alternating current (AC) system. The AC system can be run by an onboard electric generator or an electrical supply cord. These systems should have circuit breakers installed to protect against electrical overload (RV One Superstores, 2019). These ranges are manufacturer based and may vary.

Heating and cooking energy can also be supplied to the appliances by onboard LPG tanks ranging from 5 to 80 gallons, based on the manufacturer. All LPG tanks have a pressure relief valve to vent the tank at a predesignated pressure. When an LPG tank heats due to flame impingement, the pressure inside the tank increases. When this pressure is greater than the relief valve setting, the gas is vented. This venting continues until the pressure inside the tank is less than the pressure relief setting. Once the pressure is reduced, the relief valve closes. If the fire continues to heat the tank, the valve remains open and continues to expel all of the tank's contents. These flammable vapors are heavier than air and seek the lowest point while awaiting an ignition source.

Motor home interiors are furnished with the same type of furniture, carpeting, and appliances as in a house. These furnishings contain plastics, synthetic fabrics, foam rubber, and particleboard. The use of these materials increases the fire load per square foot.

The structural parts of motor homes consist of lightweight aluminum that stretches over the framing. Plywood or particleboard is used inside the walls and floors and polystyrene or polyurethane foam is added between layers of board. This combination offers little fire resistance and once ignited, there is little to stop the entire structure from being consumed and collapsing. The products of combustion are also toxic, producing formaldehydes and a variety of cyanogens. All of these can be fatal when inhaled.

Passengers in motor homes often do not use seat belts. When an accident occurs, these individuals are tossed inside the home. Firefighters may find entire families injured. **FIGURE 9-6** shows examples of Class A, B, and C motor homes.

A.

B.

C.

FIGURE 9-6 **A.** Class A motor home with extension side. **B.** Class B motor home. **C.** Class C motor home.

FIGURE 9-7 Bus fueled by CNG.

Fire Tactics and Strategy for Motor Homes, Buses, and Recreational Vehicles

There are many different kinds of buses, such as school buses, public transportation buses, group charter buses, and commercial bus lines. Fires in buses are not much different from fires in large vans or trucks. Because of the possibility of people being trapped inside the vehicle by a fuel-fed fire or vehicle accident, a quick fire attack and rescue are needed.

While most buses are powered by diesel engines, many public buses are powered with natural gas or methane. Methane is the simplest hydrocarbon. It is odorless, colorless, and tasteless and is the primary part of natural gas. It is a desirable fuel when compared to gasoline. It provides a high heat value and its combustion produces less volatile compounds compared to that of gasoline. For these reasons, many cities use CNG/LPG to fuel city buses (**FIGURE 9-7**).

Methane is lighter than air, so the fuel storage location on top of the bus is relatively safe. Any leakage will rise into the atmosphere.

Trucks

Trucks are vehicles that carry cargo. The cargo varies in size and shape as well as hazard. It can be explosive or toxic and may not always have a placard or warning label. Trucks carry goods over long distances and many have sleeping compartments. Compartments can be exposed to fire from inside (e.g., smoking) or the cab or cargo area.

Most international goods are shipped in containers. These can be easily unloaded from a ship and transferred quickly to a truck for ground transport. See Chapter 10 for information on international placard requirements.

Challenges Encountered in Truck Fires

Several potential fire and safety hazards exist with the variety of goods carried by trucks. Firefighters must use caution when extinguishing these fires until the cargo can be identified.

Even if a nonhazardous cargo is burning in a large trailer, accessing the fire can be dangerous. Many trailers have only two double doors on the rear and no side access. This means firefighters are limited in how they approach a cargo fire inside the trailer (**FIGURE 9-8**).

Saddle tanks are typically found on large trucks (**FIGURE 9-9**). These are large fuel tanks located on the side of the cab. They are usually under the driver and passenger doors. They range from 50 to 250 gallons (189.3 to 946.4 liters) each. The saddle tanks are vulnerable because they are mounted to the outside of the body frame and are exposed to damage from vehicle accidents and road debris. Leaking fuel tanks can present a serious fire problem, so flowing fuel fires must be dealt with quickly because they can spread rapidly. Such fires are a serious life hazard to firefighters working in or around flowing fuel.

The air supply tank for the braking system is located behind the fuel tank. Collision from the side could split the fuel tank and open the air brake system, which would automatically activate the braking system. This would cause the vehicle to come to an abrupt stop. The air brake system uses air to keep the brake disengaged. Releasing the air in this scenario would prevent the vehicle from proceeding and create more problems.

FIGURE 9-8 Cargo box on a tractor trailer.

FIGURE 9-9 Exposed fuel tank.

Large tractor trailers may have air skirts to reduce air resistance as the truck moves. They can hide the saddle tanks and their proximity to the battery. Knowledge of trucks and their design is helpful to firefighters. A walk-through at a truck dealership can help firefighters to locate and understand vehicle fuel and electrical systems.

Some trucks have air suspension systems. If this system fails, air pressure is released, causing the trailer to drop several inches. Firefighters rescuing a victim under a trailer with air suspension should set cribbing in place before starting any rescue work. **Cribbing** uses blocks of wood or plastic to lift a vehicle's frame off the wheels and stabilize it. **FIGURE 9-10** shows cribbing used to hold and stabilize a vehicle.

TIP

Firefighters must wear full protective clothing and SCBA when attacking and overhauling a confined cargo area to protect against toxic gases and smoke.

Truck Brake Fires

Firefighters occasionally encounter a truck with its brakes on fire. The brakes can become overheated when applied. This most often happens when a heavy vehicle is descending a steep decline. Once heated, the

FIGURE 9-10 Cribbing used to hold and stabilize a lifted vehicle.

metal brake drums can crack if cooled too quickly. Firefighters should apply fog spray in short bursts to slowly cool the drums.

Fire Tactics and Strategy for Trucks

In most cases, the cargo area will be the focus of a truck or trailer fire. Always check for placards. Do not assume that placards will be posted or that a truck has proper placards. Also, trucks have a **bill of lading**, or documents that indicate the amount and type of cargo or freight being transported. With the advancement of technology, some delivery companies now rely on handheld computer-based devices for the bill of lading information.

In the late 1980s, firefighters responded to a fire at a construction site. They were told that explosives might be on site but were not told where. Firefighters started suppression activities, unaware that a burning truck trailer (without warning placards) contained explosives (USFA, 1988). Firefighters were unaware that the truck trailer contained a mixture of ammonium nitrate and fuel oil with aluminum pellets. When the truck was exposed to fire, it exploded, killing six firefighters.

Firefighters should be aware that once a transport container is on private property, the DOT hazardous materials placards may be removed. During the size-up of the incident, firefighters should obtain as much information as possible about the materials stored on site. If such information is not available, then caution should be taken accordingly.

Fires in Railed Equipment

A fire department may have knowledge of its local building target hazards and have excellent prefire plans. Unfortunately, some of the most dangerous problems for a department are the freight trains that pass daily through the jurisdiction. Any fire department that covers an area with train tracks might respond to a rail incident. Trains, like ships, can carry large amounts of people or cargo. In some cases, cargo may be hazardous. Light rail and subway systems can have a high life hazard when involved in fire because of the large numbers of people they transport.

Railcar Construction

The U.S. DOT regulates the rail transport of hazardous goods. The Federal Railroad Administration (FRA) enforces the regulations. The Association of American Railroads (AAR) oversees the design and construction of the containers that transport hazardous materials (AAR, 2022).

Rail transport is a business with special procedures, regulations, and equipment. Firefighters should visit and preplan any rail facilities in their jurisdiction. Fire departments with rail traffic running through their area should arrange drills and seminars with rail officials.

Locomotives

Locomotives are the power for the train. The diesel-electric locomotive is a generator that creates electricity to power the train. Cross-country trains carry up to 5,000 gallons (18.93 kL) of fuel for the engines. A leaking tank creates a fire problem.

There is a chance of fire in the engine itself. Most of the hazard is the electricity from the generators driven by the engines. Newer locomotives control the electrical power for the train with switches and circuit breakers.

Placards

The size and weight of trains also make them dangerous. These provide inertia energy. Once the train is in motion, it requires a long distance before it can stop. Trains have a **consist** or **waybill**, which are documents that identify hazardous cargo onboard. The train's engineer usually keeps these documents in the cab.

Boxcars

Boxcars are railroad cars that are enclosed and used to carry general freight. They get their name from their appearance because they look like boxes with wheels. They carry a variety of goods that may or may not be hazardous. These goods have varying levels of flammability. Boxcars protect the cargo from weather and contamination. They are made of wood, which adds fuel if the car is on fire. Some boxcars may contain a refrigeration system. The compressor is electric and run by a diesel generator attached to the boxcar. The fuel for the generator is stored in a fuel tank attached to the car.

Flatcars

Flatcars are railcars that are open to the elements. They come in different sizes depending on the type of cargo they carry. Some flatcars are made of wood, so the main fire concern is the wooden car and the cargo.

Intermodal Railcars

Intermodal railcars are flatcars with an intermodal container. The container allows it to be transferred quickly between a ship, truck, or rail carrier, without any handling of the freight itself when changing modes.

FIGURE 9-11 An IMO 101 tank. Like the totes, these are bulk tanks that can carry a large quantity of product. They are placed on ships and then delivered locally by a truck, although trains can also be used.

Courtesy of UBH International Ltd.

For example, an intermodal container could be shipped by sea and then transferred by crane to a truck. It could be loaded back onto a truck for delivery without ever off-loading the product.

The hazards with intermodal equipment relate to the cargo. The cargo can contain hazardous materials, such as flammable or pressurized gases, greases, and liquids. See the placards and labeling requirements in **FIGURE 9-11**. An intermodal tank is commonly referred to as an IM or IMO.

Gondola Cars

Gondola cars are railcars that have flat bottoms and four walls. They may have a cover for the cargo onboard. Some are made of wood, but most often they are made of steel. Those constructed of wood can be a fire problem with some goods, but in most cases the only fire concern is the hazard of the cargo itself.

Hopper Cars

Hopper cars are a type of railroad freight car that is generally made of metal with sides and ends that are fixed. These cars transport dry bulk materials, such as fertilizers, chemicals, salt, flour, and grains. The main hazard of a hopper car is the cargo itself.

Passenger Railcars

Passenger railcars are railcars that carry people. They may also provide specific services for passengers, such as riding, sleeping, dining, and storage. Hazards with passenger cars begin with the number of people on board. Unlike freight trains, which have a small crew, passenger trains carry several hundred people. Older passenger cars have flammable interiors and are subject to fires from smoking and other ordinary combustibles.

Passenger cars also have cooling systems, air-conditioning systems, and electrical systems. All may cause a fire.

Tank Railcars

Tank railcars are tanks mounted on railroad frames with wheels. They are designed to transport liquid products and vary widely in size and capacity. Even though tank cars can carry products that are as safe as water, a burning tank car fueled by a flammable liquid is common. Tank cars can carry a few hundred gallons to as much as 45,000 gallons (over 170 kL) of product. They may have one or more compartments. They may be nonpressurized and carry gasoline or pressurized and carry LPG. Tank cars may also be connected by piping, thus forming a tank train. Further, several 1-ton containers of chlorine are commonly transported on a flat car, called a multiunit tank car.

The product and the type of tank car play an important role in determining the fire suppression strategy and tactics. For example, if flame is touching a railcar containing LPG and there is not enough water to reduce the internal tank pressure, firefighters should evacuate. They should take a defensive or nonattack mode. **FIGURE 9-12** shows an example of a defensive approach to a tank car fire.

Electric Locomotives

Electric locomotives run on electricity. They use a **third rail** or overhead wire that carries from 25,000 to 50,000 volts. Firefighters must ensure that no one touches or crosses over the third rail because that person could get electrocuted.

Subway Rail Vehicles

A subway fire can be disastrous. In an urban area, hundreds of people could find themselves trapped on an underground train. The intense heat and dense smoke in a confined space pose a serious life threat to passengers and crew members. Tunnels are lit with small emergency lights. If the third rail is energized, there is a serious life threat to firefighters working in the tunnel.

Reaching these fires is challenging, and in some cases they are difficult to extinguish. Hose lines must be stretched by hand unless the subway has a built-in standpipe system. Breathing equipment is always needed, and supplying fresh air bottles complicates an already serious firefight.

FIGURE 9-12 A defensive approach using unstaffed hose streams on a tank car.

Fire Tactics and Strategy for Railed Transportation Equipment

Fires and other emergencies in railed vehicles offer unique fire behavior and safety problems. They combine high-voltage electrical systems with diesel fuel/electric engines to run the train. Large amounts of fuel are needed because many trains travel long distances.

Rail systems have many fire and safety requirements. They haul a wide variety of cargo and have many types of vehicles and units. Cargo can contain hazardous materials that, if mixed, can explode. Firefighters should identify the cargo and review the recommendations for evacuation.

If firefighters can fight the fire safely, they should approach downwind and carry enough water. Fire crews should review attack recommendations for tank cars carrying flammable gases. These cars are prone to boiling liquid/expanding vapor explosions (BLEVEs) and require special precautions. Firefighters must wear protective equipment with SCBA.

Aircraft

Fires and safety issues on aircraft differ from fires on other transport equipment and structure fires. The differences are the speed at which these fires develop and the intensity of the heat. According to the National Transportation Safety Board, speed is of concern because 80% of aviation accidents happen due to human error. Takeoff and landing are considered the most dangerous time periods. There is also no escape route when the aircraft is in the air. Once on the ground, passengers find themselves in a long, narrow tube with multiple exits. These exits are indicated in the preflight instructions that the flight attendants give passengers. **FIGURE 9-13** shows a fire attack with aircraft firefighting trucks.

FIGURE 9-13 Aircraft firefighting trucks make a coordinated attack.

The **Federal Aviation Administration (FAA)** is an agency of the DOT that oversees all aspects of civil aviation in the United States. The FAA controls the amount and type of firefighting systems at airports using an indexing system. Under federal regulations, aircraft are sorted by length into indexes. They range from Index A, where the aircraft is less than 90 ft (27.43 m), to Index E, where the aircraft are longer than 200 ft (60.96 m). The FAA requires fire protection equipment based on the frequency of aircraft landings, type of aircraft using the runways, and the aircraft index. The regulations can be found in the *Code of Federal Regulations*, title 14, part 139 and a recent update by the FAA in advisory circular 150/5210-6B. Responders should know that each plane must have an air bill for each flight. The **air bill** identifies hazardous materials shipping by air. The captain is responsible for these papers, which can be found in the cockpit.

Aircraft Fuel

Aircraft fuels are identified by their ease of ignition. They have a wide range of flammability and are subject to changing altitudes, temperatures, outgassing of dissolved oxygen, and movement of the fuel.

There are a variety of jet fuels, but the primary one used is Jet A fuel. It is a kerosene-grade fuel, which, because it has a higher ignition temperature than gasoline, is considered relatively safe during fueling. While its ignition temperature is higher than for gasoline, once ignited, this kerosene fuel is just as dangerous as gasoline.

Aqueous film-forming foam (AFFF) is particularly suited to fires in aircraft fuel spills. It readily supplies a surface film over kerosene because of its fluidity. AFFF flows smoothly and helps to cover the fuel spill. This type of foam is compatible with dry chemical agents, which also may be needed on aircraft fires. Firefighters must be aware that AFFFs drain rapidly and provide less burnback resistance compared to protein-based foams.

Aircraft fuels may also be released as a mist after an aircraft accident. The fuel mist is exposed to hot metal, disrupted electrical circuits, and other ignition sources. If the fuel is in a mist form, firefighters must prepare for an explosion or rapidly spreading fire. In some cases, foam can be used to suppress vapors and provide an escape for passengers. To date, there are no new developments in anti-misting additives.

Another concern with aviation fuels is the need to make sure the aircraft is electrically grounded. If the aircraft is not grounded, a static spark could ignite available vapors. Most airport runways have electrical grounding posts to prevent explosions from sparks.

Hydraulic Systems and Fluids

Like other vehicles, aircraft have hydraulic systems. This means that hydraulic fluids are operating under pressure. Always approach hydraulic cylinders with extreme caution. When released, fluid under pressure will spray as a mist. An ignition source can light the mist (vapor cloud), immediately resulting in fire or an explosion.

Oxygen Systems

Many jets have oxygen/air systems that activate if the cabin pressure drops. These systems can increase the flammability of materials in the passenger compartment, including the passengers' clothing. Firefighters should know how these systems operate and how to shut them down.

Electrical Systems

Most jets have special electrical units working at 24 volts. These systems provide energy for several of the electrical power-driven motors. These motors drive the wings, flaps, rudder, and other aircraft systems. Some jets have water-powered systems as backup for the aircraft control systems. Aircraft fire and safety drills at an airport can help firefighters to understand the aircraft systems.

De-icing/Anti-icing Fluids

Anti-icing fluids keep ice off the wings and tail portion of the aircraft. These fluids are a combination of ethylene glycol, propylene glycol, alcohol, and glycerin. Some aircraft engines may also have alcohol/water systems. While not as great a hazard as other aircraft fuels, alcohol will burn with an almost invisible blue flame and requires great amounts of water to dilute the fuel. In most aviation settings, the amounts of alcohol are limited.

Pressurized Cylinders

An aircraft has different pressurized cylinders. While oxygen cylinders have pressure relief valves, others may not. Pressurized cylinders are used for hydraulic fluids, fire extinguishing systems, rain repellant systems, pneumatic systems, and other compressed gases. All of these cylinders have been known to explode during aircraft firefighting operations.

Tire, Rim, and Wheel Assemblies

Large aircraft tires may have pressures in excess of 200 psi. They are usually filled with nitrogen, an inert gas, to protect the tire from the heat generated during takeoffs and landings.

Because of the high pressures, these tires can explode with the force of a bomb when overpressurized, overheated, or damaged. Firefighters must use caution when close to these tires while fighting a fire.

Escape Slides

The larger jets use escape slides, which deploy and inflate automatically within seconds from the time the emergency exit opens. Firefighters arriving at a crash scene must assist passengers as they use these chutes and move them quickly away from the aircraft.

Military Aircraft

Firefighters responding to military aircraft fires must be aware that explosives are used in some military aircraft. These devices eject the pilot seat and canopy with great force. Firefighters should take a prefire tour of a military base to familiarize themselves with systems found on military aircraft.

Military aircraft have a wide variety of armaments in and on the aircraft. These aircraft may also transport hazardous materials. Special training and experience are crucial in dealing with military aircraft fires and incidents. This training is beneficial should a military aircraft land at a civilian airport.

Crash Scene Security

Aircraft emergencies attract the attention of the media and public almost immediately. First responders must establish an area around the crash for emergency operations. This area also preserves evidence at the scene. There are two perimeter zones around an aircraft crash. The inner security perimeter is at least 300 ft (91.44 m) from the outer limits of the wreckage and debris path. This provides room for first responders to perform firefighting, rescue, and medical functions. They can also do any crime scene or crash scene investigation work required. This inner security zone is surrounded by an outer security zone that contains spectators and the press. The second zone should extend from the inner zone to at least another 300 ft (91.44 m). When first established, both areas must be large enough to open the space for all activities and keep the public safe. The areas can be narrowed later when the incident is under control.

Regulations for Aircraft

The FAA has one set of regulations that are adhered to worldwide. It has maintained that the most critical issue associated with fire safety is the rate of heat release. As the rate increases, the time for a safe exit from the vehicle decreases rapidly. The FAA subdivides aircraft incidents into three types of fire scenarios. The first fire scenario includes fires that occur after an air crash. The aircraft fuel ignites, resulting in a fire in the fuselage or passenger area. The second and third fire situations involve fires while an aircraft is in flight. These fires may occur in the passenger cabin or cockpit. They can spread rapidly once detected. The third situation involves fire that remains undetected for a while, such as when a fire is in a hidden area (e.g., a restroom or cargo space).

The FAA routinely conducts tests using actual aircraft for a variety of fast, in-flight, and post-crash fire scenarios. In each test, scientists assess the release of heat, smoke, and gases. The results show that improved fire-resistant materials are needed. The criterion was that the material in the passenger compartment and cockpit would allow passengers at least 5 minutes to escape from the aircraft. Passenger exit could only start after the aircraft landed either by crashing or pilot action. As a result of these tests, some aircraft now have heat sensors with extinguishing systems in the cargo areas.

Aircraft Engines (On the Ground)

Fires in aircraft engines on the ground are generally not serious because the fire attack can be made by ground units. There is a difference, however, depending on whether the fire is in a piston-driven engine or a turbine engine. If the fire is within the engine **nacelle**, the covering or cowling that protects the aircraft's engine, fire crews should use the onboard extinguishing system. If this fails, firefighters can use hose lines with fog nozzles to provide enough cooling to control the fire.

Jet (Turbine) Powered Aircraft Engine Fires

Fires in the combustion chamber of jet engines are best controlled if the engine can be kept from turning over. Observe extreme caution, however, when working around a running jet engine. Personnel should never stand within 25 ft (7.6 m) of the front or side, or to the rear of the engine outlets. The suction of some engines is strong enough to draw in a 200-lb (90.7-kg) person.

Personnel should also stand clear of the turbine. In the event of an engine coming apart, this area will be the path of flying metal parts. People should stay at least 150 ft (45.7 m) from the exhaust outlet. Exhaust temperatures can reach 3,000°F (1,648.9°C) at the outlet.

Fires that are outside the combustion chamber but within the engine nacelle should be put out by the pilot or crew using the aircraft's built-in extinguishing system. If this fails, the crew should try carbon dioxide or dry chemicals. These extinguishers should not be used if magnesium or titanium metals are involved. Under these circumstances, it is best to allow the fire to burn itself out. Foam or water spray should be used to keep the nacelle and surrounding exposed parts of the aircraft cool.

Wheel Fires

Potential wheel fires should be approached with caution. The fire apparatus should be parked within effective firefighting distance from the aircraft but never to the side of the aircraft or in line with the wheel's axle. If the tire should explode, tire debris is thrown to the sides but not to the front or rear. If it is necessary for firefighters to approach the aircraft, it should be done from the front or rear.

Smoke around the brake drums and a tire does not mean that the wheel is on fire because overheating of brakes occurs often. If the brakes are overheated, they should be allowed to cool by air using only a smoke ejector to help reduce the cooling time. Carbon dioxide extinguishers are frequently used for cooling wheels and can be used if there are actual flames in the wheel. However, a dry chemical extinguisher is preferred because it is less likely to chill the metal in the wheel parts.

Water should be used only when a dry chemical extinguisher is not available; however, rapid reduction of heat on the wheels may result in an explosion. Firefighters should protect themselves from a possible explosion by using the fire apparatus as a shield. The water should be applied in a fine spray and in short bursts of 5 to 10 seconds. At least 30 seconds should elapse between bursts. Water should be used only as long as flames are visible.

Strategy and Tactics for Aircraft

Aircraft fires and accidents present unique problems because passengers are inside an aluminum tube surrounded by flammable fuel. Once released, the fuel can leak into the passenger compartment. Fire can engulf the entire aircraft in flames.

The size-up of aircraft fires on land must consider the passengers' escape and extinguishment of the fire. If the fire cannot be put out quickly, there must be an escape route. Fire crews can confine the fire using an aircraft rapid intervention vehicle. An aircraft **rapid intervention vehicle (RIV)** is a large pumper used for aircraft firefighting. It carries large amounts of water and foam. Some of these vehicles have two engines. One engine propels the vehicle quickly, while the second engine powers the pump. This allows for a fast-moving attack while pumping foam or water.

The RIVs have a monitor (foam application device), which is operated from the cab. This lets firefighters discharge foam while advancing on the fire (**FIGURE 9-14**).

FIGURE 9-14 Aircraft rescue firefighting vehicle with firefighter in specialized protective clothing.

The foam provides a cooling and smothering action and blankets the vapors from the leaking fuel to prevent ignition. This application of foam can confine the fire to allow passengers to escape. Some units carry large, dry powder extinguishers for use on magnesium and other combustible metal fires.

Specialized training and equipment are needed for aircraft firefighting. Special tools penetrate the aircraft to apply foam inside the fuselage when needed. Firefighters must wear protective clothing that can withstand the high temperatures of aviation fuels. Technology is offering improvements over the foil/silver reflective personal protective equipment currently being used. This material tends to crack over time.

Boats

There is a possibility of a small boat fire in any fire department's jurisdiction that has a body of water. In most cases, small boats are brought to the water on a trailer, but some boats may be docked in a boat marina. Boat marinas will vary in size from those docking only a few boats to those storing a large number. The method of storage consists of one or more floating docks extending into the body of water.

In the marina, small boats are protected from the weather during nonuse by canvas covers. These large covers present potential fire problems because fire can rapidly spread if one boat catches fire. Firefighters must preplan for this type of runaway fire to reduce the risk of a fire destroying all of the boats in the marina.

Getting water to a boat fire may be a problem if the boat is tied up to a dock far from shoreline and water is unavailable. During the prefire planning, firefighters should develop alternative methods to obtain water.

Larger boat marinas provide fire hydrants adjacent to the walkways. The hydrants should be tested periodically. At other marinas, firefighters should draft the supply of hose lines. A **draft** is the process of moving or drawing water away from a static source of water by a pump. If engines cannot be positioned close enough to draft, a siphon ejector may be used.

Fires in the cabin or superstructure area should be attacked using foam or water spray streams. A single 1 1/2-in. (38-mm) or 1 3/4-in. (44-mm) line will normally be sufficient for a fire in a small boat. Two lines will be required if the fire is large or in a small yacht. A fire in a small yacht should be attacked from both sides to avoid pushing the fire onto an adjacent docked boat.

Most fires on small boats start near the engine or in the bottom of the boat. The **bilge** is the bottom area of a boat or ship. This area can trap flammable liquids or vapors, which can catch on fire when the engine starts. The fire usually starts with an explosion. A carbon dioxide or dry chemical extinguisher is very effective if the fire is well contained under the floor decking and has not progressed.

Water in the form of fog or spray streams is also effective; however, it could sink the boat and spread burning fuel onto the adjacent waterway. If too much water is used, crews may need to pump out (or dewater) the boat to keep it from sinking. Firefighters should always think about protecting the boats docked nearby.

Regardless of the methods used to extinguish the initial fire, there will probably be burning materials in the furnishings inside the boat. These items need to be thoroughly overhauled. There is also the possibility of unburned gasoline in the bilge, which will need to be siphoned into a container. It may be necessary to request a flammable-liquid vacuum truck if the spill is extensive. Fire crews need to disconnect the wiring from the batteries and shorelines before beginning their work.

Occasionally, a small boat on fire will be set adrift before the arrival of the fire department. This is to keep it from damaging adjacent boats. In such cases, the boat should be secured before attacking the fire with hose lines. If this is not done, hose stream application may drive it into adjacent exposures.

Ship Fires

In the United States, ship-related regulations are established by the federal government and enforced by the U.S. Coast Guard (USCG). There may also be local and state laws, but in most cases, these are per federal regulations.

For foreign waters, there is a treaty called the **Safety of Life at Sea (SOLAS)**. This treaty contains the minimum fire and safety standards for ships on international voyages. Because the United States signed the SOLAS treaty, U.S. ships using foreign waters must comply with its provisions.

Another agency concerned with fires and safety regulations at sea is the International Maritime Organization. The **International Maritime Organization (IMO)** is an agency of the United Nations that specializes in maritime issues. It must uphold the SOLAS treaty and has issued several fire test methods (IMO resolutions) cited in the SOLAS treaty.

Fires in the Hold of a Ship

Ship fires are greatly affected by the conductivity of the steel construction. The transfer of heat by the steel can spread fire among compartments, making fires more difficult to control than most structure fires. Other problems include the generation of smoke and heat. Ship cargo holds are confined spaces where smoke and heat can build. Without proper ventilation, this combination of intense heat and thick, toxic smoke makes putting out ship fires dangerous and difficult for firefighters. Ships have a **bulkhead**, which is the main wall or supporting structure. Some bulkheads are watertight and fire resistive. They can be built to prevent the spread of water and fire in case of damage to the ship.

As with all fires, size-up requires the first-arriving firefighter to find out what is burning and the location and extent of the fire. Cargo ships have a **dangerous cargo manifest**, or documents used to identify the hazardous cargo of a marine vessel. Firefighters must find out if any hazardous materials are on board. If so, they need to find in what cargo holds they are stored and what hazards they present.

Firefighters must determine if the ship has a carbon dioxide or steam fire extinguishing system. They must also find out if the system is operable. The system can be used after a review of the cargo manifest, the extent of the fire, and the ability of the system to contain the fire. Always check the compatibility of the extinguishing system with the materials on fire.

If available and compatible with the cargo, foam of medium or high expansion (20 to 1,000 times) may be used. It can fill holds of ships where a fire is difficult or impossible to reach. Foam reduces the fire's access to convection currents and air. The water content of the foam also cools and lessens the available oxygen by steam displacement.

Remember that oxidizers provide oxygen when heated. Therefore, do not put out a fire involving an oxidizer by filling the hold with carbon dioxide or excluding oxygen by introducing steam. Students should review the report of the fire and explosion of the *S.S. Grand Camp* in Texas City, Texas, in 1947.

One problem with using water to put out fires on a ship is the possibility of capsizing the ship. The cargo may absorb the added water and upset the stability of the ship. The ship's officer is responsible for the ship's stability. It is vital for firefighters to keep in close communication with the ship's officers if the stability of the ship is threatened.

Tanker Ships

Tanker ships present a fire problem that differs from the cargo ship because cargo areas contain flammable liquids. Firefighters suffocate these fires. Generally, one of two types of fire extinguishing systems is installed. The two systems are a carbon dioxide or a steam system. Both work on flammable liquids if all cargo tank openings can be closed to reduce the inflow of additional air. Once the ship's fire extinguishing system is in service, firefighters need to verify that the system is working properly.

For tanker ship fires, firefighters should know what material is on fire. Foam is often the only effective agent available. Fire crews can use water sparingly to precool the liquid before applying the foam. If materials such as asphalt or tar are involved in the fire, crews should not apply foam. This could result in a steam explosion.

Fire Tactics and Strategy for Boats and Ships

Marine fires have some of the same hazards as other vehicles, which complicate firefighting activities. Like aircraft, safe evacuation is crucial. Boat and ship fires can move quickly and produce a great deal of deadly smoke. Ships carry a wide variety of cargo, some of which may be hazardous. The electrical, fuel, and ventilation systems can malfunction, creating a fire or other safety problems. Just as in a steel or concrete structure, radio communications can be challenging due to the amount of metal the ship holds.

Small boats are made of wood, fiberglass, or aluminum, all of which can burn or create safety problems for those onboard. Larger ships are made of steel and have large cargo spaces that are poorly ventilated. The steel transmits heat, which can spread the fire.

In sizing up these fires, firefighters should look at the cargo manifest. The manifest reveals the nature and problems of the cargo and the type and condition of the onboard firefighting system, if one is available. Firefighters need protective clothing with SCBA and an extra supply of air bottles for a large fire. Prefire planning and close coordination with the harbormaster are recommended to prepare firefighters for marine fires. In extreme situations, it may be better to sink the boat or ship in shallow water and refloat it after the fire is out. **FIGURE 9-15** shows a cargo ship in port.

FIGURE 9-15 A cargo ship in port.

CASE STUDY
CONCLUSION

1. **What classification of fire would this be?**

   This fire involved a flammable liquid and would be a Class B fire.

2. **What class extinguishing agent would be best to extinguish this fire?**

   For this fire, a Class B foam would be used because it is carbon repellent and would create a foam barrier over the flammable liquid. This would prevent vapors from mixing with oxygen and reigniting.

3. **If you had access to the tanker, how would you determine the name of the product being carried in the tanks?**

   If you had access to the tanker, you would check the bill of lading to determine the name of the product.

WRAP-UP

SUMMARY

- The National Highway Traffic Safety Administration developed a small-scale regulatory fire test in 1969 to test the flammability of vehicle interiors.
- An alternative fuel vehicle is powered by a fuel or energy source other than petroleum.
  - These fuels include compressed natural gas (CNG), liquefied petroleum gas (LPG), battery powered (electrical), hydrogen, ethanol, and biodiesel, and a combination of electricity and gasoline-powered vehicles (hybrid).
- To extinguish a fire in a passenger vehicle, the fire attack should be from the downwind side and upslope.
- Motor homes are divided into three categories: Class A, B, or C. Motor homes have the same fire hazards as stationary homes.
- Truck compartments can be exposed to fire from inside (e.g., smoking) or the cab or cargo area.
- Railed equipment includes railcars, boxcars, flatcars, intermodal railcars, gondola cars, hopper cars, passenger railcars, tank railcars, and subway rail vehicles.
- The differences between fires and safety issues on aircraft and other vehicle fires are the speed at which these fires develop and the intensity of the heat generated.
- Firefighters must preplan for runaway boat fires to reduce the risk of destruction of all boats in the marina.
- In the United States, ship-related regulations are established by the federal government and enforced by the U.S. Coast Guard (USCG). The International Maritime Organization (IMO) is another organization concerned with fires and safety regulations at sea.
- Ship fires are greatly affected by the conductivity of the steel construction.
- Cargo ships have a dangerous cargo manifest, or documents used to identify the hazardous cargo of a marine vessel.

KEY TERMS

air bill A document used to identify the hazardous materials shipping by air.

alternative fuel vehicle A vehicle powered by a fuel or energy source other than petroleum.

bilge The bottom area of a boat or ship.

bill of lading Documents that indicate the amount and type of cargo or freight being transported.

boxcar A railroad car that is enclosed and used to carry general freight.

bulkhead A main wall or supporting structure made of steel.

compressed natural gas (CNG) Natural gas that is compressed and contained within a pressurized container.

consist Document used to identify the hazardous cargo of a rail transportation vehicle; also called a waybill.

cribbing Blocks of wood or prefabricated plastic used to lift a vehicle's frame off the wheels and stabilize it.

dangerous cargo manifest Documents used to identify the hazardous cargo of a marine vessel.

draft The process of moving or drawing water away from a static source of water by a pump.

Federal Aviation Administration (FAA) An agency of the DOT that oversees all aspects of civil aviation in the United States.

flatcar A railcar that is open to the elements.

gondola car A railcar with a flat bottom and four walls that can carry a variety of goods.

hopper car A type of railroad freight car used to transport loose bulk commodities.

hybrid electric vehicle (HEV) Vehicle run by a variety of fuels and powering systems; often powered by an internal combustion engine and a lithium-ion battery.

inflator A pressurized container that releases cool/hot pressurized gases that fill the appropriate impact curtain or bag.

intermodal railcar A flatcar with an intermodal container that can use multiple modes of transportation (rail, ship, and truck), without any handling of the freight itself when changing modes.

International Maritime Organization (IMO) An agency of the United Nations that specializes in maritime issues; responsible for maintaining the SOLAS treaty.

lithium-ion battery A rechargeable battery made from lithium iron phosphate, lithium manganese oxide, and lithium nickel manganese oxide.

nacelle The covering or cowling that protects the engine of the aircraft.

National Highway Traffic Safety Administration (NHTSA) An agency within the DOT that works to reduce injuries, deaths, and economic losses due to motor vehicle accidents; developed a small-scale regulatory fire test for vehicle interiors that is still used.

passenger railcar Railcar that carries people.

rapid intervention vehicle (RIV) A large pumper used for aircraft firefighting.

Safety of Life at Sea (SOLAS) An international treaty containing the minimum fire and safety standards for ships on international voyages.

side impact curtain (SIC) A vehicle passenger restraint system that deploys more quickly and forcefully from the side because of the shorter distance to the passenger.

supplemental restraint system (SRS) An additional passenger restraint system, which may be added to the vehicle roof, dash, or roof pillar.

tank railcar A tank mounted on a railroad frame with wheels designed to transport a variety of liquid products.

thermal runaway event An event in which a lithium-ion battery self-heats uncontrollably.

third rail The rail used in subways and the wire for overhead electric train power systems to provide electricity to the train.

REVIEW QUESTIONS

1. Why are transportation fires a greater concern today than in the past?
2. Why are today's passenger car fires more difficult to extinguish?
3. What types of fuels may firefighters encounter in hybrid electric vehicles?
4. What are some of the fire problems that may be encountered with fires involving motor homes, buses, and recreational vehicles?
5. Why are ship fires difficult to fight? Discuss some of the fire behavior problems that may be encountered on a ship.

DISCUSSION QUESTIONS

1. What extinguishing methods for vehicle fires are available for firefighters today?
2. What are four types of cargo manifests? Identify the form of transportation to which they apply.
3. What are the serious issues that separate hydrocarbon-fueled vehicle fires from those in electric vehicles?
4. What is the best method of water application on a pressurized rail tank car?

APPLYING THE CONCEPTS

A train of 100 cars is proceeding down a grade. At a sharp curve, the train derails and 30 cars plunge down a ravine toward an active river. Three tank cars and two hopper cars are found in the water.

At the derailment site, a fire has started. The fire is impinging on four high-pressure tank cars and six boxcars. The engineer and conductor are at the derailment site when you arrive and are cooperative.

1. What are the first issues of operational importance that you would address?
2. What information can the conductor provide for you that will help firefighters devise a suppression strategy?
3. What additional resources/participating partners would need to be contacted?

REFERENCES

Ahrens, Marty. *Vehicle Fires.* National Fire Protection Association. March 2020. Accessed March 24, 2023. https://www.nfpa.org/-/media/Files/News-and-Research/Fire-statistics-and-reports/US-Fire-Problem/osvehiclefires.pdf.

All About Automotive. "Supplemental Restraint System." February 2, 2018. Accessed March 24, 2023. https://allaboutautomotive.com/2018/02/02/supplemental-restraint-system/.

Association of American Railroads. *Hazardous Materials Regulations of the Department of Transportation.* Tariff No. BOE-6000 B. Washington, DC: Bureau of Explosives, 1981.

Association of American Railroads. "Hazardous Materials & Freight Rail Tank Car Regulations." n.d. Accessed March 24, 2023. https://www.aar.org/article/freight-rail-hazmat-regulations/.

Bensen, Jackie. "Burning Hybrid SUV Took Fire Crews 3 Hours to Extinguish." NBC4 Washington. May 4, 2022. Accessed March 28, 2023. https://www.nbcwashington.com/news/local/burning-hybrid-suv-took-fire-crews-3-hours-to-extinguish/3042420/.

Burnett, John. "Fire Safety Concerns for Rail Rapid Transit Systems," *Fire Safety Journal* 8 no. 1 (1998): 103–106.

Federal Emergency Management Agency and Charles Jennings. *Gasoline Tanker Incidents in Chicago, Illinois and Fairfax County, Virginia: Case Studies in Hazardous Materials Planning,* Technical Report 032. 1990. https://www.interfire.org/res_file/pdf/Tr-032.pdf.

Gesell, Laurence E. *The Administration of Public Airports,* 5th ed. Chandler, AZ: Coast Aire Publications, 2007.

Gustin, Bill. "New Fire Tactics for New Car Fires," *Fire Engineering* 149 no. 5 (March 1996): 62–67. https://www.fireengineering.com/firefighting/new-fire-tactics-for-new-car-fires/

Gustin, Bill. "Expect the Unexpected in Vehicle Fires," *Fire Engineering* 150 no. 8 (August 1997): 87–94. https://www.fireengineering.com/firefighting/expect-the-unexpected-in-vehicle-fires/.

National Fire Protection Association. *The Research Advisory Council on Fire and Transportation Vehicles.* Quincy, MA: National Fire Protection Research Foundation, 2004.

National Fire Protection Association. *Fire Protection Handbook,* 20th ed. Quincy, MA: National Fire Protection Association, 2008.

National Highway Traffic Safety Administration. *Test Procedures for Evaluating Flammability of Interior Materials.* January 25–27, 2017. Accessed March 24, 2023. https://www.nhtsa.gov/sites/nhtsa.gov/files/documents/2017saebhennessey.pdf.

RV One Superstores. "RV Electrical Systems Explained." December 6, 2019. Accessed April 14, 2023. https://rvonestaugustine.rvone.com/2019/12/06/rv-electrical-systems-explained/.

Shaw, Ron. "New Auto Safety Technology, Part 1." *Fire Engineering* 157 no. 5 (May 2004): 78–82.

Smith, D. A. "Some Aspects of Fire Safety Design on Railways." *Fire and Materials* 8 no. 1 (March 1984): 6–9. https://onlinelibrary.wiley.com/doi/abs/10.1002/fam.810080103.

Stang, John. "Washington Wants to Plug in to the Next Thing in Fuel: Hydrogen." Crosscut. May 18, 2022. Accessed March 25, 2023. https://crosscut.com/environment/2022/05/washington-wants-plug-next-thing-fuel-hydrogen.

St. Louis, Ed, and Steve Wilder. "Train Disasters Test the Fire Service: Tragedy on the City of New Orleans," *Fire Engineering* 156 no. 6 (June 1999): 61–71.

U.S. Coast Guard. *Amendments to the 1994 Safety at Sea (SOLAS) Treaty.* Navigation and Vessel Inspection Circular 4-04, 2004.

U.S. Department of Energy. "Alternative Fuels Data Center." n.d. Accessed March 24, 2023. https://afdc.energy.gov/fuels/.

U.S. Department of Transportation. *Safety First.* n.d. Accessed March 24, 2023. https://www.transportation.gov/briefing-room/safetyfirst/national-highway-traffic-safety-administration.

U.S. Department of Transportation. *Emergency Response Guidebook.* Washington, DC: Government Printing Office, 2020.

U.S. Environmental Protection Agency. "Green Vehicle Guide: A Glimpse into Hydrogen & Transportation." n.d. Accessed March 24, 2023. https://www.epa.gov/greenvehicles/glimpse-hydrogen-transportation.

U.S. Fire Administration. *Transportation Fires, The Research Advisory Council on Fire and Transportation Vehicles.* Quincy, MA: National Fire Protection Research Foundation, 2004. https://www.nhtsa.gov/press-releases/early-estimate-2021-traffic-fatalities.

CHAPTER

10

Hazardous Materials and Warning Systems

LEARNING OBJECTIVES

After studying this chapter, you will be able to:

- Describe the U.S. Department of Transportation hazardous materials warning system, including its advantages and disadvantages.
- Explain the types of information available to first responders in the *Emergency Response Guidebook*.
- Describe the National Fire Protection Association standard 704 warning system, including its advantages and disadvantages.
- Explain the requirements, purposes, and value of Safety Data Sheets to first responders.
- Define the responsibilities and duties of state and local emergency planning committees and how their plans and documents inform first responders.
- Identify other sources of information that can be valuable to firefighters.
- Discuss the four types of radiation and their penetrating powers.
- Explain the issues that make weapons of mass destruction incidents complex and the reasons for the development of the National Incident Management System.

CASE STUDY

Highway Gasoline Tanker Fire

On a major U.S. highway, a tanker carrying 1,000 gallons of gasoline, a flammable liquid, swerved to avoid a car stopped in the middle lane of traffic. The tanker hit the concrete center divide barrier and burst into flames. It was 11 PM. The temperature was 75°F, and there was a light, 5-mph northwesterly wind. Humidity was at 45%. Three engines and a truck responded to put out the fire. The entire highway was closed in the direction of travel.

1. What are your tactical options?
2. What would happen if the fire crew lets the fire burn itself out?
3. What would happen if the fire crew puts out the fire with a Class B foam?

Introduction

This chapter provides an overview of systems that warn first responders of hazardous materials, and as defined in the Federal Superfund Program, hazardous substances. We discuss laws that control hazardous materials and rules put in place to protect those likely to encounter hazardous materials.

Both private and governmental agencies have warning and information systems. This text will look at the two major systems used by first responders. It is important to stress how vital these warning systems are to first responders and the need for their continued improvement.

The final topics in the chapter are newer threats to first responders: terrorism, weapons of mass destruction, active shooter scenarios, and riots. Their impact on how emergencies, planning, and preparedness are managed and organized has become a vital part of the fire service. In today's world, the fire service needs to be aware of current events. This allows for an informed response and trained fire personnel.

Hazardous Materials and First Responders

After World War II, new types of emergencies occurred in the United States. These incidents were linked to the behavior of chemicals, petroleum, and nuclear products that appeared as products of convenience. The building blocks for these products were hazardous materials. A **hazardous material** is a material in any form or quantity that poses an unreasonable risk to safety, health, and property when transported in trade. The products came to the attention of first responders when they were misused or involved in unintended fires or accidents. Today, many hazardous materials are part of our culture. They have made it possible to enjoy many of the conveniences of modern living. Removing these hazardous materials from our society is not possible, so it is vital that their usage is as safe as possible. We must effectively control their handling, transport, and storage.

In the 1940s, few hazardous products were available to the public. After World War II, mass-produced homes and products led to the use of chemical-based items. Polyvinyl chloride plastics (PVCs) and acrylonitrile butadiene styrene (ABS) were used in the production of furniture, transport, and clothing. These thermoplastic polymers are impact resistant and chemical resistant, and do not conduct electricity. They are also very lightweight and stronger than existing plastics. As the popularity of these products grew, demand for them increased significantly. With the increase in demand, problems began to surface. It became apparent that manufacturers needed to control the handling of these substances.

At the national level, the government addressed the problem of abandoned hazardous wastes. This concern resulted in Congress enacting the Resource Conservation and Recovery Act of 1976. This act was followed by the Comprehensive Environmental Response, Compensation, and Liability Act (CERCLA) in 1980 and the Superfund Amendments and Reauthorization Act (SARA) in 1986. These three legislative documents were signed by Presidents Carter and Reagan. They form the basis of the Superfund program, which funds organizations that clean up hazardous substances. A **hazardous substance** is any substance named under the Clean Water Act and CERCLA as posing a threat to waterways and the environment when released. For the Occupational Safety and Health Administration (OSHA) and the Environmental Protection Agency (EPA), hazardous substances refer to any chemical that is a physical or health hazard.

Other legislation and federal actions resulted in the National Oil and Hazardous Substances Pollution Contingency Plan (NCP). This plan helps local communities to prevent and mitigate harm from hazardous materials releases. Local first responders have been assigned duties under this plan, so all local agencies should review the plan and its requirements. The **National Response Team (NRT)** was formed in 1993 to provide better federal agency coordination during emergencies. It consists of 15 federal agencies that help states with their emergency response planning and during major hazardous materials incidents.

As co-chairs of the NRT, the EPA and the U.S. Coast Guard play major roles in environmental protection. These two agencies protect U.S. waterways. The EPA protects inland waters, while the Coast Guard oversees coastal waters and larger waterways, such as Lake Michigan. Together, these agencies run a federal program for preventing, mitigating, and responding to releases of hazardous substances that threaten human health and the environment.

During the 1990s, there was a rise in acts and threats of terrorism in the United States. In response to these acts, in 1998, President Clinton issued presidential decision directive 62 (PDD 62). The purpose of this act was to meet threats of terrorism in the 21st century with a systematic approach.

The directive defined the roles of agencies involved in counterterrorism. A nationwide effort was made to locate potential targets and develop response and consequence plans. This legislation was especially important to the fire service because of funding provisions for training, mitigation, and enforcement of regulations of hazardous materials incidents. The directive called for the review of personal protective equipment (PPE) and training for first responders from the 120 largest U.S. cities. First responders were trained in the hazards, response procedures, and equipment needed to respond to an emergency involving a chemical or biological agent.

The directive includes a requirement to notify first responders of an existing hazard stored on a property or being transported. This legislation requires that Safety Data Sheets be on site where the materials are stored. **Safety Data Sheets (SDS)** are worksheets that manufacturers and marketers are required to prepare. These worksheets provide workers with detailed information about the properties of a commercial product that contains hazardous substances. This is vital information for first responders. SDS are explained in detail in the Emergency Planning and Community Right-to-Know Act section.

Two Major Warning Systems

Two systems warn first responders of hazardous materials. The first one is the hazardous material placard used in transportation. The second is the color-coded hazardous warning system for fixed facilities, such as warehouses, manufacturing plants, and research and development institutions.

The two organizations that provide warning systems are:

- U.S. Department of Transportation (DOT), and
- National Fire Protection Association (NFPA).

U.S. Department of Transportation System

The **U.S. Department of Transportation (DOT)** is the federal department responsible for the regulations for the safe transport of hazardous materials. The DOT also trains and inspects carriers and shippers of hazardous materials to ensure that they are in compliance. It has a system to alert the public of vehicles transporting hazardous materials on the highways. This system divides materials into nine major classifications of hazards. It then subdivides the materials to identify a substance.

The DOT's placard system requires a placard on all four sides of the tank, trailer, or intermodal carrier. The tractor is only required to have a placard on the front of the vehicle. This is to ensure that the tractor trailer can be identified from any side as carrying hazardous materials. Also, should the trailer be separated from the tractor, it can still be identified as carrying hazardous materials. Once the trailer is delivered to private property, and the tractor is unhitched from the trailer, the placards may be removed from the tractor.

The DOT placard system applies to all forms of transportation of hazardous materials. These are roadways, railways, air, and water. Placards are found on all road and rail transports. Aircraft and ships are not required to have a placard on their exteriors. On ships, however, intermodal containers and bulk containers are required to have placards on all four sides.

A nationwide standardized warning system was needed because these vehicles transport materials across state lines. The DOT, with the United Nations, developed an international classification system that identifies hazardous materials in transit using color-coded warning labels with symbols. These labels also use a numeric system to identify the hazardous material being transported.

The DOT publishes the ***Emergency Response Guidebook* (ERG)**, which is a manual to help first responders handle hazardous materials. It contains the chemical names and commonly used names for most toxic chemicals. This manual is provided to every first response emergency apparatus nationwide and is also distributed in Canada and to some South American countries free of charge. It can be accessed through an online application for mobile phones and tablet use.

The book is divided into color-coded sections. In general, white provides information on placards and other types of vehicle identification. Yellow identifies chemicals and their four-digit United Nations (UN) numbers. Blue lists chemicals in alphabetical order. Orange outlines how to deal with specific chemicals, and Green describes hazards to humans if the chemical is inhaled and its reaction with water.

The ERG's cross-reference system is helpful. For example, if information from one of the identification methods is known, then other methods are cross-referenced for verification and more information.

TIP

The ERG provides a quick reference to keep first responders safe; however, the ERG is *not* designed to provide specific information on *all* hazardous materials.

TIP

The ERG's recommendations should serve *only* during the initial operation. Once operations are under way, first responders should obtain specific information. Fire or police dispatch centers should have emergency contact numbers on hand.

The DOT label system combines four different ways to warn the first responder. The placard has a picture of the hazard at the top of the triangle that shows the hazard. The placard background color identifies the hazard class; this is also the easiest way to identify the hazard. In the middle of the placard, the hazard is identified in English or with the UN classification number. At the bottom of the placard, the hazard class and division number are shown.

As previously mentioned, the DOT system divides hazardous materials into nine major classes. Each of the nine classes is subcategorized for specific material descriptions. For example, an explosive that is Class 1.1 is more dangerous than a Class 1.6 explosive, which is not likely to explode. The nine classes identified by the DOT system are listed in **TABLE 10-1**. **FIGURE 10-1** shows the DOT's nine transportation classifications of hazardous materials.

DOT System Advantages and Disadvantages

The ERG is a guide that can help first responders quickly identify hazards of the material(s) involved in an emergency. It helps them protect themselves and the public.

Although the ERG does not provide specific information on every hazardous material, first responders can check the bill of lading or the SDS. One of these should be attached to the transporting agency's documentation. Once the specific materials have been identified, the first responder can look up in the ERG recommendations for handling, mitigation, and evacuation procedures.

A helpful feature of the ERG is the recommended isolation and evacuation distances. Response guidelines have special directions for specific products. The ERG includes 24-hour emergency response phone numbers and other contact information. First responders should remember that exceptions to the regulations may allow materials to ship without labels, placards, or markings. Such exceptions are routine in the shipping industry.

Another advantage is the DOT placard system. DOT placards have an easy recognition feature for the hazard of the product. Each placard has a background color, and these colors represent the hazard. For example, red means flammability, yellow means oxidizers, and blue means dangerous when wet. This provides a visible warning without first responders having to walk up to the vehicle. The placard can be seen with binoculars and the naked eye.

The DOT placard system can be manipulated to represent products not being carried. An example would be using a radioactive placard on a trailer carrying large-screen TVs. This may be an effort to protect the product being transported. Placards can also be removed before the product is removed from the trailer. This is common when the trailer is parked on private property. The DOT placards only apply to vehicles in transit on public roads.

National Fire Protection Association System

The National Fire Protection Association (NFPA) is an organization that reduces the impact of fire and other hazards. One of the functions of the NFPA is creating

TABLE 10-1 DOT Nine Classes of Hazards

Hazard Class	Product Example
Class 1—Explosives	
1.1 Explosives with a mass explosion hazard	Black powder
1.2 Explosives with a projection hazard	Detonating cord
1.3 Explosives with significant fire hazard	Propellant explosives
1.4 Explosives with no significant blast hazard	Practice ammunition
1.5 Very insensitive explosives; blasting agents	Prilled[a] ammonium nitrate
1.6 Extremely insensitive detonating substances	Fertilizer–fuel oil mixtures
Class 2—Gases	
2.1 Flammable gas	Propane
2.2 Nonflammable gas	Anhydrous[b] ammonia
2.3 Poisonous gas by inhalation	Phosgene
2.4 Corrosive gases (Canada)	Hydrogen bromide; anhydrous
Class 3—Flammable liquids, combustible liquids	
No divisions	Gasoline, kerosene
Class 4—Flammable solids, spontaneously combustible materials, and dangerous when wet materials	
4.1 Flammable solids	Magnesium
4.2 Spontaneously combustible materials	Phosphorus
4.3 Dangerous when wet materials	Calcium carbide
Class 5—Oxidizers and organic peroxides	
5.1 Oxidizers	Ammonium nitrate
5.2 Organic peroxides	Ethyl ketone peroxide
Class 6—Poisonous materials	
6.1 Poisonous	Arsenic
6.2 Infectious substances (etiological agent)	Rabies, human immunodeficiency virus (HIV), hepatitis B
Class 7—Radioactive materials	Radioactive isotopes
No divisions	
Class 8—Corrosive	
No divisions	Caustic soda

Continues.

TABLE 10-1 DOT Nine Classes of Hazards CONTINUED

Hazard Class	Product Example
Class 9-Miscellaneous hazardous materials	
9.1 Miscellaneous (Canada only)	Polychlorinated biphenyls (PCBs), molten sulfur
9.2 Environmental hazardous substances (Canada only)	PCBs, asbestos
9.3 Dangerous wastes (Canada only)	Fumaric acid

[a]Prilled is the process to convert a material into a granular, free-flowing form.
[b]Anhydrous indicates that all of the water has been removed.
Data from U.S. Department of Transportation 2020. *2020 Emergency Response Guidebook*. https://www.phmsa.dot.gov/sites/phmsa.dot.gov/files/2020-08/ERG2020-WEB.pdf.

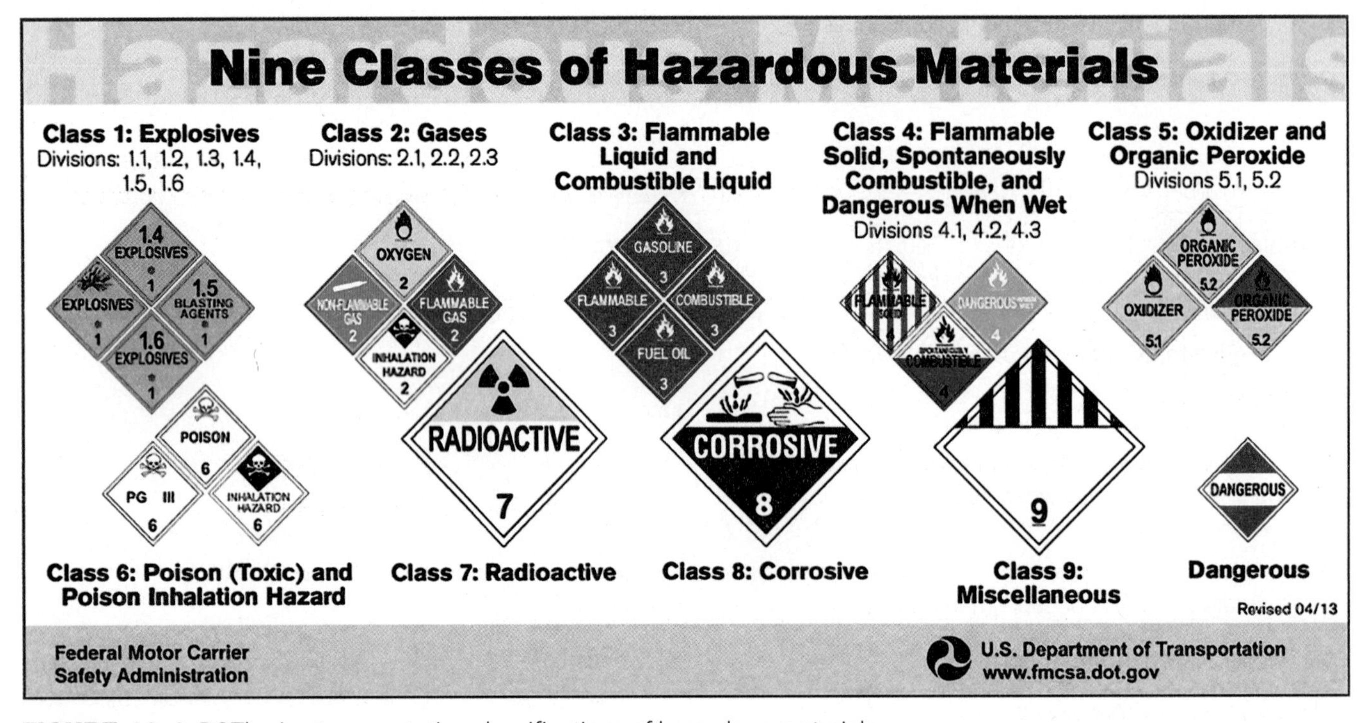

FIGURE 10-1 DOT's nine transportation classifications of hazardous materials.

Courtesy of the U.S. Department of Transportation.

codes and standards. The NFPA has made 300 codes and standards covering many buildings, processes, services, and equipment in the United States. The **National Fire Protection Association (NFPA) 704 system** is the warning method used for hazardous materials stored in buildings. The NFPA 704 standard for placards is vital to warn first-responding units.

The NFPA 704 placard system uses a color and numbering scheme to warn emergency responders of hazardous materials. They could be stored in tanks, small containers, or fixed facilities, or handled on the premises. The system does *not* provide the first responder with specific information about the substance, but similar to the DOT requirement, the building owner must have an SDS on hand.

The placards are diamond-shaped figures divided into four quadrants. Each quadrant is color coded to correspond with one of three hazards. Blue is for health hazards, red is for flammability hazards, and yellow is for chemical reactivity hazards. The quadrants are numbered 0 to 4 to show the degree of the hazard (with 4 as the greatest hazard). The fourth quadrant warns of special hazards, such as water reactivity (W), or that the material is an oxidizer (OXY). Other hazard signs include the trifoil, which indicates a radiation hazard, ALK for alkalis, and CORR for corrosives. **FIGURE 10-2** shows us that the material is a serious health hazard but does not present a fire hazard or a serious reactivity hazard. The material is also identified as an oxidizer.

A.

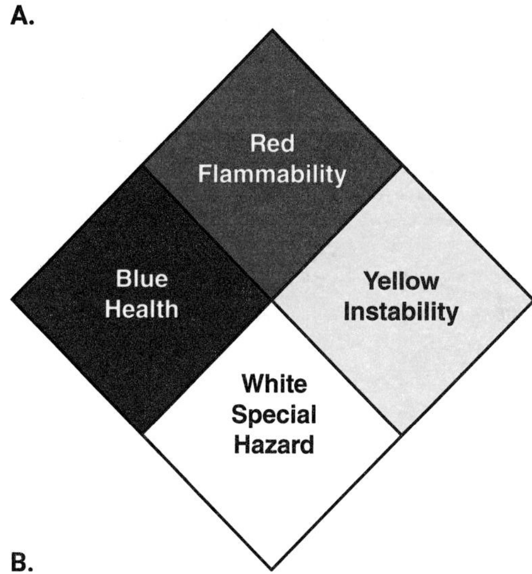

B.

FIGURE 10-2 NFPA placard. Identification of the four sections of the NFPA Diamond.

NFPA 704 System Advantages and Disadvantages

Like the DOT warning system, the NFPA placard system also offers a visual cue using a color-coded placard shown on the building. A disadvantage is that many plant facilities have more than one hazardous product on site. While each chemical may have an NFPA placard on its carton, the placard placed on the building only indicates the highest hazard of each chemical. For example, a building used to store chlorine, ammonia, methyl ethyl ketone, and acetylene has the following information.

- Chlorine is rated individually as 4 (Health), 0 (Flammability), and 0 (Reactivity) or 4-0-0.
- Ammonia (anhydrous) is rated individually as 3 (Health), 1 (Flammability), and 0 (Reactivity) or 3-1-0.
- Methyl ethyl ketone is rated individually as 1 (Health), 3 (Flammability), and 0 (Reactivity) or 1-3-0.
- Acetylene is rated individually as 0 (Health), 4 (Flammability), and 2 (Reactivity) or 0-4-2.

In this case, the NFPA placard on the front door would read 4 (Health), 4 (Flammability), and 2 (Reactivity). The numbers are not added, but the most dangerous one (the highest number) of each hazard is used to create a new number. Although this is not a distinct disadvantage, it can lead to confusion.

NFPA Placard Identification System

As mentioned earlier, the rating levels for the NFPA placard system are from 0, which equals no hazard, to 4, which equals an immediate hazard. The following is an example of how each rating is assigned to the material. It should be noted that there are various rating definitions.

Flammability

4 Flammable gases, volatile liquids, pyrophoric materials

3 Ignites at room temperatures

2 Ignites when slightly heated

1 Needs to be preheated to burn

0 Will not burn

Reactivity

4 Can detonate or explode at normal conditions

3 Can detonate or explode if strong initiating source is used

2 Violent chemical change if temperature and pressure are elevated

1 Unstable if heated

0 Normally stable

Health

4 Severe health hazard

3 Serious health hazard

2 Moderate hazard

1 Slight hazard

0 No hazard

FIGURE 10-3 shows an NFPA 704 placard on a warehouse door. Remember that the placard system is not chemical specific. No chemical identification system can reflect the effects of a combination of several chemicals.

FIGURE 10-3 NFPA 704 placard on a warehouse door.

Reproduced from National Fire Protection Association 2022. *Standard System for the Identification of the Hazards of Materials for Emergency Response*, 2022 Edition. Quincy, MA: National Fire Protection Association.

Another weakness of this system is that it is impossible to determine the exact chemicals present. It is also impossible to know the quantity of chemicals in the building or where they are located. While this situation may not provide first responders with all of the information, it is a place to start. To find out prefire planning information, the Emergency Planning and Community Right-to-Know requirements (discussed next) inform first responders.

Emergency Planning and Community Right-to-Know Act

The 1986 **Emergency Planning and Community Right-to-Know Act (EPCRA)** helps communities plan for chemical incidents. It requires businesses to report the use, storage, and releases of hazardous materials to the government. EPCRA also provides the public and local governments with news of potential chemical hazards in their communities. Each state governor appoints a State Emergency Response Commission. The **State Emergency Response Commission (SERC)** coordinates emergency response plans to chemical emergencies. The SERCs then set up local emergency planning districts and Local Emergency Planning Committees. **Local Emergency Planning Committees (LEPCs)** develop and maintain local emergency response plans to chemical emergencies. The plan outlines actions to take and evacuation plans for each district. The SERC supervises and coordinates the activities of the LEPC and processes requests for information.

The LEPC must include local officials, such as police, fire, civil defense, public health, transportation, and environmental professionals. It must also include representatives of buildings subject to the emergency plans, community groups, and the media. The LEPCs develop an emergency response plan for their jurisdiction, ensure an annual review, and inform the community about the chemicals stored in their area.

The four major area requirements to protect the community and first responders from hazardous materials releases are:

- emergency planning;
- emergency release notification;
- hazardous chemical storage reporting; and
- toxic chemical release inventory.

The requirement calls for an emergency plan to handle a hazardous materials incident. Included in the plan are emergency response procedures, equipment and facilities, evacuation plans (if needed), and training for first responders.

The law requires the facilities housing the hazardous materials to warn the public of a release. Facility managers do this by dialing 911. Specific information must be given to the 911 centers or operators. The law also requires the buildings to maintain SDS and to record how much hazardous material was released. The SDS mandate is used in addition to both the DOT and NFPA warning systems. It provides more in-depth information for product end users, handlers, and first responders. **TABLE 10-2** details the information requirements for the SDS.

SDS required by OSHA are depicted in **FIGURE 10-4**.

Other Sources of Information

Additional information sources, including U.S. government departments, private agencies in cooperation with the U.S. government, and the military, contribute vital information for the handling and management of hazardous materials incidents. We will identify these agencies and explain their functions. Each has a specific area of expertise as well as regulations and procedures to follow when handling hazardous materials.

Chemical Transportation Emergency Center (CHEMTREC)

CHEMTREC is a national call center that guides emergency responders at a chemical incident. This group

TABLE 10-2 Requirements for SDS Forms[a]

Identification	Product identifier, manufacturer or distributor name, address, phone number, emergency phone number, recommended use, and restrictions on use.
Hazard(s) identification	All hazards regarding the chemical and required label elements.
Composition/Information on ingredients	Information on chemical ingredients and trade secret claims.
First aid measures	Important symptoms/effects (acute, delayed) and required treatment.
Firefighting measures	Suitable extinguishing techniques, equipment, and chemical hazards from fire.
Accidental release measures	Emergency procedures, protective equipment, and proper methods of containment and clean-up.
Handling and storage	Precautions for safe handling and storage, including what not to transport with this material (incompatibilities).
Exposure controls/personal protection	OSHA's permissible exposure limits (PELs), threshold limit values (TLVs), appropriate engineering controls, and personal protective equipment (PPE).
Physical and chemical properties	The chemical's characteristics.
Stability and reactivity	Details stability and possible hazardous reactions that the chemical may produce.
Toxicological information	Details how contact with the chemical may occur (inhalation, ingestion, or absorption contact), symptoms, acute and chronic effects, and numerical measures of toxicity.

[a]Note: Since other agencies regulate this information, OSHA does not enforce Sections 12 through 15 (ecological information, disposal considerations, transport information, and regulatory information) (29 CFR 1910.1200(g)(2)).

Reproduced from Occupational Safety and Health Administration (OSHA). n.d. OSHA Quick Card. Accessed April 12, 2023. https://www.osha.gov/sites/default/files/publications/OSHA3493QuickCardSafetyDataSheet.pdf.

provides critical information and can also provide response teams if necessary. In some cases, the SDS will not have enough specific information on a chemical or compound or how to deal with it during an emergency. Fortunately, there are some resources on the internet that can offer instant answers.

Wireless Information System for Emergency Responders (WISER)

The **Wireless Information System for Emergency Responders (WISER)** helps first responders in hazardous materials incidents. Developed by the National Library of Medicine (NLM), WISER helps first responders identify a chemical, its properties, and the effects of exposure. In addition to many other features, it supports a fully searchable version of the *ERG 2020.*

Unfortunately, in February 2023, the NLM discontinued WISER. The users who have installed the app on a mobile device or downloaded the WISER system to their personal computer still have access to the system, but the data are no longer being updated.

For users in need of a system similar to WISER, the NLM recommends, among other systems, the **Chemical Companion**, which is also known as the Emergency Response Decision Support System (ERDSS). The system is a reference for 2,000 common chemicals, exposure guidelines, and hazardous concentration levels. It also provides isolation and protective action distances.

This system is available as a tablet app and downloadable software. Mobile apps are currently being developed.

Environmental Protection Agency (EPA)

The goal of the **Environmental Protection Agency (EPA)**, a government agency, is to protect human life and the environment. The EPA assumes the part of the

SAFETY DATA SHEET

Section 1. Identification

Product Name: Synonyms:	**Ammonia, Anhydrous** Ammonia
CAS REGISTRY NO:	7664-41-7
Supplier:	Tanner Industries, Inc. 735 Davisville Road, Third Floor Southampton, PA 18966
Website:	www.tannerind.com
Telephone (General): **Corporate Emergency Telephone Number:** **Emergency Telephone Number:**	215-322-1238 **800-643-6226** **Chemtrec: 800-424-9300**
Recommended Use:	Various Industrial / Agricultural

Section 2. Hazard(s) Identification

Hazard: Acute Toxicity, Corrosive, Gases Under Pressure, Flammable Gas, Acute Aquatic Toxicity

Classification:
Acute Toxicity, Inhalation (Category 4)
Skin Corrosion / Irritation (Category 1B)
Serious Eye Damage / Irritation (Category 1)
Gases Under Pressure (Liquefied gas)
Flammable Gases (Category 2)
Acute Aquatic Toxicity (Category 1)

Note: (1 - Most Severe / 4 - Least Severe)

Pictogram:

Signal word: **Danger**

Hazard statements:
Harmful if inhaled.
Causes severe skin burns and serious eye damage.
Flammable gas.
Contains gas under pressure; may explode if heated.
Very toxic to aquatic life.

Precautionary statements:
Avoid breathing gas/vapors.
Use only outdoors or in well-ventilated area.
Wear protective gloves, protective clothing, eye protection, face protection.
Keep away from heat, sparks, open flames and other ignition sources. No smoking.

FIGURE 10-4 Example of the first page of an SDS form for anhydrous ammonia.

Courtesy of Tanner Industries, Inc., Southhampton, PA.

lead agency for carrying out Title III reporting and training requirements. The EPA is also responsible for hazardous waste sites and Superfund waste site clean-up. It serves as the chair for the 15 federal agencies that comprise the NRT. As discussed, the NRT is available to local communities to assist and coordinate between federal, state, and local agencies.

National Oceanic and Atmospheric Administration (NOAA)

The **National Oceanic and Atmospheric Administration (NOAA)** is an organization that works with the EPA, the U.S. Department of Commerce, and other organizations to develop the Chemical Reactivity Worksheet. The Worksheet is a software program that first responders can use to find out the chemical reactivity of thousands of chemicals. It also provides the compatibility of materials used to absorb chemicals and the safety of construction materials in chemical processes.

U.S. Coast Guard (USCG)

The **U.S. Coast Guard (USCG)** is the branch of the U.S. military in charge of enforcing maritime laws. It works in concert with the Department of Homeland Security. The U.S. Coast Guard is also responsible for boating safety and inspection of seagoing vessels.

The U.S. Coast Guard offers a chemical and hazardous response information system. This online information database contains physical data and emergency procedures in a searchable format. Local fire departments have the responsibility of notifying the U.S. Coast Guard when a hazardous materials release will impact the coastal waters. **FIGURE 10-5** shows the approved warning label for materials that pollute water.

FIGURE 10-5 Marine pollutant indicator.

Courtesy of the National Oceanic & Atmospheric Administration.

National Institute of Occupational Safety and Health (NIOSH)

The **National Institute of Occupational Safety and Health (NIOSH)** is a federal agency under the U.S. Department of Health and Human Services that is responsible for checking the toxicity of workspaces. It also oversees all other matters relating to safe industrial practices. NIOSH publishes *Pocket Guide to Chemical Hazards,* a resource for employers, workers, and health specialists to learn about chemicals and their hazards. The guide is available in print, online, PDF, and mobile app. NIOSH also has a personal protective technology (PPT) program that researches standards for respirators, protective gear, and sensors for hazardous materials. The agency also publishes the Permeation Calculator, a software program for analyzing chemical permeation data.

Department of Homeland Security (DHS)

The **Department of Homeland Security (DHS)** protects Americans from terrorist threats. It protects U.S. territory from terrorist attacks and responds to natural disasters. It is improving technology to detect explosives and other weapons, prevent cyber attacks on infrastructure, and share information with other agencies.

The DHS also oversees offices that deter terrorism. In 2018, the DHS formed the Countering Weapons of Mass Destruction (CWMD) office. The job of this office is to prevent terrorist attacks from weapons of mass destruction against the United States by communicating with similar agencies.

Department of Defense (DOD)

The **Department of Defense (DOD)** combats biological and chemical terrorism. It is a federal department that supplies staff, training, and equipment for a full range of hazardous substances. DOD laboratories and bases can be sources of guidance, equipment, and goods for use in local emergencies.

Department of Energy (DOE) and Nuclear Regulatory Commission (NRC)

In the past, the Department of Energy and the Nuclear Regulatory Commission trained state and local emergency personnel. These personnel would train for an emergency at a DOE or an NRC-licensed facility. The **Department of Energy (DOE)** is responsible for reducing radioactive waste from the U.S. nuclear weapons program and nuclear reactors. The **Nuclear Regulatory Commission (NRC)** is a group within the DOE, which furnishes licenses to facilities that use nuclear products.

Nuclear Radiation

Radiation is energy being given off through electromagnetic waves. These waves are commonly referred to as light. Nuclear radiation is the energy given off from fast-moving nuclear particles. These are in three forms: alpha, beta, and gamma radiation. Nuclear radiation is successfully used in the form of x-rays and as a source of power. There is a hazard to life when this radiation is used for weapons.

Nuclear Radiation Releases

Since the 1950s, use of nuclear materials has increased worldwide. In addition to nuclear fuels used for power, nuclear materials are used for medical purposes, industrial controls, and household items (e.g., smoke detectors). They are also used for basic research and military systems. The chances of accidental releases are increasing as more radioactive materials are transported and handled.

The increase in terrorist acts has also made the threat of a release of radioactive materials a possibility. Because firefighters are often the first responders, they need to train to handle radiological incidents. Fire crews also need protection against exposure to radioactive materials, radioactive waves, and contamination.

Radioactive Materials

There are many different forms of radioactive materials. They can be the emissions from a damaged energy plant, the result of a nuclear weapon's release, or an accident in a medical or research laboratory. They can also be used by terrorists to threaten for concessions to demands or to cause death and injuries.

Radioactive materials go through spontaneous transformation and release radiant energy or atomic particles. Firefighters are primarily concerned with three forms of radiation: alpha, beta, and gamma. **Alpha radiation** is radiation that gives off alpha particles. This is the least dangerous type of radiation that firefighters may encounter. Alpha radiation is least harmful as an external radiation hazard and most harmful as internal radiation. Alpha particles have low penetrating power and seldom penetrate the skin. Full PPE will provide the needed protection from alpha radiation.

Beta radiation gives off beta particles. Beta particles are smaller than alpha particles but have a higher energy level and more penetrating power. They can travel through the air from 10 to 100 ft (3 to 30 m). Large amounts of beta radiation can seriously damage skin tissue. Beta particles enter the body through damaged skin, ingestion, or inhalation. They will cause organ damage like that caused by gamma particles. Full PPE including SCBA is required to protect against beta radiation.

Gamma radiation is one of the most dangerous forms of penetrating radiation. Gamma rays have 10,000 times the power of alpha particles and 100 times the power of beta particles. They pass easily through the body and strike organs, and in doing so, cause extensive internal damage.

Gamma radiation consists of rays similar to x-rays. Protection from external radiation can be achieved three ways: time, distance, and shielding. Time refers to the amount of time one is exposed to the radiation. Distance measures how far from the radioactive source one is standing. Shielding describes the type of protective clothing being worn.

Neutron radiation is a product of nuclear fission. This type of radiation is not often encountered by firefighters. Like gamma radiation, neutron radiation will penetrate PPE.

As a person's distance from the radiation source doubles, the amount of radiation decreases in proportion to the square of the distance. In other words, doubling the distance from the source reduces the radiation exposure by one-fourth. Also, shorter exposures to radiation lead to smaller doses of the radiation. First responders must reduce the amount of time they are exposed. One method might be to rotate assignments to limit the time an individual is exposed.

Certain dense materials, such as lead, concrete, earth, and water, will stop some of the radiation rays. The thickness of the shielding needed depends on the type of the material giving off the radiation, the amount of radiation, and the distance from the source. **FIGURE 10-6** shows the stopping power of various materials. **FIGURE 10-7** shows an example of radiation protective clothing for firefighters.

Measuring Radiation

We measure absorbed radiation doses with rads. A **radiation-absorbed dose (rad)** is a unit to describe the amount of radiation absorbed by an object or person (absorbed dose). The rad reflects how much energy radioactive sources deposit in materials as they pass through them.

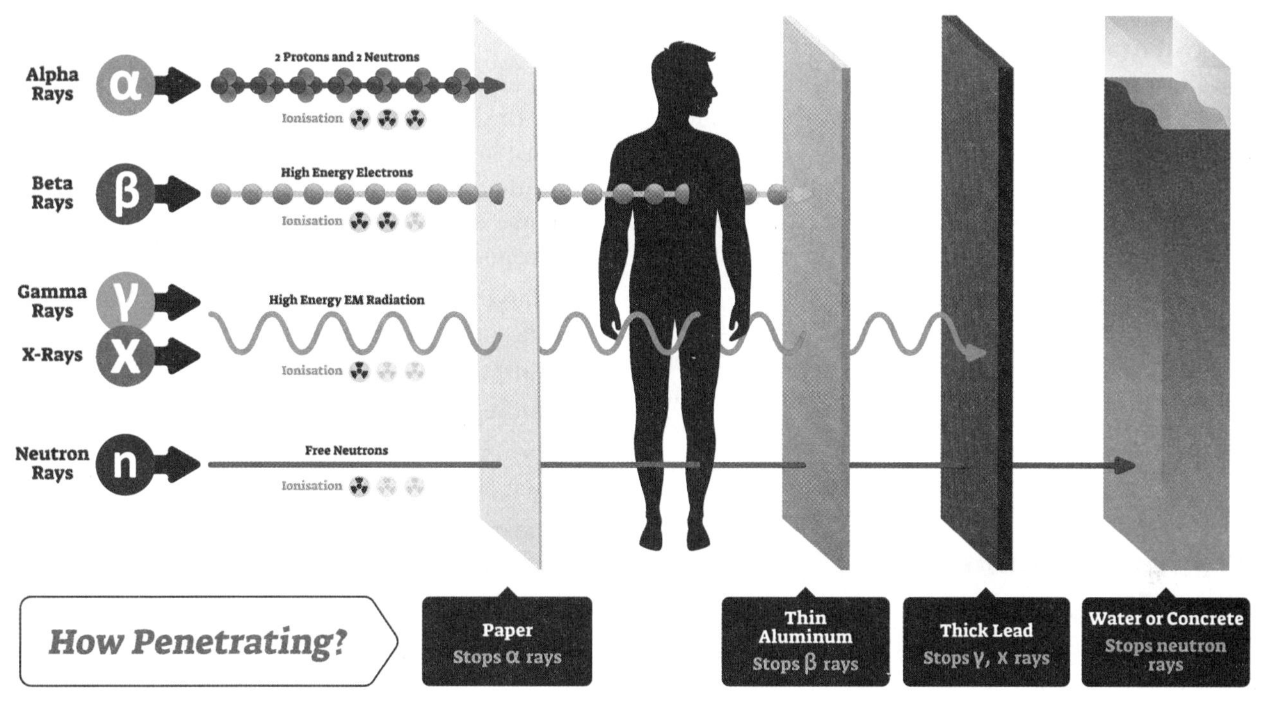

FIGURE 10-6 Radiation penetration.

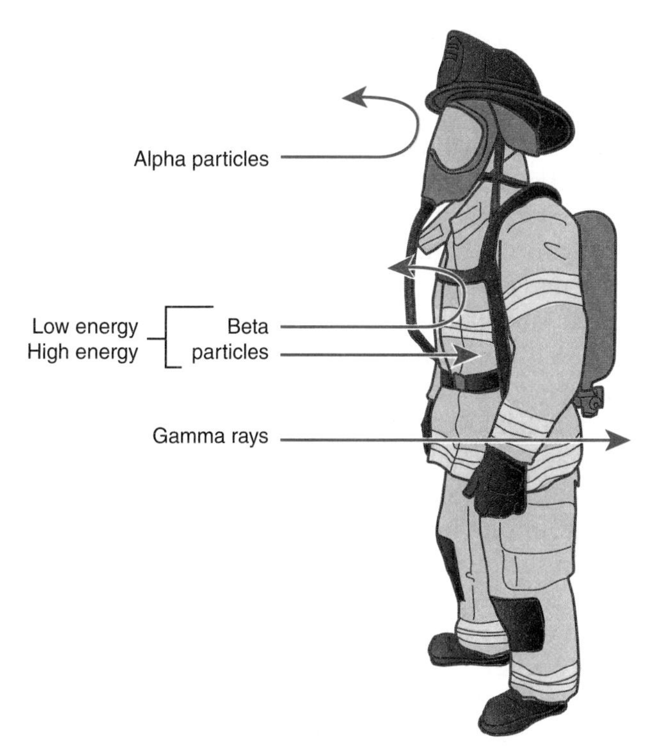

FIGURE 10-7 Radiation protective clothing for firefighters.

To assess the risk of radiation, the absorbed dose of energy in rads is multiplied by the relative biological effectiveness. The **relative biological effectiveness (RBE)** measures how damaging a given type of particle is compared to an equivalent dose of x-rays.

The **roentgen equivalent man (rem)** is a unit of radiation dosage (e.g., from x-rays) applied to humans. It is defined as the dosage in rads that will cause the same amount of biological injury as one rad of x-rays or gamma rays. This is the important dosage measurement for first responders because there is a maximum amount of rem that people can be exposed to without serious health consequences.

Almost any radioactive material can be used to make a radiological dispersal device. A **radiological dispersal device (RDD)** is a device, such as an explosive, used to disperse radiological products. One such product is fissile material, which is the waste from the fuel of nuclear reactors. It can also come from radiological medical, industrial, and research waste. To cause the most damage, the most radioactive fissile material would be the active ingredient in the RDD.

An RDD would have a great impact on a target population. An uninformed public may add to the terrorist's goals. Such a public reaction would be more likely to initiate widespread panic, and people would move away from the target area. Unless the public has an understanding of nuclear, biological, and chemical (NBC) terrorism, any use of RDDs is likely to invoke the fear of nuclear war.

The disadvantage of RDDs over nuclear weapons is the use of explosive materials. If a terrorist uses radioactive waste, which is different than fissile material, that person is likely to be contaminated by radiation. The first responder who handles an RDD is also more likely to become tainted because radioactive wastes transfer from devices to hands. First responders must decide if they can safely enter contaminated areas. They should measure the radiation and wear protective gear while entering the zone.

Radiation Exposures to First Responders

The exposure to radiation by first responders is usually allowed to exceed that of people who work with radioactive materials. It is acceptable as long as certain limits are not exceeded. The EPA Protective Action Guidelines recommend that in an emergency for saving lives or protection of a large population, entry to a radioactive zone is permitted as long as the whole-body exposure does not exceed 25 rem. Emergencies include search and rescue or situations where lives are at stake.

Terrorism

Terrorism is any act by an individual or group to create fear and injure civilians, frighten government leaders, or alter government actions by violent means. In today's times, this can also include active shooter incidents, hostage taking, and riots.

Since the early 1960s, there have been terrorist acts in Northern Ireland and the Middle East. Most Americans were not personally affected because these acts took place overseas. In the late 1970s and early 1980s, terrorist acts against Americans stationed overseas occurred more often. During this same time, more acts of violence took place within U.S. borders. The situation in the United States has escalated since the early 1990s. The 1993 bombing of the World Trade Center, the 1995 bombing of the Alfred P. Murrah Federal Building, and the 2001 World Trade Center and Pentagon air attacks all signal that first responders and the American public face a grave threat.

During the same period, terrorists began to use a variety of materials to attack their targets. The devices that spread these materials are called **weapons of mass destruction (WMD)**. Weapons of mass destruction use hazardous materials, nuclear radiation, biological agents, chemicals, explosives, and agricultural agents to kill or injure numerous people. First responders must be trained, equipped, and prepared to handle these attacks efficiently and safely.

WMD attacks are more difficult than the usual incident response. There are four areas of difficulty:

1. A large number of people needing immediate assistance at the same time
2. Multiuse of first responders, some of a technical nature, conducted simultaneously
3. The immediate involvement of federal and state agencies
4. The overwhelming response of the national media

During the World Trade Center attack in 2001, people needed urgent medical care and rescue at the same time from responding emergency agencies. This incident signaled the need to provide multiple emergency functions. The fire service was needed for both evacuation and fire attack to save trapped people. The emergency medical service (EMS) was very active with the number of people injured. EMS also had to arrange for the expected deluge of patients at the hospitals. Law enforcement was also busy, controlling traffic and crowds and assisting in evacuations. This incident required agencies to carry out multiple functions at the same time and in a way that made the most of limited resources.

Soon, government agencies arrived at the scene. The **Federal Bureau of Investigation (FBI)**, the Department of Justice's investigative branch, wanted to investigate the crime while the incident was still active. The state of New York called for disaster aid from the Federal Emergency Management Agency. At the same time, New York City police and firefighters, assisted by the New York State National Guard, were at the site of the fallen World Trade Center buildings performing clean-up and recovery activities. This need for local, state, and federal agencies to manage many tasks at once in one location while responding to a major terrorist incident resulted in the 2003 presidential directive 5 (HSPD-5). The directive to the newly formed DHS (which was created in response to the attacks) was to create a National Incident Management System (NIMS).

The DHS protects U.S. territory from terrorist attacks. It also responds to natural disasters. HSPD-5 was followed by directives HSPD-7 and HSPD-8, which had conditions for NIMS. NIMS could not be a new Incident Command System (ICS). It had to combine all parts of preparing for and managing incidents. This system has the following components:

- Command and management
- Preparedness resource management
- Communications and information management
- Supporting technologies and ongoing management and maintenance

ICS is the command and management part of NIMS. NIMS manages incidents that cover more than one jurisdiction. Combined efforts need to have a whole and coordinated plan under NIMS. NIMS receives direction from a unified command. The **unified command (UC)** is a team system that allows all agencies responding to the incident to adopt one set of objectives and strategies.

Immediately after 9/11, the news media overwhelmed the incident, seeking answers that were not yet available. At a WMD incident such as this, the news media may have a national impact. The public could lose confidence in the government. For this reason, news releases must be carefully crafted, accurate, and complete.

All of these factors make a WMD incident challenging for first responders. It can be handled successfully with careful preincident planning, joint drills among agencies, and training on specialized equipment.

Active Shooter Scenarios

Following the Columbine school shooting in April 1999, emergency responders changed the policies and procedures for response and what actions to take to defuse such a situation. Law enforcement now immediately engages with the shooter or shooters to draw their attention away from possible victims. Today, Active Shooter/Hostile Event Response (ASHER) teams, which consist of police and fire personnel, and Advanced Law Enforcement Rapid Response Training (ALERRT) are in place to ensure an immediate evacuation of wounded individuals. Fire and emergency medical services work in the warm zone to triage, treat, and remove those who have been wounded. The injured are then moved to a safe area for transport to hospitals.

Riots

A **riot** is an action taken by a group of people to cause fear and disorder in a community. Riots may be caused by political unrest, fear of an injustice, or actions taken by an individual or individuals that many find offensive. The rioters will often cause large-scale violence, involving fires, explosions, and traffic blockades.

Such actions are often referred to as urban terrorist activity. The resulting fear and destruction may cause great personal fear and an unsafe feeling for personal protection.

CASE STUDY

CONCLUSION

1. **What are your tactical options?**

   Your tactical options are direct attack with enough Class B foam and defensive attack by protecting the exposures and letting the product burn out completely.

2. **What would happen if the fire crew lets the fire burn itself out?**

   Letting the fire burn itself out can result in various scenarios. One of these scenarios is that there would be less contaminated runoff.

3. **What would happen if the fire crew puts out the fire with a Class B foam?**

   It is essential to have enough Class B foam to extinguish the fire. Large Class B fires may spread once the product is released from the vessel, allowing for the flammable liquid to travel beyond the original location.

WRAP-UP

SUMMARY

- After World War II, new types of emergencies occurred in the United States. These incidents were linked to the behavior of chemicals, petroleum, and nuclear products that appeared as products of convenience.
- Three acts form the basis of the Superfund program:
  - The Resource Conservation and Recovery Act of 1976
  - The Comprehensive Environmental Response, Compensation, and Liability Act (CERCLA) of 1980
  - The Superfund Amendments and Reauthorization Act (SARA) of 1986
- Two agencies warn first responders of hazardous materials using placard systems: the Department of Transportation and the National Fire Protection Association.
- The 1986 Emergency Planning and Community Right-to-Know Act (EPCRA) helps communities plan for chemical incidents.
  - The State Emergency Response Commission (SERC) coordinates emergency response plans.
  - The Local Emergency Planning Committees (LEPCs) develop and maintain local emergency response plans.
- Other information sources for first responders include:
  - Chemical Transportation Emergency Center (CHEMTREC)
  - Wireless Information System for Emergency Responders (WISER)
  - Chemical Companion (also known as the Emergency Response Decision Support System [ERDSS])
  - Environmental Protection Agency (EPA)
  - National Oceanic and Atmospheric Administration (NOAA)
  - U.S. Coast Guard (USCG)
  - National Institute of Occupational Safety and Health (NIOSH)
  - Department of Homeland Security (DHS)
  - Department of Defense (DOD)
  - Department of Energy (DOE)
  - Nuclear Regulatory Commission (NRC)
- Firefighters are primarily concerned with three forms of radiation: alpha, beta, and gamma.
- Almost any radioactive material can be used to make a radiological dispersal device.
- Terrorism is any act by an individual or group to create fear and injure civilians, frighten government leaders, or alter government actions by violent means.
- The reasons WMD attacks are more difficult than the usual incident response are:
  - A large number of people needing immediate assistance at the same time
  - Multiuse of first responders, conducted simultaneously
  - The immediate involvement of federal and state agencies
  - The overwhelming response of the national media

KEY TERMS

alpha radiation Radiation that gives off alpha particles.

beta radiation Radiation that gives off beta particles.

Chemical Companion A system that references 2,000 common chemicals, exposure guidelines, hazardous concentration levels, and isolation and protective action distances; also known as the Emergency Response Decision Support System (ERDSS).

CHEMTREC A national call center that guides emergency responders at a chemical incident.

Department of Defense (DOD) A federal department that supplies staff, training, and equipment for a full range of hazardous substances.

Department of Energy (DOE) The federal department that is responsible for reducing radioactive waste from the U.S. nuclear weapons program and nuclear reactors.

Department of Homeland Security (DHS) A federal department that protects U.S. territory from terrorist attacks and responds to natural disasters.

Department of Transportation (DOT) A federal department responsible for the regulations for the safe transport of hazardous materials.

Emergency Planning and Community Right-to-Know Act (EPCRA) An act that helps communities plan for chemical incidents and informs the public and local governments of potential chemical hazards.

Environmental Protection Agency (EPA) A government agency whose purpose is to protect human life and the environment.

***Emergency Response Guidebook* (ERG)** A manual that contains the chemical names and commonly used names for most toxic chemicals; meant to help first responders handle hazardous materials.

Federal Bureau of Investigation (FBI) Under the direction of the Department of Justice, this federal agency is responsible for the criminal investigation of acts of terrorism.

gamma radiation One of the most dangerous forms of penetrating radiation.

hazardous material A material in any form or quantity that poses an unreasonable risk to safety, health, and property when transported in trade.

hazardous substance Any substance named under the Clean Water Act and the Comprehensive Environmental Response, Compensation, and Liability Act (CERCLA) as posing a threat to waterways and the environment when released.

Local Emergency Planning Committee (LEPC) A committee that develops and maintains local emergency response plans to chemical emergencies.

National Fire Protection Association (NFPA) 704 system The warning method used for hazardous materials stored in buildings.

National Institute of Occupational Safety and Health (NIOSH) A federal agency under the U.S. Department of Health and Human Services that is responsible for checking the toxicity of workspaces and all other matters relating to safe industrial practices.

National Oceanic and Atmospheric Administration (NOAA) An organization that works with the EPA, the U.S. Department of Commerce, and other organizations to develop the Chemical Reactivity Worksheet.

National Response Team (NRT) A team of 15 federal agencies that help states with their emergency response planning and during major hazardous materials incidents.

neutron radiation A product of nuclear fission; one of the most dangerous forms of radiation.

Nuclear Regulatory Commission (NRC) A group within the Department of Energy, which furnishes licenses to facilities that use nuclear products.

radiation-absorbed dose (rad) A unit to describe the amount of radiation absorbed by an object or person (absorbed dose).

radiological dispersal device (RDD) A device, such as an explosive, used to disperse radiological products (e.g., fissile material).

relative biological effectiveness (RBE) A measure of how damaging a given type of particle is compared to an equivalent dose of x-rays.

riot An action taken by a group of people to cause fear and disorder in a community.

roentgen equivalent man (rem) A unit of radiation dosage (e.g., from x-rays) applied to humans.

Safety Data Sheet (SDS) A required worksheet that manufacturers and marketers prepare to provide workers with detailed information about the properties of a commercial product that contains hazardous substances.

State Emergency Response Commission (SERC) Appointed by the governor, this commission coordinates emergency response plans, which are developed by the LEPCs.

terrorism Any act by an individual or group to create fear and injure civilians, frighten the government, or alter government actions by violent means.

unified command (UC) A team system that allows all agencies responding to the incident to adopt one set of objectives and strategies.

U.S. Coast Guard (USCG) The branch of the U.S. military in charge of enforcing maritime laws.

weapon of mass destruction (WMD) Weapons that use hazardous materials, nuclear radiation, biological agents, chemicals, explosives, or agricultural agents to kill or injure numerous people.

WISER A system that helps first responders in hazardous materials incidents by identifying a chemical, its properties, and the effects of exposure; shut down as of February 2023.

REVIEW QUESTIONS

1. What are the two types of hazardous materials warning systems and their benefits to first responders?
2. What are the four types of radiation? Give examples of how to protect responders for each type.
3. What are the potential benefits of the LEPC for first responders?
4. What is an RDD is and how it will affect an emergency response?

DISCUSSION QUESTIONS

1. How does a hazardous material affect living conditions and environment changes?
2. What are the benefits of the DOT *Emergency Response Guidebook* (ERG)?
3. How does the DOT placard system differ from the NFPA 704 placard system?

APPLYING THE CONCEPTS

You are dispatched to a structure fire at a known chemical production facility. While en route, you receive updated information about one of the production fabrication stations as being the point of origin.

On your arrival, you are met outside by a member of the company's in-house emergency response team. They confirm the information you have already received.

1. On your way to this facility, what resource could you access to assist you on arrival?
2. On arrival at the facility, what are some details you would like to know to assist you in your action plan?
3. How might the on-site emergency response team assist you?

REFERENCES

Adobe. "How to Read the Emergency Response Guidebook Color Meanings." n.d. Accessed March 29, 2023. https://www.adobe.com/sign/hub/how-to/how-to-use-emergency-response-guidebook.

American Institute of Chemical Engineers. "Chemical Reactivity Worksheet." Updated July 12, 2019. Accessed March 29, 2023. https://www.aiche.org/ccps/resources/chemical-reactivity-worksheet.

Angle, James S. *Occupational Safety and Health in the Emergency Services*, 2nd ed. Clifton Park, NY: Delmar Cengage Learning, 2005.

Britannica.com. "Rem." n.d. Accessed March 29, 2023. https://www.britannica.com/science/rem-unit-of-measurement.

Centers for Disease Control and Prevention, National Institute for Occupational Safety and Health (NIOSH). "NIOSH Pocket Guide to Chemical Hazards." Last reviewed February 18, 2020. Accessed March 29, 2023. https://www.cdc.gov/niosh/npg/default.html.

Centers for Disease Control and Prevention, National Institute for Occupational Safety and Health (NIOSH). "Permeation Calculator." August 2017. Accessed March 29, 2023. https://www.cdc.gov/niosh/docs/2007-143c/default.html.

Chemical Companion. "ERDSS Chemical Companion." n.d. Accessed April 12, 2023. https://www.chemicalcompanion.org/.

Fitch, J. Patrick, Ellen Raber, and Dennis R. Imbro. "Technology Challenges in Responding to Biological or Chemical Attacks in the Civilian Sector." *Science* 302 no. 5649 (2003): 1350–1354.

Gordon, John C., and Steen Bech-Nielsen. "Biological Terrorism: A Direct Threat to Our Livestock Industry," *Military Medicine* 151 no. 7 (July 1986): 357–363.

Hawley, Chris. *Hazardous Materials Incidents*, 3rd ed. Sebastopol, CA: O'Reilly Media, 2017.

International Association of Fire Chiefs. "IAFC Survey," *Fire Engineering* 158 no. 1 (2005).

Mangold, Tom, and Jeff Goldberg. *Plague Wars*. New York, NY: St. Martin's Press, 2001.

National Library of Medicine. "WISER." 2023. Accessed March 29, 2023. https://www.nlm.nih.gov/wiser/index.html.

National Oceanic and Atmospheric Administration. "Updated Software to Better Protect Responders to Chemical Hazards, Spills." April 3, 2013. Accessed March 29, 2023. https://www.noaa.gov/noaa-updated-software-better-protect-responders-chemical-hazards-spills.

Nuclear, Biological & Chemical Warfare. "NBC Links." Placke & Associates, n.d. Accessed March 29, 2023. http://www.nbc-links.com/.

Occupational Safety and Health Administration. *OSHA Brief: Hazard Communication Standard: Safety Data Sheets.* n.d. Accessed March 29, 2023. https://www.osha.gov/sites/default/files/publications/OSHA3514.pdf.

Randall, Stephen. "Chemical Manufacturers Association Creates MSDS Data Base," *Fire Engineering* 148 no. 6 (1995).

U.S. Department of Agriculture. *Lesson 4: Unified Command.* n.d. Accessed March 29, 2023. https://www.usda.gov/sites/default/files/documents/ICS300Lesson04.pdf.

U.S. Department of Environmental Protection. "Superfund." Last updated April 18, 2023. Accessed May 16, 2023. https://www.epa.gov/superfund.

U.S. Department of Health and Human Services. "Chemical, Biological and Respiratory Protection," workshop sponsored by NIOSH-DOD-OSHA. February 2000.

U.S. Department of Homeland Security. "Countering Weapons of Mass Destruction Office." n.d. Accessed March 29, 2023. https://www.dhs.gov/countering-weapons-mass-destruction-office.

U.S. Department of Homeland Security. "Preventing Terrorism Overview." n.d. Accessed March 29, 2023. https://www.dhs.gov/preventing-terrorism-overview.

U.S. Department of Justice, Office of the Inspector General. "The Federal Bureau of Investigation's Efforts to Improve the Sharing of Intelligence and Other Information." December 2003. Accessed March 29, 2023. https://oig.justice.gov/reports/FBI/a0410/background.htm#:~:text=In%20May%201998%2C%20the%20President,consequence%20management%20after%20terrorist%20acts.

U.S. Department of Justice, Office of Justice Programs. "EPA's Role in Counter-Terrorism Activities." February 1998. https://www.ojp.gov/ncjrs/virtual-library/abstracts/epas-role-counter-terrorism-activities.

U.S. Department of Labor and Occupational Safety and Health Administration. "Clarification of the Definition of a Hazardous Chemical and the Requirements for Material Safety Data Sheets." May 15, 1997. Accessed March 29, 2023. https://www.osha.gov/laws-regs/standardinterpretations/1997-05-15-1#:~:text=A%20hazardous%20chemical%2C%20as%20defined,29%20CFR%201910.1200(d).

U.S. Department of Labor and Occupational Safety and Health Administration. "Safety and Health Topics." http://www.osha.gov/SLTC/radiation/index.html and http://www.osha.gov/SLTC/biologicalagents/index.html.

U.S. Department of Transportation. *Emergency Response Guidebook.* Washington, DC: Government Printing Office, 2020. https://www.phmsa.dot.gov/sites/phmsa.dot.gov/files/2020-08/ERG2020-WEB.pdf.

U.S. Department of Veterans Affairs. "VA Technical Reference Model v 23.3: Chemical Companion: Emergency Response Decision Support System (ERDSS)." Last updated June 30, 2020. Accessed April 12, 2023. https://www.oit.va.gov/Services/TRM/ToolPage.aspx?tid=10839.

U.S. Environmental Protection Agency. "Definitions of Hazardous Chemical and OSHA's MSDS Requirement for Determining Applicability of EPCRA 311/312." n.d. Accessed March 29, 2023. https://www.epa.gov/epcra/definition-hazardous-chemical-and-oshas-msds-requirement-determining-applicability-epcra.

U.S. Environmental Protection Agency. "Emergency Planning and Community Right-to-Know Act (EPCRA)." n.d. Accessed March 29, 2023. https://www.epa.gov/epcra.

U.S. Environmental Protection Agency. "What Is Superfund?" n.d. Accessed March 29, 2023. https://www.epa.gov/superfund/what-superfund.

U.S. Environmental Protection Agency. *PAG Manual: Protective Action Guides and Planning Guidance for Radiological Incidents.* January 2017. Accessed April 12, 2023. https://www.epa.gov/sites/default/files/2017-01/documents/epa_pag_manual_final_revisions_01-11-2017_cover_disclaimer_8.pdf.

U.S. Government Accountability Office. "How Does Homeland Security Combat Weapons of Mass Destruction and How Could Their Efforts Improve?" May 3, 2022. Accessed March 29, 2023. https://www.gao.gov/blog/how-does-homeland-security-combat-weapons-mass-destruction%2C-and-how-could-their-efforts-improve.

U.S. National Response Team. "About NRT." n.d. Accessed March 29, 2023. https://www.nrt.org/NRT/About.aspx.

U.S. Nuclear Regulatory Commission. "Rad (RADIATION ABSORBED DOSE)." n.d. Accessed March 29, 2023. https://www.nrc.gov/reading-rm/basic-ref/glossary/rad-radiation-absorbed-dose.html.

Walsh, Donald W., Hank T. Christen, Christian E. Callsen, Geoffrey T. Miller, Paul M. Maniscalco, Graydon C. Lord, and Neal J. Dolan. *National Incident Management System: Principles and Practice,* 2nd ed. Burlington, MA: Jones & Bartlett Learning, 2012.

Appendix A

Fire Behavior and Combustion Course Content Correlation Guide

The National Fire Science Curriculum Advisory Committee identified ten desired outcomes, involving eleven content areas for the course. This text addresses each desired outcome within the eleven content areas.

Fire Behavior and Combustion Core Course Desired Outcomes

1. Identify physical properties of the three states of matter.
2. Categorize the components of fire.
3. Recall the physical and chemical properties of fire.
4. Describe and apply the process of burning.
5. Define and use basic terms and concepts associated with the chemistry and dynamics of fire.
6. Explain the effect and dangers of air movement on the combustion process.
7. Discuss various materials and their relationship to fires as fuel.
8. Demonstrate knowledge of the characteristics of water as a fire suppression agent.
9. Articulate other suppression agents and strategies.
10. Compare other methods and techniques of fire extinguishments (U.S. Fire Administration, 2019).

FESHE Content Area Comparison

The following table provides a comparison of the eleven FESHE content areas in this text.

Fire and Emergency Services Higher Education (FESHE) Course Correlation Grid		
Name:	Fire Behavior and Combustion	*Fire Behavior and Combustion* Chapter Reference
Course Description:	This course explores the theories and fundamentals of how and why fires start, spread, and how they are controlled.	
Prerequisite:	None	
Course Outline:	**I.** Introduction	
	A. Matter and Energy	1, 2
	B. The Atom and Its Parts	2
	C. Chemical Symbols	2
	D. Molecules	2
	E. Energy and Work	1
	F. Forms of Energy	3
	G. Transformation of Energy	3
	H. Laws of Energy	3
	II. Units of Measurements	
	A. International (SI) Systems of Measurement	1, Appendix B
	B. English Units of Measurement	1, Appendix B
	III. Chemical Reactions	
	A. Physical States of Matter	2
	B. Compounds and Mixtures	2
	C. Solutions and Solvents	2
	D. Process of Reactions	2

Fire and Emergency Services Higher Education (FESHE) Course Correlation Grid

	IV. Fire and the Physical World	
	A. Characteristics of Fire	3
	B. Characteristics of Solids	2
	C. Characteristics of Liquids	2
	D. Characteristics of Gases	2
	V. Heat and Its Effects	
	A. Production and Measurement of Heat	1, 2, 4
	B. Different Kinds of Heat	3
	VI. Properties of Solid Materials	
	A. Common Combustible Solids	3
	B. Plastic and Polymers	2
	C. Combustible Metals	3, 4
	D. Combustible Dust	2
	VII. Common Flammable Liquids and Gases	
	A. General Properties of Gases	2
	B. The Gas Laws	2
	C. Classification of Gases	2
	D. Compressed Gases	2, 4
	VIII. Fire Behavior	
	A. Stages of Fire	3
	B. Fire Phenomena	3
	1. Flashover	3
	2. Backdraft	3
	3. Roll-Over	3
	4. Flameover	3
	5. Heat Flow	3
	C. Fire Plumes	3
	IX. Fire Extinguishment	
	A. The Combustion Process	3
	B. The Character of Flame	3
	C. Fire Extinguishment	3, 4
	X. Extinguishing Agents	
	A. Water	4
	B. Foams and Wetting Agents	4
	C. Inert Gas Extinguishing Agents	4
	D. Halogenated Extinguishing Agents	4
	E. Dry Chemical Extinguishing Agents	4
	F. Dry Powder Extinguishing Agents	4
	XI. Hazards by Classification Types	
	A. Hazards of Explosives	10
	B. Hazards of Compressed and Liquified Gases	4
	C. Hazards of Flammable and Combustible Liquids	3, 4
	D. Hazards of Flammable Solids	3, 4

Fire and Emergency Services Higher Education (FESHE) Course Correlation Grid		
	E. Hazards of Oxidizing Agents	4
	F. Hazards of Poisons	10
	G. Hazards of Radioactive Substances	10
	H. Hazards of Corrosives	10

Modified from U.S. Fire Administration. 2019. *National Fire Academy FESHE Model Curriculum Associate's (Core)*. https://www.usfa.fema.gov/downloads/pdf/nfa/higher_ed/associate_curriculum_core.pdf

Appendix B

Reference Tables

Standard conversions					
To Change	**To**	**Multiply***	**To Change**	**To**	**Multiply***
Length			**Mass**		
Inches	Feet	0.08333	Ounces	Pounds	0.06250
Inches	Millimeters	25.40	Pounds	Ounces	16
Feet	Inches	12	Long tons	Pounds	2,240
Feet	Yards	0.3333	Short tons	Pounds	2,000
Yards	Feet	3	Short tons	Long tons	0.8928
Area			**Volume**		
Square inches	Square feet	0.006944	Inches of mercury	Pounds per square inch	0.4912
Square feet	Square inches	144	Ounces per square inch	Inches of mercury	0.1273
Square feet	Square yards	0.1111	Pounds per square inch	Inches of mercury	2.036
Square yards	Square feet	9	Pounds per square inch	Atmospheres	0.06805
Volume			Atmospheres	Pounds per square inch	14.70
Cubic inches	Cubic feet	0.0005787	Atmospheres	Inches of mercury	29.92
Cubic feet	Cubic inches	1,728			
Cubic feet	Cubic yards	0.03704			
Cubic yards	Cubic feet	27			
Cubic inches	Gallons	0.004329			

**Values within this table are rounded up to the fourth nearest significant figure.*

Conversion factors for water
1 U.S. gallon = 8.345 pounds
1 U.S. gallon = 0.1337 cubic feet
1 U.S. gallon = 231 cubic inches
1 U.S. gallon = 0.8327 Imperial gallon
1 U.S. gallon = 3.785 liters
1 Imperial gallon = 10.00 pounds
1 Imperial gallon = 0.1605 cubic feet
1 Imperial gallon = 277.4 cubic inches
1 Imperial gallon = 1.201 U.S. gallons
1 Imperial gallon = 4.546 liters
1 liter = 2.204 pounds
1 liter = 0.03532 cubic feet
1 liter = 61.02 cubic inches
1 liter = 0.2642 U.S. gallon
1 liter = 0.2200 Imperial gallon
1 cubic foot of water = 62.43 pounds
1 cubic foot of water = 1,728 cubic inches
1 cubic foot of water = 7.481 U.S. gallons
1 cubic foot of water = 6.229 Imperial gallons
1 cubic foot of water = 28.32 liters
1 pound of water = 0.01602 cubic feet
1 pound of water = 27.712 cubic inches
1 pound of water = 0.1198 U.S. gallon
1 pound of water = 0.1000 Imperial gallon
1 pound of water = 0.4536 liter
1 inch of water = 0.0361 pound per square inch
1 inch of water = 0.07348 inch of mercury
1 inch of water = 0.5780 ounce per square inch
1 inch of water = 5.197 pounds per square foot
1 foot of water = 0.4335 pound per square inch
1 foot of water = 62.43 pounds per square foot
1 foot of water = 0.8827 inch of mercury
1 pound per square inch = 2.307 feet of water
1 inch of mercury = 13.60 inches of water
1 inch of mercury = 1.133 feet of water
1 ounce per square inch = 1.730 inches of water
1 pound per square inch = 27.68 inches of water

Conversion of inches to millimeters

in.	mm	in.	mm	in.	mm	in.	mm
1	25.4	26	660.4	51	1,295.4	76	1,930.4
2	50.8	27	685.8	52	1,320.8	77	1,955.8
3	76.2	28	711.2	53	1,346.2	78	1,981.2
4	101.6	29	736.6	54	1,371.6	79	2,006.6
5	127.0	30	762.0	55	1,397.0	80	2,032.0
6	152.4	31	787.4	56	1,422.4	81	2,057.4
7	177.8	32	812.8	57	1,447.8	82	2,082.8
8	203.2	33	838.2	58	1,473.2	83	2,108.2
9	228.6	34	863.6	59	1,498.6	84	2,133.6
10	254.0	35	889.0	60	1,524.0	85	2,159.0
11	279.4	36	914.4	61	1,549.4	86	2,184.4
12	304.8	37	939.8	62	1,574.8	87	2,209.8
13	330.2	38	965.2	63	1,600.2	88	2,235.2
14	355.6	39	990.6	64	1,625.6	89	2,260.6
15	381.0	40	1,016.0	65	1,651.0	90	2,286.0
16	406.4	41	1,041.4	66	1,676.4	91	2,311.4
17	431.8	42	1,066.8	67	1,701.8	92	2,336.8
18	457.2	43	1,092.2	68	1,727.2	93	2,362.2
19	482.6	44	1,117.6	69	1,752.6	94	2,387.6
20	508.0	45	1,143.0	70	1,778.0	95	2,413.0
21	533.4	46	1,168.4	71	1,803.4	96	2,438.4
22	558.8	47	1,193.8	72	1,828.8	97	2,463.8
23	584.2	48	1,219.2	73	1,854.2	98	2,489.2
24	609.6	49	1,244.6	74	1,879.6	99	2,514.6
25	635.0	50	1,270.0	75	1,905.0	100	2,540.0

The above table is exact on the basis: 1 in. = 25.4 mm

Conversion of millimeters to inches							
mm	in.	mm	in.	mm	in.	mm	in.
1	0.039370	26	1.023622	51	2.007874	76	2.992126
2	0.078740	27	1.062992	52	2.047244	77	3.031496
3	0.118110	28	1.102362	53	2.086614	78	3.070866
4	0.157480	29	1.141732	54	2.125984	79	3.110236
5	0.196850	30	1.181102	55	2.165354	80	3.149606
6	0.236220	31	1.220472	56	2.204724	81	3.188976
7	0.275591	32	1.259843	57	2.244094	82	3.228346
8	0.314961	33	1.299213	58	2.283465	83	3.267717
9	0.354331	34	1 338583	59	2.322835	84	3.307087
10	0.393701	35	1.377953	60	2.362205	85	3.346457
11	0.433071	36	1.417323	61	2.401575	86	3.385827
12	0.472441	37	1.456693	62	2.440945	87	3.425197
13	0.511811	38	1.496063	63	2.480315	88	3.464567
14	0.551181	39	1.535433	64	2.519685	89	3.503937
15	0.590551	40	1.574803	65	2.559055	90	3.543307
16	0.629921	41	1.614173	66	2.598425	91	3.582677
17	0.669291	42	1.653543	67	2.637795	92	3.622047
18	0.708661	43	1.692913	68	2.677165	93	3.661417
19	0.748031	44	1.732283	69	2.716535	94	3.700787
20	0.787402	45	1.771654	70	2.755906	95	3.740157
21	0.826772	46	1.811024	71	2.795276	96	3.779528
22	0.866142	47	1.850394	72	2.834646	97	3.818898
23	0.905512	48	1.889764	73	2.874016	98	3.858268
24	0.944882	49	1.929134	74	2.913386	99	3.897638
25	0.984252	50	1.968504	75	2.952756	100	3.937008

The above table is approximate on the basis of 1 in. = 25.4 mm, 1/25.4 = 0.039370078740+

Metric equivalents	
Length **U.S. to Metric**	**Metric to U.S.**
1 inch = 2.540 centimeters	1 millimeter = 0.03937 inch
1 foot = 0.3048 meter	1 centimeter = 0.3937 inch
1 yard = 0.9144 meter	1 meter = 3.281 feet or 1.094 yards
1 mile = 1.609 kilometers	1 kilometer = 0.6214 mile
Area	
1 inch^2 = 6.452 centimeter^2	1 millimeter^2 = 0.001550 inch^2
1 foot^2 = 0.09290 meter^2	1 centimeter^2 = 0.1550 inch^2
1 yard^2 = 0.8361 meter^2	1 meter^2 = 10.76 foot^2 or 1.196 yard^2
1 acre^2 = 4,047 meter^2	1 kilometer^2 = 0.3861 mile^2 or 247.1 acre^2
Volume	
1 inch^3 = 16.39 centimeter^3	1 centimeter^3 = 0.06102 inch^3
1 foot^3 = 0.02832 meter^3	1 meter^3 = 35.32 foot^3 or 1.308 yard^3
1 yard^3 = 0.7646 meter^3	1 liter = 0.2642 gallon
1 quart = 0.9464 liter	1 liter = 1.057 quart
1 gallon = 0.003785 meter^3	1 meter^3 = 264.2 gallons
Weight	
1 ounce = 28.35 grams	1 gram = 0.03527 ounce
1 pound = 0.4536 kilogram	1 kilogram = 2.205 pounds
1 ton = 0.9072 metric ton	1 metric ton = 1.102 tons
Velocity	
1 foot/second = 0.3048 meter/second	1 meter/second = 3.281 feet/second
1 mile/hour = 0.4470 meter/second	1 kilometer/hour = 0.6214 mile/hour
1 mile/hour = 1.609 kilometers/hour	
Acceleration	
1 inch/second^2= 0.02540 meter/second^2	1 meter/second^2 = 3.281 feet/second^2
1 foot/second^2 = 0.3048 meter/second^2	
Force	

N (newton) = basic unit of force, kg-m/s^2. A mass of one kilogram (1 kg) exerts a gravitational force of 9.8 N (theoretically 9.80665 N) at mean sea level.

Temperature Conversion: Celsius (°C) to Fahrenheit (°F)	
°C = (°F − 32.00°F)/1.800	°F = (1.800 × °C) + 32.00°F
Rankine (Fahrenheit Absolute) = Temp. °F + 459.67 deg.	Kelvin (Celsius Absolute) = Temp. °C + 273.15 deg.
Freezing point of water: Celsius = 0 deg.; Fahrenheit = 32 deg.	Boiling point of water: Celsius = 100 deg.; Fahrenheit = 212 deg.
Absolute zero: Celsius = −273.15 deg.; Fahrenheit = −459.67 deg.	

Appendix C

Wildland—10 Standard Fire Orders and 18 Watch Out Situations

Tracing back over the years, it is found that 90% of fire fatalities and injuries on wildland fires can be directly attributed to a violation of the standard fire orders. Both the standard orders and the obvious or "watch out" situations can be printed in card form, and the card can then be carried with the firefighter in a pocket or helmet.

Standard Fire Orders

- Keep informed of fire weather conditions and forecasts.
- Know what your fire is doing at all times.
- Base all actions in current and expected fire behavior.
- Identify escape routes and safety zones, and make them known.
- Post a lookout when there is possible danger.
- Be alert, keep calm, think clearly, and act decisively.
- Maintain prompt communications with your forces, your supervisor, and adjoining forces.
- Give clear instructions and be sure they are understood.
- Maintain control of your personnel at all times.
- Fight fire aggressively, but provide for safety first.

National Wildfire Coordinating Group, 2022

The following "watch out" situations were developed after a detailed study of many fire accidents, injuries, and deaths. These situations are a warning sign for firefighters to be extra careful as they are finding themselves in a situation where others have suffered.

Watch Out Situations

- Fire not scouted and sized up.
- In country not seen in daylight.
- Safety zones and escape routes not identified.
- Unfamiliar with weather and local factors influencing fire behavior.
- Uninformed on strategy, tactics, and hazards.
- Instructions and assignments not clear.
- No communication link with crew members or supervisor.
- Constructing line without safe anchor point.
- Building fireline downhill with fire below.
- Attempting frontal assault on fire.
- Unburned fuel between you and fire.
- Cannot see main fire; not in contact with someone who can.
- On a hillside where rolling material can ignite fuel below.
- Weather becoming hotter and drier.
- Wind increases and/or changes direction.
- Getting frequent spot fires across line.
- Terrain and fuels make escape to safety zones difficult.
- Taking a nap near fireline.

National Wildfire Coordinating Group, 2023

REFERENCES

National Wildfire Coordinating Group. 2022. "10 Standard Firefighting Orders." Last modified/reviewed March 2022. https://www.nwcg.gov/committee/6mfs/10-standard-firefighting-orders

National Wildfire Coordinating Group. 2023. "18 Watch Out Situations, PMS 118." Last modified/review January 25, 2023. https://www.nwcg.gov/publications/pms118

Acronyms

ABS Acrylonitrile budadiene styrene

AC Alternating current

ADA Americans with Disabilities Act

AFFF Aqueous film-forming foam

ALERRT Advanced Law Enforcement Rapid Response Training

ASHER Active Shooter/Hostile Event Response

ASME American Society of Mechanical Engineers

ASTM American Society for Testing and Materials

BLEVE Boiling liquid/expanding vapor explosion

BP Boiling point

BTU British thermal unit

CAFS Compressed air foam systems

CDC Centers for Disease Control and Prevention

CERCLA Comprehensive Environmental Response, Compensation, and Liability Act

CHEMTREC Chemical Transportation Emergency Center

CISM Critical incident stress management

CMA Chemical Manufacturers Association

CNG Compressed natural gas

CWMD Countering Weapons of Mass Destruction

DC Direct current

DHS Department of Homeland Security

DOD Department of Defense

DOE Department of Energy

DOJ Department of Justice

DOT Department of Transportation

EEOC Equal Employment Opportunity Commission

EMS Emergency Medical Services

EPA Environmental Protection Agency

EPCRA Emergency Planning and Community Right-to-Know Act

ERDSS Emergency Response Decision Support System

ERG *Emergency Response Guidebook*

ERT Emergency response technology

FAA Federal Aviation Administration

FARS Firefighter air replenishment system

FBI Federal Bureau of Investigation

FCEV Fuel cell electric vehicle

FDC Fire department connection

FDNY New York City Fire Department

FEMA Federal Emergency Management Agency

FMZ Fire management zone

FOG Field operations guide

FRA Federal Railroad Administration

GPS Global positioning system

HEV Hybrid electric vehicle

HIV Human immunodeficiency virus

HRR Heat release rate

HSPD Homeland Security presidential directive

HVAC Heating, ventilation, and air conditioning

IAP Incident action plan

IAQ Indoor air quality

IBC International Building Code

IC Incident commander

ICS Incident Command System

IDLH Immediately dangerous to life and health

IMO International Maritime Organization

ISO Insurance Services Office

IUPAC International Union of Applied Chemistry

JFO Joint field office

JOC Joint operations center

LEL Lower explosive limit

LEPC Local Emergency Planning Committee

LNG Liquefied natural gas

MBE Management by exception

MBO Management by objectives

MCI Mass-casualty incident

MDT Mobile data terminal

MOU Memorandum of Understanding

NBC Nuclear, biological, and chemical

NCP National Oil and Hazardous Substances Pollution Contingency Plan

NDPO National Domestic Preparedness Office

NFIRS National Fire Incident Reporting System

NFPA National Fire Protection Association

NHTSA National Highway Traffic Safety Administration

NIMS National Incident Management System

NIOSH National Institute of Occupational Safety and Health

NIST National Institute of Standards and Technology

NLM National Library of Medicine

NOAA National Oceanic and Atmospheric Administration

NRC Nuclear Regulatory Commission

NRT National Response Team

NWCG National Wildfire Coordination Group

OSHA Occupational Safety and Health Administration

PAR Personnel accountability report

PAS Personnel accountability system

PASS Personal alert safety system

PCB Polychlorinated biphenyl

PEL Permissible exposure limits

PPT Personal protective technology

PPE Personal protective equipment

PPV Positive-pressure ventilation

PVC Polyvinyl chloride plastic

rad Radiation-absorbed dose

RBE Relative biological effectiveness

RDD Radiological dispersal device

RECEO-VS Rescue, exposures, confinement, extinguishment, overhaul, and ventilation and salvage

rem Roentgen equivalent man

RIC Rapid intervention crew

RIT Rapid intervention team

RIV Rapid intervention vehicle

SARA Superfund Amendment and Reauthorization Act

SCBA Self-contained breathing apparatus

SDS Safety data sheet

SERC State Emergency Response Commission

SIC Side impact curtain

SIPS Side impact protection system

SLICE-RS Size up–locate the fire–identify the flow path–cool from safe location–extinguish–rescue–salvage

SOLAS Safety of Life at Sea (treaty)

SOP Standard operating procedure

SRS Supplemental restraint system

THOR Tactical Hazardous Operations Robot

TIC Thermal imaging camera

TLV Threshold limit value

UC Unified command

UEL Upper explosive limit

UL Underwriters Laboratories

UN United Nations

USCG U.S. Coast Guard

USFA U.S. Fire Administration

VD Vapor density

VP Vapor pressure

VEIS Vent-Enter-Isolate-Search

WISER Wireless Information System for Emergency Responders

WMD Weapons of mass destruction

Glossary

A

absolute pressure The measurement of pressure exerted on a surface, including pressure from the atmosphere, measured in pounds per square inch absolute.

absolute zero The temperature at which there is no movement of the molecules.

acid A chemical that releases hydrogen ions when dissolved in water.

acidity The amount of acid in a substance.

aerial fuel Fuel that includes all green and dead materials in the upper forest canopy.

air bill A document used to identify the hazardous materials shipping by air.

alkalinity A measure of water's ability to neutralize or cancel out acids.

alpha radiation Radiation that gives off alpha particles.

alternative fuel Any energy source other than the hydrocarbon fuels.

alternative fuel vehicle A vehicle powered by a fuel or energy source other than petroleum.

America Burning The 1973 presidential committee report on the U.S. fire service.

***America Burning Recommissioned, America at* Risk** The 2000 revisit of both the original *America Burning* report (1973) and the *America Burning Revisited* report (1987).

America Burning Revisited The 1987 revisit of the *America Burning* report to review the progress on the recommended improvements.

anchor point A safe location, usually a barrier to fire spread, where fire crews can start building a fire line.

aqueous film-forming foam (AFFF) A foam created by combining water and perfluorocarboxylic acid that is used on flammable liquid fires, such as those in fuel.

area ignition The accumulation of heated gases from individual fires that ignite either simultaneously or in quick succession.

aspect The direction a slope faces to the sun; also called exposure.

atmospheric stability The air's tendency to either rise and form clouds and storms (instability) or sink and create clear skies (stability).

atom The smallest unit of an element that takes part in a chemical reaction; made up of a neutron, a proton, and a cloud of orbiting electrons.

atomic number The number of protons in an element.

atomic weight The average number of protons and neutrons in atoms of a chemical element; sometimes referred to as atomic mass.

auto-exposure The lapping of fire from one floor to an upper floor on the outside of the building, sometimes through windows.

autoignition temperature The temperature at which a material will ignite in the absence of any outside source of heat; also referred to as the spontaneous ignition temperature.

automatic aid A contract between agencies to provide resources to any incident on dispatch.

B

backdraft A sudden reignition of room contents once the oxygen has been used up; the introduction of oxygen then creates an immediate explosion.

backfire A fire set to burn the area between the control line and the fire's edge to remove fuel in advance of the fire, to change the direction of the fire, and/or to slow the fire's progress.

backpressure The pressure applied in the opposite direction of the water flowing from a nozzle; also called nozzle reaction pressure.

balloon frame construction method An outdated method in which the wood studs run from the foundation to the roof and the floors are nailed to the studs.

base A substance that will react with acids in aqueous solution to form salts, while releasing heat.

below grade A floor lower than ground level.

beta radiation Radiation that gives off beta particles.

bilge The bottom area of a boat or ship.

bill of lading Documents that indicate the amount and type of cargo or freight being transported.

black zone The fuel or vegetation that has been burned; it is considered a safe area because the fuel has been removed.

boiling liquid/expanding vapor explosion (BLEVE) The result of the explosive release of vessel pressure, portions of the tank, and the burning vapor cloud of gas with the accompanying radiant heat.

boiling point (BP) The temperature at which a liquid will convert to a gas at a vapor pressure equal to or greater than atmospheric pressure.

boil over The expulsion of a tank's contents by the expansion of water vapor that has been trapped under the oil and heated by the burning oil and metal sides of the tank.

bottom The rearmost area of a fire, used as the anchor or starting point of the fire suppression activities; also called a toe.

boxcar A railroad car that is enclosed and used to carry general freight.

Boyle's law A theory that states the more a gas is compressed, the more the gas becomes difficult to compress further.

British thermal unit (BTU) Unit of measure for the heat energy needed to raise the temperature of 1 pound of water by 1 degree Fahrenheit.

brush Dense vegetation that consists of shrubs or small trees.

bulkhead A main wall or supporting structure made of steel.

burn back resistance The ability to prevent any flame from breaking through the foam barrier.

burning index An estimate of the potential difficulty to contain the fire as it relates to the flame length at the most rapidly spreading part of a fire's edge.

bypass valve A tap that allows a water supply to travel around the pump and provide fire protection.

C

carbon dioxide A nonflammable liquid asphyxiant gas.

carbon monoxide (CO) A fire gas product that consists of a single atom of carbon and a single atom of oxygen.

catalyst A substance that speeds up a chemical reaction but is not changed or used up by the reaction.

cellulose material A complex carbohydrate of plant cell walls used to make paper or rayon.

cellulosic material Material made by changing cellulose chemically.

Celsius scale A temperature scale based on water freezing at 0° and boiling at 100°; also known as centigrade.

central core construction Method of construction where the elevators, stairs, restrooms, and support systems are located in the center of the building; also called center core construction.

Charles's law A theory that states a gas will expand or contract in direct proportion to an increase or decrease in temperature.

chemical bond The attractive force that binds two atoms in a combination that is stable at room temperature, but becomes unstable at a high temperature.

chemical change Molecules of a material are changed by a process.

Chemical Companion A system that references 2,000 common chemicals, exposure guidelines, hazardous concentration levels, and isolation and protective action distances; also known as the Emergency Response Decision Support System (ERDSS).

CHEMTREC A national call center that guides emergency responders at a chemical incident.

chimney A topographic feature that has three walls and a steep, narrow chute.

churning the air Pulling air into the building from the outside and blowing it out again.

circumference The length around a circle.

Class A fire Fire involving ordinary cellulosic materials.

Class A foam Foam used on a Class A fire.

Class B fire Fire involving flammable liquids and gases.

Class B foam Foam used on a Class B fire.

Class C fire Fire involving energized electrical equipment or wires.

Class D fire Fire involving combustible metals.

Class K fire Fire involving liquid cooking materials.

clean agent suppression system Fire suppression system that uses nontoxic agents.

closed cup test A lid is placed on a cup to confine the vapors above the cup. The sample in the cup is heated and stirred, and an ignition source is introduced at random times to find the flash point of the sample.

cockloft A void space, about 3 feet deep, between the ceiling area and the underside of the roof.

cold trailing A method to determine whether a fire is still burning by inspecting and feeling with gloved hands to detect any heat source.

collapse zone The border around a structure that is usually 1.5 times the height of the building where debris would fall during a collapse.

combination attack method A method that takes advantage of the direct and indirect attack methods by combining both methods.

combustion A chemical reaction between a fuel and oxygen that creates smoke, heat, and light.

compartmentation The process of dividing large structures, such as high-rise buildings, into compartments or units to provide safe areas and prevent the spread of fire.

compound A substance formed from two or more elements joined with a fixed ratio.

compressed air foam system (CAFS) A fire suppression system that injects air into the foam solution with an air compressor.

compressed gas Any material that, when in a closed container, has an absolute pressure of more than 40 psi at 70°F (21°C) or an absolute pressure exceeding 104 psi at 130°F (54°C), or both.

compressed natural gas (CNG) Natural gas that is compressed and contained within a pressurized container.

conduction The transfer of heat energy from the hot to the cold side of a medium through collisions of molecules to molecules or atoms to atoms.

conflagration A fire with building-to-building flame spread over a great distance.

consist Document used to identify the hazardous cargo of a rail transportation vehicle; also called a waybill.

control line All natural and created barriers used to stop a fire from spreading; this is where you begin the backfire.

control valve A valve used to control the flow of water in sprinkler and standpipe systems.

convection The transfer of heat by circulating air currents.

convection column The ascending column of gases, smoke, and debris produced by a fire.

convergence cluster behavior People who feel threatened gather as a group to feel safe.

covalent bond The sharing of electron pairs by combined atom(s).

cribbing Blocks of wood or prefabricated plastic used to lift a vehicle's frame off the wheels and stabilize it.

crown fire Fire that has traveled from the ground to the forest canopy and spread through it.

cryogenic Material with boiling points of no greater than −150°F (−101°C) that are transported, stored, and used as liquids.

cybernetic building system Control system that combines building services, such as energy control (HVAC systems), fire and security systems, building transport systems, earthquake detection, and real-time checks of a building.

D

dangerous cargo manifest Documents used to identify the hazardous cargo of a marine vessel.

day of transition The first day when offshore winds subside and cool, moist onshore flows begin gradually returning.

dead load The full weight of the building and its static parts.

debriefing A voluntary group discussion that is used when there is no urgent need to discuss the incident; part of critical incident stress management.

decay stage The stage when the fire has used up all of the fuel and combustion will cease.

decision-making model A five-step process used to solve problems.

decomposition temperature The temperature to which the material must be heated for it to decompose.

defensive attack mode A fire tactic used outside a building to put out a fire.

defusing The process of talking about the experience that occurs immediately after the incident; part of critical incident stress management.

deluge system A dry pipe system that protects areas that are being consumed by a fast-spreading fire; all sprinkler heads are activated at the same time.

Department of Defense (DOD) A federal department that supplies staff, training, and equipment for a full range of hazardous substances.

Department of Energy (DOE) The federal department that is responsible for reducing radioactive waste from the U.S. nuclear weapons program and nuclear reactors.

Department of Homeland Security (DHS) A federal department that protects U.S. territory from terrorist attacks and responds to natural disasters.

Department of Transportation (DOT) A federal department responsible for the regulations for the safe transport of hazardous materials.

diameter The distance from one side of a circle to the other.

diffusive flaming A category of flaming combustion characterized by yellow flames that are part of the combustion process.

direct attack method A method of fire suppression in which personnel and resources work at or very close to the edge of the burning fire.

draft The process of moving or drawing water away from a static source of water by a pump.

drain time The time needed for the water to drain away from the foam solution.

drip loop A loop formed by the electrical supply lines; also called a meter head.

dry pipe A system that has air in the sprinkler piping until it is activated; each sprinkler head is activated by heat.

E

electron A very light particle with a negative electrical charge, a number of which surround the nucleus of most atoms.

element The simplest form of matter.

Emergency Planning and Community Right-to-Know Act (EPCRA) An act that helps communities plan for chemical incidents and informs the public and local governments of potential chemical hazards.

***Emergency Response Guidebook* (ERG)** A manual that contains the chemical names and commonly used names for most toxic chemicals; meant to help first responders handle hazardous materials.

endothermic A type of reaction in which heat (energy) is absorbed when the reaction takes place.

entrainment The gathered or captured cooler air that replaces the rising heated air surrounding the point of combustion.

envelopment action A fire suppression technique that suppresses a fire at many points and in many directions simultaneously.

Environmental Protection Agency (EPA) A government agency whose purpose is to protect human life and the environment.

evaporation The process by which molecules in a liquid state (e.g., water) spontaneously become gaseous (e.g., water vapor).

exothermic A type of reaction that will release or give off heat.

expansion ratio The volume of a substance's liquid form compared to the volume of its gas form.

explosive/flammability range The range of concentrations of gases or materials (dusts) in the air that permit the material to burn.

exposure A property threatened by radiant heat from a fire in another structure or an outside fire.

eyebrow dormer A concrete extension over the top of openings, such as windows, doors, and balconies.

F

Fahrenheit scale A temperature scale based on water freezing at 32° and boiling at 212°.

families Groups of elements in the periodic table with like properties.

Federal Aviation Administration (FAA) An agency of the DOT that oversees all aspects of civil aviation in the United States.

Federal Bureau of Investigation (FBI) Under the direction of the Department of Justice, this federal agency is responsible for the criminal investigation of acts of terrorism.

field operations guide (FOG) A system that classifies units into strike teams and task forces; available in print, online, or as an app.

finger A narrow strip of fire that has separated from the main body of the fire.

fire A rapid oxidation process that involves heat, light, and smoke in varying intensities.

fire damper A device that, when turned on, closes the air flow in the HVAC system ducting.

fire department connections (FDCs) Fixtures, either stand-alone or part of the building's outside wall, that allow firefighters to supply additional water to the fire suppression sprinkler system or the standpipe system.

fire department standpipe connection Where the fire pumper connects to pressurize the building standpipe system.

firefighter air replenishment systems (FARS) Standpipes for air that remove the need for firefighters to leave a burning building to refill their SCBAs.

fire flow The water supply needed to put the fire out.

fire line A fuel break that is created when firefighters clear an area twice as wide as the fuel is tall.

fire load The total amount of fuel that might be involved in a fire, measured by the amount of heat given off when that fuel burns; expressed in BTUs.

fire management zone (FMZ) A zone within an engine company's area where similar hazards are grouped because they have roughly equal needed fire flow and number of hazards.

fire modeling A simulation used to figure out a fire's outcome.

fire point The lowest temperature of a liquid sufficient to produce vapor that will ignite from an outside ignition source and sustain combustion.

fire pump A specially designed pump that increases the water pressure serving a fire protection system.

fire retardant Any substance that is used to slow or stop the spread of fire or to decrease a fire's intensity.

fire tetrahedron A four-sided model showing the heat, fuel, oxygen, and chemical reaction necessary for combustion.

fire tornado A vortex comparable to a standard land-based tornado.

fire triangle A fire requires the presence of heat, fuel, and oxygen, which is depicted in a model.

first-in jurisdiction The area where a fire department has a duty to respond.

flame over The flames that travel through unburned gases in the upper portions of a confined area during the fire's development; also called roll-over.

flaming combustion An exothermic or heat-releasing chemical reaction with flames between a substance and oxygen; can be broken into two categories: premixed flaming and diffusive flaming.

flammable gas A gas that is flammable at atmospheric temperature and pressure within a mixture with air of 13% or less (by amount or volume) or that has a flammability range with air of more than 12%.

flanking action A technique in which the fire is attacked on both flanks at the same time.

flash fuel Fuel such as grass, leaves, ferns, tree moss, and some types of slash that ignites readily and is consumed rapidly when dry.

flashover When all the contents of a room or enclosed compartment reach their ignition temperature at the same time, resulting in an explosive fire.

flashover fire A fire that happens when all the contents of a room or enclosed space reach their ignition temperature and explode in fire.

flash point The lowest temperature of a liquid at which vapors are produced and will ignite when an outside ignition source is present.

flatcar A railcar that is open to the elements.

fluid mechanics The study of fluids at rest or in motion.

foehn wind A dry wind with a strong downward component; locally called Santa Ana, North, Mono, Chinook, or East Wind.

free radical A fragment of a molecule that has at least one unpaired electron.

friction loss The pressure lost by fluids while they are moving through pipes, hose lines, or other limited spaces.

fuel ladder The progression of the fire from the lighter to heavier fuels.

fuel loading The amount of fuel available to burn in a given area.

fully developed stage The stage of a fire where there is maximum generation of heat and flames; all fuel and oxygen are used up.

G

gamma radiation One of the most dangerous forms of penetrating radiation.

global positioning system (GPS) A navigation system that uses satellites in space to map locations on Earth.

gondola car A railcar with a flat bottom and four walls that can carry a variety of goods.

gravity wind Wind that occurs when air is pushed over high elevations and flows downhill.

green area The unburned area next to vegetation fires.

group A vertical column of elements in the periodic table, each with the same number of valence electrons, resulting in similar properties and reactivity.

growth stage The stage where the fire increases fuel consumption and heat generation.

gypsum board Board made with a core of gypsum that is primarily used in inside finishes for walls and ceilings; also called drywall or plasterboard.

H

halogenated compound Compound formed from any chemical reaction in which one or more halogenated atoms are combined with an existing compound.

halogenated extinguishing agent An extinguishing agent that has one or more halogen atoms added to a hydrocarbon after removing the hydrogen atom from the hydrocarbon molecule.

hazardous material A material in any form or quantity that poses an unreasonable risk to safety, health, and property when transported in trade.

hazardous substance Any substance named under the Clean Water Act and the Comprehensive Environmental Response, Compensation, and Liability Act (CERCLA) as posing a threat to waterways and the environment when released.

head The outermost portion of the fire that is moving from the rear.

heat All of the energy, both kinetic and potential, within the molecules.

heating, ventilation, and air conditioning (HVAC) system A central system used to heat and cool large buildings.

heavy fuel Fuel over 6 ft tall.

heavy timber construction Masonry; however, the inside columns and beams are solid or heavy wood and are built without hidden spaces; also called Type IV or mill construction.

high-rise building A building where the floor of an occupiable story is greater than 75 ft (22.9 m) above the lowest level of fire department vehicle access.

hopper car A type of railroad freight car used to transport loose bulk commodities.

hybrid electric vehicle (HEV) Vehicle run by a variety of fuels and powering systems; often powered by an internal combustion engine and a lithium-ion battery.

hydrocarbon product An organic compound, such as benzene or methane, that contains only carbon and hydrogen.

I

incident action plan (IAP) A plan for the firefighters battling the fire during the next operational period.

Incident Command System (ICS) A management system used on the emergency scene to keep order.

incipient stage The point at which the four parts of the fire tetrahedron come together and the materials reach their ignition temperature and fire starts; also called the ignition stage.

indirect attack method A method that consists of constructing control lines or creating a backfire at a distance ahead of the fire.

inflator A pressurized container that releases cool/hot pressurized gases that fill the appropriate impact curtain or bag.

inorganic A term that describes matter composed mainly of earth materials, such as rocks, soil, air, water, and minerals at or below the Earth's surface.

insolubility A term describing materials that form a separate layer in another substance and that will either float or sink, depending on their specific gravity.

Insurance Services Office (ISO) An agency funded by the insurance companies to apply a rating schedule to cities and fire departments and set the rate for fire insurance costs.

intermodal railcar A flatcar with an intermodal container that can use multiple modes of transportation (rail, ship, and truck), without any handling of the freight itself when changing modes.

International Maritime Organization (IMO) An agency of the United Nations that specializes in maritime issues; responsible for maintaining the SOLAS treaty.

inversion layer A layer of comparatively warm air overlaying cool air.

ionic bond A bond formed between two charged atoms, or ions, to create a molecule.

island An area inside the fire that has not been burned out.

L

ladder fuel Fuel that grows and spreads from the ground surface into higher vegetation.

latent heat of vaporization The amount of energy (enthalpy) that must be added to a liquid (water) to change it into a gas (steam); also known as the enthalpy of vaporization or evaporation.

leeward Facing the direction toward which the wind is blowing.

light fuel Fuel less than 2 ft tall.

lightweight concrete A light-in-weight concrete mixture that is blown onto metal and provides a layer of material that resists fire; also called blown-on concrete.

lightweight construction Construction that uses lumber that is a smaller dimension; often 2 × 4 wooden studs, rafters, and ceiling joists are found in these structures.

liquefied petroleum gas (LPG) A term given [illegible] propane gases that have been pressurized an[illegible] in a tank.

lithium-ion battery A rechargeable battery made [illegible] ium iron phosphate, lithium manganese oxide, a[illegible] ium nickel manganese oxide.

live load Items inside a structure that are not attached t[illegible] structure or permanent.

Local Emergency Planning Committee (LEPC) A committee that develops and maintains local emergency response plans to chemical emergencies.

louvering A method of roof ventilation that reduces the firefighters' exposure to smoke and heat and initially prevents cutting the supporting roof joists.

lower explosive limit (LEL) The lowest concentration of a substance (by percentage) in air that will burn.

M

mass A measure of the amount of matter objects contain.

matter Anything that occupies space and has mass or something that occupies space and can be felt by one or more of the senses.

medium fuel Fuel 2 to 6 ft tall.

mixture A combination of substances held together by physical rather than chemical means.

molecule The smallest unit of an element or compound that keeps the chemical traits of the original material.

monitor An industrial device used to deliver large amounts of water for firefighting purposes.

monomers Gaseous or liquid small molecules that are the building blocks for polymers; also called monomer molecules.

mutual aid A contract between agencies to provide resources on request.

N

nacelle The covering or cowling that protects the engine of the aircraft.

National Fire Protection Association (NFPA) 704 system The warning method used for hazardous materials stored in buildings.

National Highway Traffic Safety Administration (NHTSA) An agency within the DOT that works to reduce injuries, deaths, and economic losses due to motor vehicle accidents; developed a small-scale regulatory fire test for vehicle interiors that is still used.

National Incident Management System (NIMS) A management system adopted by FEMA that combines resources from public and private agencies.

National Institute of Occupational Safety and Health (NIOSH) A federal agency under the U.S. Department of Health and Human Services that is responsible for checking the toxicity of workspaces and all other matters relating to safe industrial practices.

National Oceanic and Atmospheric Administration (NOAA) An organization that works with the EPA, the U.S. Department of Commerce, and other organizations to develop the Chemical Reactivity Worksheet.

National Response Team (NRT) A team of 15 federal agencies that help states with their emergency response planning and during major hazardous materials incidents.

an atom cause it to gain an extra overall negative charge; also known

ntilation A method of forced ventila- and smoke out of a structure.

with nearly the same mass as the proton, ly neutral; it is part of the nucleus of all at- the most common isotope of hydrogen.

ation A product of nuclear fission; one of the ngerous forms of radiation.

k mode A fire tactic used when a fire attack is too gerous or suppression activities are prevented.

ar Regulatory Commission (NRC) A group within the Department of Energy, which furnishes licenses to facilities that use nuclear products.

O

occupancy The building code term that provides standards to match a building's use and those who will use it with features to address fire hazards and life safety concerns.

octet rule A rule that refers to the preference of atoms to have eight electrons in the valence shell.

offensive attack mode Firefighting that makes a direct attack on a fire.

open cup test A test that measures the release of the vapors in terms of the pressure being exerted at a specific temperature; the flammable vapors or molecules driven off the flammable liquid are not trapped by a lid.

ordinary construction A type of building construction in which the outside walls are made of masonry materials.

organic A term that refers to matter in substances that were once living organisms.

overhaul The process of searching the fire scene for hidden fires or sparks that may rekindle; also helps find the origin and cause of the fire.

oxidizer or oxidizing agent Substance that presents special hazards because it reacts chemically with many combustible organic materials, such as oils, greases, solvents, paper, cloth, and wood; for example, halogens are powerful oxidizers.

P

parallel action A fire suppression technique in which the fire line is constructed parallel to and just far enough away from the fire edge to allow personnel and equipment to work effectively.

particulate The unburned product of combustion one can see in smoke.

passenger railcar Railcar that carries people.

period A horizontal row of elements in the periodic table that identifies how many orbits revolve around the nucleus of the atom.

personal alert safety system (PASS) A small device that is sensitive to motion; worn with SCBA when firefighters enter an immediately dangerous to life and health (IDLH) area.

personnel accountability report (PAR) A roll call taken of all firefighters working on the fire ground.

personnel accountability system (PAS) A tracking system to follow the entry and exit of crew members into the working area during an incident.

pH A measure of a substance's ability to react as an acid (low pH) or as an alkali (high pH).

physical change Molecules of a material are not changed by a process and the molecules remain intact.

physical property A trait that is constant to the compound; for example, a ball or chain configuration.

pike pole A device with a sharp pointed head and a curved hook; used to pry open scuttle holes and sound the roof.

piloted ignition temperature The temperature of a liquid fuel at which it will self-ignite when heated.

pincer action A fire suppression technique in which crews move along both flanks of the fire to a point where flanking forces create a pinching action near the head of the fire.

platform construction method A construction method in which the floors are built separately from the outer walls; this means that the ceiling and floor serve as a fire block.

plenum space A space above a suspended ceiling or hallway that is kept under negative pressure for return air.

pocket An area on the edge of a fire that has not been burned.

polar solvent A substance that allows firefighters' foam to be used on alcohol-based fires without breaking down the foaming agent.

polymerization The process of reacting (linking or rearranging) monomer molecules in a chemical reaction to create a polymer; the reaction can be explosive or violent.

positive ion An atom that is missing electrons; also known as a cation.

positive-pressure ventilation A process that uses mechanical fans to blow air into a structure to remove smoke and gases.

postfire conference A meeting or meetings held after an incident to talk about what fire crews saw, what actions they took, and what they learned to improve their emergency response on future incidents.

prefire planning The process of studying and preparing for firefighting activities at the scene of a hazard or structure.

premixed flaming A category of flaming combustion in which a gaseous fuel mixes with air before catching fire.

pressure relief valve Used on compressed gas cylinders to release pressure build-up within a cylinder.

proton A positively charged particle that is the nucleus of all atoms, including the most common isotope of hydrogen.

pyrolysis The process of breaking down a solid fuel with heat into gaseous parts.

R

radiation Energy that travels across a space and does not need an intervening medium, such as a solid, gas, or fluid.

radiation-absorbed dose (rad) A unit to describe the amount of radiation absorbed by an object or person (absorbed dose).

radiological dispersal device (RDD) A device, such as an explosive, used to disperse radiological products (e.g., fissile material).

radius The length from the middle of a circle to the outside of the circle.

rapid intervention team or rapid intervention crew (RIT/RIC) A group of people whose purpose is to make a rapid response to reports of firefighters who become trapped or confused in the building.

rapid intervention vehicle (RIV) A large pumper used for aircraft firefighting.

rear The portion at the edge of a fire opposite the head.

RECEO-VS An acronym used to develop strategies and tactics on the fire ground.

red flag warning An alert issued by the National Weather Service when relative humidity is below 15%, which causes fuel to become very dry and makes conditions favorable for a wildland fire.

redox A chemical reaction that involves the loss of electrons, which speeds up the oxidation process, and a gain of electrons, which slows the oxidation process; shortened term for reduction–oxidation.

rehabilitation system A group of activities for the health and safety of responders at emergency incidents.

rekindle A fire thought to be out that reignites after the fire department has left.

relative biological effectiveness (RBE) A measure of how damaging a given type of particle is compared to an equivalent dose of x-rays.

relative humidity The amount of water vapor in the air to the maximum amount that the air can hold; it is presented as a percentage.

ridge A raised land mass that divides the terrain.

riot An action taken by a group of people to cause fear and disorder in a community.

risk assessment The evaluation or comparison of risks to develop successful approaches to an incident.

roentgen equivalent man (rem) A unit of radiation dosage (e.g., from x-rays) applied to humans.

S

saddle The low topography between two high points.

Safety Data Sheet (SDS) A required worksheet that manufacturers and marketers prepare to provide workers with detailed information about the properties of a commercial product that contains hazardous substances.

Safety of Life at Sea (SOLAS) An international treaty containing the minimum fire and safety standards for ships on international voyages.

salvage operation The protection of property while fighting a fire.

saponification The process of chemically converting the fatty acid in cooking oil or grease to soap or foam.

scuttle hole An opening that allows entry into the attic area.

self-contained breathing apparatus (SCBA) A pressurized tank on a backpack that provides breathable air within a fully enclosed mask.

self-oxidizing material Material that has extra oxygen, which supports the process of combustion by making the fire stronger.

shock-sensitive A material that will react violently when struck or compressed.

Siamese connection A hose fitting with a double hose line junction that allows two hoses to connect to one line.

side impact curtain (SIC) A vehicle passenger restraint system that deploys more quickly and forcefully from the side because of the shorter distance to the passenger.

size-up An ongoing review by firefighters to identify the problems at an incident.

SLICE-RS An acronym used to develop strategies and tactics on the fire ground.

slippery water Water that has polymers, plasticlike substances, to not only reduce friction loss in the hose but also increase the amount of water that can be moved through a hose line.

smoke ejector Mechanical smoke fan that draws heated air and smoke outside of a structure.

smoldering combustion The absence of flame and the presence of hot materials on the surface where oxygen diffuses into the fuel; also called glowing combustion.

soffit A false space under stairways and projecting roof eaves or the false space above cabinets in kitchens or bathrooms.

solubility The maximum amount of the substance that can be dissolved in another substance.

sounding the roof Using a pike pole, vent hook, or ax to tap the roof ahead to see if it is solid enough to support the weight of firefighters.

span of control The number of people or resources that one supervisor can manage during an incident.

specific gravity The density of the product divided by the density of water or air; water and air are the standards that have been given the value of 1.0.

spontaneous combustion A situation in which a material self-heats to its piloted ignition temperature before catching fire.

spot fire Fire that jumps over the control line and burns the vegetation ahead of the main fire front.

stack effect The natural movement of air within a tall building caused by the temperature difference between the outdoor air and indoor air; also called chimney effect.

stairwell support procedure A system used to move resources to the fire attack area when the elevators are unsafe.

standard operating procedure (SOP) Specific instructions on how to complete a task or assignment; also called standard operational guidelines (SOG).

standpipe A manual firefighting system with piping and hose connections inside buildings.

State Emergency Response Commission (SERC) Appointed by the governor, this commission coordinates emergency response plans, which are developed by the LEPCs.

stoichiometric An ideal burning situation or a condition where there is perfect balance between the fuel, oxygen, and end products that results in the almost complete consumption of all materials with little or no energy waste.

stoichiometry of reaction The proportions of fuel, oxygen, and end products.

strategy A general plan to meet incident goals.

stratification location The point in a high-rise building where heated air has risen and is the same temperature and weight as the surrounding air; at this location, air moves horizontally.

structural integrity The engineering field that ensures a building is designed and built for safe use under normal conditions and in an emergency.

sublimation A direct change from a solid to a gas without changing into a liquid.

supplemental restraint system (SRS) An additional passenger restraint system, which may be added to the vehicle roof, dash, or roof pillar.

surface tension The tendency of molecules to be attracted to each other at the surface of a liquid.

surfactant A soap material that reduces the surface tension of water, allowing the water to go through materials more easily.

swamper A worker on a bulldozer crew who pulls the winch line, helps maintain equipment, and generally assists with suppression work on a fire.

T

tactic An action needed to complete the strategy or plan.

tandem action A direct attack method with the attacking forces working in tandem (one unit following the other).

tank railcar A tank mounted on a railroad frame with wheels designed to transport a variety of liquid products.

target hazard Structure or area that poses the greatest threats to life and property.

temperature A measure of the average molecular movement or the heat's degree of intensity.

temperature inversion Atmospheric condition in which air temperature increases with height; cooler air is found close to the ground and the warmer air is above the cooler air.

terrorism Any act by an individual or group to create fear and injure civilians, frighten the government, or alter government actions by violent means.

thermal imaging A tool that uses infrared waves to find heat given off by a substance or body.

thermal imaging device A device that discerns between objects or areas with different temperatures and gives a visual display of those areas.

thermal imbalance A condition in which turbulent circulation of steam and smoke in the fire area makes it hard for firefighters to see and breathe.

thermal layer zone Layering of air and fire gases based on their temperatures.

thermal runaway event An event in which a lithium-ion battery self-heats uncontrollably.

thermodynamics The relationship between heat and other forms of energy.

third rail The rail used in subways and the wire for overhead electric train power systems to provide electricity to the train.

time lag The time it takes for the moisture content of fuel and the surrounding air to equalize; usually expressed in hours.

transitional mode The process of shifting from an offensive attack mode to a defensive attack mode, or vice versa.

transom A small window in an older building, such as a hotel, at the top of the ceiling, usually over the room entrance.

U

ultrafine water mist Water dispensed under high pressure and sent through very fine nozzle outlets that create nearly microsized droplets.

unified command (UC) A team system that allows all agencies responding to the incident to adopt one set of objectives and strategies.

upper explosive limit (UEL) The highest concentration of a substance (by percentage) in air that will burn.

U.S. Coast Guard (USCG) The branch of the U.S. military in charge of enforcing maritime laws.

utility chase A channel used for electrical, telephone, and plumbing lines and pipes for services in a building.

V

valence electrons The electrons in the outermost shell of an atom that interact in the bonding process with other atoms.

vapor density (VD) The mass of the given vapor in comparison to the same amount of air.

vapor pressure (VP) The pressure placed on the inside of a closed container by the vapor in the space above the liquid.

vent–enter–isolate–search (VEIS) A firefighting technique that calls for venting outside the building for a safer entrance and a reduced fire flow path.

ventilation A way to remove smoke and heat from a building, allow for the entry of cooler air, and improve rescue and firefighting operations.

viscous water A combination of thickening agents that is added to water so it will cling to the surface of a fuel.

void space An area in a structure's construction that has no outside or inside entrance and is not used for anything.

W

weapon of mass destruction (WMD) Weapons that use hazardous materials, nuclear radiation, biological agents, chemicals, explosives, or agricultural agents to kill or injure numerous people.

wet pipe A system that has water in the sprinkler piping at all times, and each sprinkler head is activated by heat.

wet water An additive that reduces the surface tension of water.

white phosphorus A toxic substance made from phosphate-containing rocks that must be stored under water or oil to prevent it from catching fire.

wildland/urban interface The line, area, or zone where structures and other human developments merge with wildland or fuels from vegetation.

windward The direction from which the wind is blowing.

WISER A system that helps first responders in hazardous materials incidents by identifying a chemical, its properties, and the effects of exposure; shut down as of February 2023.

Z

zero impact time The time from dispatch to the incident to when firefighters are on scene to take action.

Index

Note: Page numbers followed "*f*" and "*t*" indicate figures and tables respectively.

D

E

F

G

H

U

V

W

Y

Z